PERSPECTIVES IN ETHOLOGY

Volume 1

PERSPECTIVES IN ETHOLOGY

Edited by
P. P. G. Bateson
Sub-Department of Animal Behaviour
University of Cambridge
Cambridge, England

and
Peter H. Klopfer
Department of Zoology
Duke University
Durham, North Carolina

Volume 1

PLENUM PRESS • NEW YORK AND LONDON

First Printing – September 1973
Second Printing – April 1975

Library of Congress Catalog Card Number 73-79427
ISBN 0-306-36601-0

© 1973 Plenum Press, New York
A Division of Plenum Publishing Corporation
227 West 17th Street, New York, N.Y. 10011

United Kingdom edition published by Plenum Press, London
A Division of Plenum Publishing Company, Ltd.
Davis House (4th Floor), 8 Scrubs Lane, Harlesden, London, NW10 6SE, England

PREFACE

In the early days of ethology, most of the major developments were in the realm of ideas and in the framework in which animal behavior was studied. Much of the evidence was anecdotal, much of the thinking intuitive. As the subject developed, theories had to be tested, language had to become more public than it had been, and quantitative descriptions had to replace the preliminary qualitative accounts. That is the way a science develops; hard-headed analysis follows soft-headed synthesis. There are limits, though, to the usefulness of this trend. The requirement to be quantitative can mean that easy measures are chosen at the expense of representing the complexly patterned nature of a phenomenon. All too easily the process of data collection becomes a trivial exercise in describing the obvious or the irrelevant.

Editors and their referees require authors to maintain high standards of evidence and avoid undue speculation—in short, to maintain professional respectability. In the main, this process is admirable and necessary, but somewhere along the line perspective is lost and a body of knowledge, with all the preconceptions and intellectual baggage that comes with it, becomes formally established. New ideas are treated as though they were subversive agents—as indeed they often are. Radical or novel theories must be fully formed if they are not to be laughed to scorn, and as new ideas peep out of the ground in the conceptual flowerbed, it is easy enough to shout "Weeds!" and dig them up by the roots. Embryonic ideas which have tremendous potential may not be easily distinguishable from embryonic nonsense. So what is to be done? We believe that occasionally the upholders of standards in a discipline—heads of departments, editors, referees, grant-giving agencies, and so on—should be rather more permissive than is their custom and should encourage the development of new theories within a subject, even though the initial attempts may be vulnerable to attack.

Feeling as we do, we could, of course, have attempted to persuade heads of departments, editors, etc., to relax their constraints on the would-be theorist. Instead, we have attempted to encourage some scientists whose work we knew about to let their hair down a little and write in a less inhibited way than they would normally if they were writing for straight journals in the field of animal behavior. Five of the articles in this volume were commissioned on topics which we suggested to the authors. The remaining three were submitted to us when word got around that we were interested in encouraging some fresh thinking. The manuscripts which arrived on our desks by the deadline were extremely diverse. Some were orthodox reviews and some were written in a style that approached blank verse. We were faced with a problem entirely of our own making. What standards should *we* apply? It is all too easy to modify and criticize in the manner of conventional editors and referees. But to do that would be to violate the principles on which the whole project was first launched. Coherence and consistency might seem obvious criteria to apply, and yet to insist on normality in these respects might emasculate an author's argument. Well, we have compromised uneasily. Where necessary, we have attempted to persuade an author to make himself intelligible to us, but left him with the option of going his own way. The formula is obviously messy, and we have no confidence that we shall please everybody. However, if we encourage some liberalizing of thought in ethology and, more tangibly, lead others to develop their ideas in print, this volume will have achieved its object.

P. P. G. Bateson
Peter H. Klopfer

ADVICE TO POTENTIAL CONTRIBUTORS

When we assembled these articles, it was with the thought that this volume might become the first in a series. The need for theoretical and speculative treatment of developments in ethology cannot be satisfied by this single present collection of essays. Furthermore, we are conscious that the range and character of material published here were very much affected by our own contacts and knowledge. Obviously, the preparation of other volumes and the time of their appearance will depend in part on the response to the present one. Nevertheless, if potential contributors to the putative series have ideas which they would like to develop, we should be grateful if they contact one of us at the addresses given below for further advice and for guidance on the preparation of manuscripts. If we do proceed, it will be with the same understanding as before, that articles will appear within a fixed and reasonably short period of time after submission. If this volume does mark the start of a series, it is our intention that it be a series that stimulates and directs new lines of research rather than chronicles the past.

Dr. P. P. G. Bateson
Sub-Department of Animal
    Behaviour
High Street, Madingley
Cambridge CB3 8AA, England

Dr. Peter H. Klopfer
Department of Zoology
Duke University
Durham, North Carolina 27706
U.S.A.

CONTENTS

Chapter 5
DESCRIBING SEQUENCES OF BEHAVIOR
by P. J. B. Slater

Chapter 6
SPECIFIC AND NONSPECIFIC FACTORS IN THE CAUSATION OF BEHAVIOR
by John C. Fentress

Chapter 7

SOCIAL DISPLAYS AND THE RECOGNITION OF INDIVIDUALS
by M. J. A. Simpson

Chapter 8
DOES THE HOLISTIC STUDY OF BEHAVIOR HAVE A FUTURE?
by Keith Nelson

Chapter 1

NATURAL RESPONSES TO SCHEDULED REWARDS[1]

John Garcia
Department of Psychology
University of Utah
Salt Lake City, Utah

J. Christopher Clarke
Department of Psychology
McMaster University
Hamilton, Ontario

and

Walter G. Hankins
Department of Psychology
University of Utah
Salt Lake City, Utah

I. ABSTRACT

We have examined the quest for a reinforcement principle to explain learned laboratory behavior. Using the ant in a maze as an example, we point out that widely divergent species can be made to appear behaviorally similar within the confines of laboratory apparatus and reward schedules. However, as an explanatory principle, reward is circular, relative, reversible, and displaceable. Responses described as freely emitted operants can be as easily viewed as respondents evoked by reward. The coping behavior of an organism in the artificial (laboratory) niche can only be understood in terms of its behavior in its natural (evolutionary) niche and its specialized receptors and neural circuits. A unitary principle stressing information is more flexible than reinforcement but also falls short of explaining all laboratory behavior.

[1]This work was supported in part by USPHS, EC 00132.

1

II. THE QUEST FOR REINFORCEMENT

At the present time, two major hypotheses concerning the general action of reinforcement are at issue in the field of experimental animal behavior. On the one hand, we have the monistic classical doctrine of *stimulus substitution* formulated by Pavlov (1927). To use the classic paradigm, he postulated that an adequate or unconditioned stimulus (US) such as food in a dog's mouth evokes a reflexive consummatory response pattern due to relatively fixed neural circuits integrating the mouth receptors with the motor response patterns of eating. This is the given, or unconditioned, response (UR). When a signal or a conditioned stimulus (CS) such as a light is presented a number of times just before the food, it will on the first trials evoke a reflexive orienting response (OR) from the animal but not the consummatory response. On subsequent trials, the animal will begin to orient, lunge toward the light, and salivate copiously. This is the conditioned response (CR), reflecting new plastic neural connections between the CS and the US–UR complex.

Confusion arose because Pavlov selected as his index of conditioning the reliable and quantifiable glandular salivary response. This is a smooth muscle "involuntary" response. However, Pavlov made it clear that the CR is a complex behavioral pattern including skeletal or "voluntary" responses such as chewing, licking, and approach movements. After conditioning, this behavioral pattern is directed at the light CS *as if it were a substitute for the food.* A dog may be actually observed to lick the light and to gulp and chew the air in response to the CS. Pavlovian scholars went on to build a system which viewed more complex behavior as a synthesis of higher-order conditioning chains and secondary signaling systems with temporal contiguity as the principal associative law.

Philosophers, as well as physiologists, have wrestled with the unsatisfactory dichotomy of "involuntary," "determined," or *evoked reactions* as opposed to "voluntary," "free," or *emitted acts.* In psychology, Thorndike (1911) developed a system of learning where coping acts are emitted by trial and error, and *those acts leading to the goal are "stamped in" or reinforced by goal achievement;* but as Seligman (1970) points out, Thorndike's early formulations emphasized that *certain stimulus conditions and certain acts naturally belong together,* a notion similar to Pavlov's. It remained for Skinner (1938) to proclaim the monistic antithesis to Pavlov's concept. He distinguished between "respondents" or *evoked reactions,* which are unimportant as far as the modification of behavior is concerned, and "operants" or *emitted acts,* which are crucial to learning.

Skinner's paradigm was a hungry pigeon in a box emitting acts such as pecking at spots, walking in circles, and stretching its neck at various frequencies. If food is delivered immediately after any of these acts, its frequency or probability of occurrence is increased or "reinforced." According to Skinner (1948), this effect is so powerful that when food is arbitrarily delivered, the pigeon will "superstitiously" repeat what it happened to be doing prior to the delivery of food, as if it were a completely economic creature convinced that the food is the fruits of its labor. Skinner went on to build his system of operant conditioning on the notion that initially an organism is emitting responses rather freely from a universe of possible responses; however, *the immediate consequence of any emitted act affects the future probability of emission of that act.* Under controlled conditions in the laboratory it is only necessary to follow some particular variant of the emitted act to shape behavior into new forms which were never present in the history of the individual or the species. The original universe of responses could thus be expanded, reconstructed, or completely changed by "experimental selection" in a manner similar to the operation of "natural selection."

Mowrer (1960) consolidated the voluntary–involuntary breach into a dualistic system proposing that Pavlovian classical conditioning is the rule for establishment of autonomic responses with temporal contiguity as the associative law, while Thorndikian or Skinnerian instrumental conditioning is the rule for skeletal responses and reinforcement is the modification principle. In this system, the organism's motivational states—characterized by events such as diffuse smooth muscle responses, endocrine reactions, and cardiovascular changes elicited by appetitive and aversive stimuli—are involuntarily attached to the CS by temporal contiguity, but the organism's free coping behavior is modified by achievement of satisfaction or relief from annoyance. Mowrer (1963) has since gone on to explore the usefulness of sin and guilt as explanatory concepts for "more complex" human behavior.

In the 1940s and 1950s, Clark Hull (1943) and his associates at Yale, aided and abetted by Kenneth Spence (1956) and his associates at Iowa, attempted to formulate a theoretical statement of how reinforcement operates in the context of an elaborate "hypotheticodeductive" theory of learning. This theory was intended to be comprehensive enough to explain learning of *all organisms in all situations.* Although global theories of this scope are no longer in fashion, the core idea on which it was based still has many supporters.

Originally, the Hull–Spence treatment of reinforcement in learning concerned those substances which the animal requires to satisfy a biological need. Food, by this account, is reinforcement for the hungry animal because food deprivation brings about a physiological deficit threatening the survival of the animal. For the same reason, escape from a painful or dangerous situation is also reinforcing. This position was later modified somewhat, when emphasis was placed on *drive reduction* rather than *need reduction. Drive is the psychological concomitant of physiological need.* The drive reduction theory was stated forcefully and directly: no learning would occur unless (1) *the animal is in some need state which produces drive* and (2) *that response which is followed closely in time by reduction in drive is learned.* This seemed to account better for experimental data which showed animals able to learn responses followed by reward long before any redress in a tissue deficit could have occurred. Ultimately, under the influence of Spence (1947), the physiological specification of need was completely scrapped, but the drive reduction notion of reinforcement was retained.

Major difficulties with the Hull–Spence theory arose out of "sensory reinforcement" experiments, which demonstrated, for example, that rats kept in the dark would press a bar or turn a wheel for a brief period of light. Other rats kept in the light would learn the same response to get into the dark (Kish, 1966). Monkeys would perform a great amount of work just for the chance to view other monkeys or humans (Butler, 1953). Sensory drives, social drives, and exploratory drives were postulated until it was obvious that *drive, divested of its biological basis, is a response-inferred gratuity,* which reflects the myriad of conditions under which an animal will increase its response parameters.

A bold attempt was then made to rally psychologists around a theory of the modification of behavior based on the *empirical law of effect.* This law goes no further than to assert that *whatever follows a response and increases its strength is a reward.* Proponents were content to avoid all discussions of the *basic reinforcement mechanisms* and to deal with an "empty" organism. Given a deprived animal, such *rewards could be completely specified in terms of their physical parameters. A reward so specified should have transsituational generality.* That is, it should operate in many situations for that deprived animal (Meehl, 1950). Everyone could agree that for a food reward, a hungry rat would run a maze, or press a lever, or learn a variety of other tricks.

This attempt to specify reward in physicalistic terms ran afoul of empirical data which indicated that *reward has a relative value for the organism.* For example, Tinklepaugh (1928) trained monkeys to open a box by showing one a piece of lettuce being placed inside. Since the monkey repeatedly opened the box and ate the lettuce, lettuce was a reward. He then showed the monkey a banana being placed in the box, and, behind a screen, he replaced the banana with a piece of lettuce. On opening the box, the monkey became very agitated and refused the lettuce, as if, when *contrasted* to the banana, the lettuce had somehow lost its reward value. Recently, a number of *behavioral contrast* studies have indicated that birds given a reward of one magnitude for key pecking will show more pecking in that situation if the quality or magnitude of rewards in other situations is reduced. Since short-term satiation effects were ruled out, a case seems to be made for describing reward in relative terms. Neither the drive reduction theory nor the empirical notion with emphasis on the physical characteristics of reward can account for these behavioral contrast data.

A major attempt at synthesis not often recognized is inherent in the ethological system of Lorenz and Tinbergen. In contrast to the tightly controlled laboratory paradigms central to the Pavlovian and Skinnerian systems and their derivatives, the ethological paradigm is a field observation in a natural setting. Tinbergen (1951) provides an example of the male three-spined stickleback during its reproductive cycle. In response to the increasing length of day and increasing temperature, this fish migrates into shallow water and establishes a territory. Physiological changes provide a general reproductive instinct, or drive, which prepares the male fish to display a *hierarchy of behavior patterns:* fighting, nest building, mating, and care of offspring. Some of these behavioral patterns are evoked or *"released" by specific stimuli*, as fighting is evoked by a red-bellied male intruding into the fish's territory. Nest building is released by other features of the territory, and ultimately some behavioral patterns follow as a sequence of previous acts, as when paternal behavior follows mating.

The general factor in motivational states is evidenced by *displacement which occurs when one behavioral pattern to a releasing stimulus is blocked, and the animal displays another behavioral pattern to another stimulus feature of the immediate situation.* Each behavioral pattern also has connected with it a specific factor or *store of action-specific energy which accumulates as a function of endogenous factors.* When this accumulation is high, the animal may emit a response to a previously inadequate stimulus or *emit the response in a vacuum where no apparent releaser is present.* Thus the dichotomy between evoked reactions and emitted acts becomes blurred.

Lorenz (1966) stresses the evolutionary aspect of behavior, noting that the species accumulates information through natural selection in much the same way an individual accumulates information from experience. During evolution, variations in behavior and structure leading to survival and reproduction are incorporated in the genome and passed on to the progeny. Thus species-specific patterns are shaped by natural selection as operant behavior is shaped by reinforcement.

This emphasis on genetically controlled behavioral patterns evoked by releasing stimuli would appear to make ethology more compatible with the Pavlovian system than with the Skinnerian system, but the ethologists differentiate between *appetitive behavior* and *consummatory behavior.* The hawk displays adjustive tactics as it searches out and pursues a pigeon. This *appetitive component is shaped and directed by sign stimuli and is in sharp contrast to the consummatory behavior evoked by the releasing stimuli, as* when the hawk seizes, tears, and eats the pigeon in a stereotypical hawklike fashion. The appetitive component is clearly Skinnerian and the consummatory component clearly Pavlovian.

In this chapter, we will examine some recent experimental findings which make it difficult to accept either the Pavlovian or the Skinnerian hypothesis as complete accounts of how reinforcement works. In fact, these data place a burden on the dualistic synthesis of both notions and call for a more broadly based complex informational notion of behavior. We will examine experimental data which call into question any conditioning notion, or for that matter any hypothesis based upon a great deal of work with a few species in a few situations, which purports to account for behavior of all creatures in all environmental niches. We will try to show that conditioning conceptions are seriously embarrassed by experimental results demonstrating (1) that animals, wherever possible, act on the basis of information extracted from the immediate experimental session, in the context of what has gone on in previous sessions and what has gone on between sessions; (2) that coping behavior comes in biologically meaningful "chunks" guided by that information; (3) that an understanding of the natural niche in which the species evolved is necessary to explain the behavior in the artificial niche created by the experimenter; and (4) that ultimately any explanation of behavioral patterns must be verified in neurological terms.

III. THE ANT IN A MAZE

An elegant example of the blending of ethological and experimental behavioral techniques is provided by Robert Fleer's (1972) study of reversal learning in the ant. Fleer observed the behavioral patterns of ants for years, virtually making one species (*Pogomyrmex californicus* Buckley) his household pets before he began his experiments. He found these ants to be ideal subjects because they were relatively slow in movement and could not climb the walls of glass containers. Their tolerance of water deprivation proved to be remarkably similar to that of rats, and they drank in a specific consummatory pattern, licking at the rate of one or two licks per second. Since ants, like rats, are familiar with tunnels, Fleer decided that a T-maze was an appropriate test instrument.

These ants become very disturbed if their exoskeleton is wet, and they will cover water droplets in their nests with debris. Therefore, he watered his animals with care, using tiny wicks suspended from the ceiling of the goal box so that he would not inadvertently splash water on his subjects. The ants had to rear up to drink, and after a few trials they learned to clamp the wick with their jaws whenever Fleer tried to remove the wick. He found it necessary to encase the wick in a smooth plastic container to prevent the ants from gripping the wick effectively. In Fleer's experiments, he never imposed a trial on an ant when it appeared emotionally upset, that is, racing around with its posterior (gaster) raised, which indicates it is very upset, or moving its jaws rhythmically, which indicates that it is slightly upset. If it was grooming its antennae, or if it was investigating a part of its home cage, it was allowed to complete the ongoing chunk of behavior. If it was moderately active and appeared alert, Fleer gently tipped its plastic home cage so that the ant slid into the start box of a tiny T-maze. He eschewed the use of tweezers to pick the ant out of the end-box, using instead an emeryboard designed to polish fingernails, placing it in front of the ant and allowing it to climb aboard.

Fleer tested the plasticity of the appetitive behavior of the thirsty ant, reasoning that this desert species needed to forage for water in its natural habitat. In the beginning, he placed the damp wick in the first arm of the T-maze and gave his ants ten trials on the first day. If an ant chose the correct arm, it was allowed to lick for 15 sec. If the ant chose the wrong arm, it received no water on that trial because it was prevented from retracing its path by a blockade placed at the choice point. All the ants learned the correct choice on the first day; the "brightest" ant made the correct choice on the last seven trials, and the "dullest" ant made the correct choice on the last four trials.

On the second day, the damp wick was placed in the opposite arm of the maze. Most of the ants went to the first arm on the initial trials, indicating that they remembered the lesson from the previous day, but by the end of ten trials they had all relocated the moisture in the second arm of the maze. On the third day, the moisture was returned to the first arm for another ten trials. This alternating procedure was continued for each daily block of ten trials. On the seventh day, Fleer rested, since six out of ten ants began demonstrating one-trial reversals, indicating what in a primate would be called a "win–stay, lose–shift" hypothesis, or perhaps a learning set (see Fig. 1).

These are fanciful notions to apply to a social insect which is assumed to have little functional plasticity and which is said to achieve its behavioral ends via structural metamorphosis, differentiation, and specialization within its group of nestmates. Nevertheless, Fleer went on to demonstrate that the ant can master a simultaneous visual discrimination problem and its reversal, and that it will display the partial reinforcement effect—that is, it will respond longer under extinction conditions (i.e., with no reward) if it has been trained under conditions in which reward was probable rather than certain. The most cogent explanation of this effect is that extinction (no reward) trials are more informative to animals trained under 100% reward than to animals trained under partial (e.g., 50%) reward reinforcement. Furthermore, when a thirsty ant, trained in the T-maze with the wick on one side 70% of the time

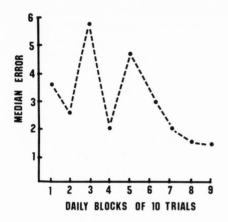

Fig. 1. Spatial reversal learning by thirsty ants in a T-maze. Water reward was alternated each day from one arm to the other; by day 7, individual ants were showing one-trial reversals. Note the variability caused by reversals on days 1–4 compared to the orderliness on days 6–9.

and 30% on the other side, was allowed to correct itself on each trial, this ant did not merely "match" its behavior to the environmental contingencies but "maximized" its chances by consistently choosing the 70% arm first. Such mastery of probability is said to be a characteristic restricted to higher species (Bitterman, 1965).

As Skinner (1959) has pointed out, behavior is remarkably lawful or consistent from species to species, once we make allowances for the particular way in which that species makes contact with its environment. He points out that given a fixed-interval schedule in which a lever press will deliver a reward, if and only if a constant temporal interval has elapsed from the last reward man, monkey, pigeon, and now probably the ant, will display the same behavioral pattern when excursions of a lever are the only aspects of behavior recorded. All species will press with increasing frequency as time passes since the last reward. The question for the student of behavior is whether it is fruitful to restrict to systematic formulations that narrow segment of the data, and relegate specific ways in which each individual animal makes its contact with the environment to the unwritten lore of the laboratory. The more fundamental laws may lie hidden in the way that men, monkeys, pigeons, and ants are constrained so that all species emit similar patterns of behavior.

IV. REWARD IS CIRCULAR

The nature of reward is as ephemeral as that of reinforcement. One of the best examples of this is the behavior of a rat in a shuttlebox where the animal is given a buzzing signal warning him that a footshock will be applied if he does not move from the first compartment he occupies into the alternative compartment. Usually, the rat is given 5 sec to move before he receives the shock. With repeated trials, the rat learns to heed the warning buzzer and avoids the shock by shuttling back and forth on cue. Normally, the buzzer turns off as the rat crosses the centerline separating the two halves of the box.

What rewards the rat's shuttling response? At first, it appears to be the electric shock which causes the rat to run, but shock does not always reinforce running; quite the opposite, many rats will freeze into immobility. If one observes a rat closely and delivers shock in gradually increasing intensity, a variety of meaningful behavioral chunks in response to shock emerge. At very low intensities, the rat is apt to pick up one paw, examine it, and lick it, a behavior more appropriate to a thorn in the paw than to an electric grid. This happens repeatedly, as the shock apparently stimulates a single sensitive spot on one paw of the rat as that paw comes into contact with the electrified grid, thus mimicking a thorn in the rat's paw.

At a higher shock intensity, one observes the running escape response that works well in the shuttlebox. The rat now acts as if the floor is hot, and his every move tends to confirm that notion. When he jerks his front paws up or leaps in the air, the shock ceases. When he contacts the floor again, he contacts the shock also and when he leaves the compartment he leaves the shock behind him. If the shock is very strong, then the rat freezes as if he has suffered a severe trauma, to which immobility is the appropriate response. Anyone who has suffered a severe electric shock can appreciate the seizure-like effect on the deep muscle and tendon receptors, which call for immobility in the case of wounds and broken bones.

Even when the shock is at the appropriate escape level, rats have a difficult time learning the "two-way avoidance" or shuttlebox problem as compared to the "one-way" avoidance task, because in the shuttlebox on the second trial the animal must seek safety in the compartment from which he has just escaped danger in the preceding trial. And so it goes: On odd-numbered trials, the first compartment is dangerous and the second compartment safe. On even-numbered trials, the opposite is true. This is a difficult problem, but most rats can learn the shuttlebox game. However, some do not. It seems that these "failures" do not formulate the appropriate rules for playing the shuttlebox game.

Even after many training trials, some rats will settle for an "escape" response; these rats will not shuttle until the shock occurs, as if they did not hear the buzzer. But if the buzzer sounds and the shock is not delivered, some of these "failure" rats will run as if they had been shocked while others will warily creep into the next compartment. It seems that these "failures" have formulated different hypotheses concerning the nature of the shuttlebox game. They act as if the shock itself shuttles from compartment to compartment. If the shock is not here, it must be there. The former time their run using buzzer duration as a precise temporal cue, while the latter need confirmation of the onset of shock as a cue to make them run. When the shock does not appear, they cautiously explore the shuttlebox as if they are trying to locate the shock. If another rat is placed in the compartment with the shocked rat, the latter is likely to attack this stimulus rat, as if attack behavior engendered by pain is released by this stimulus. In fact, rats will give up a successful avoidance habit to engage in this species-specific aggressive behavior, as if the stimulus rat is the cause of its cutaneous pain.

On the other hand, the animals which learn the shuttlebox game soon develop a smooth efficient shuttling response and then rarely, if ever, get shocked. This is the classic problem of avoidance learning. If the animal is no longer being rewarded by shock termination, why does he continue to shuttle, and how does the experimenter extinguish the shuttle habit? Katsev (1965) solved the problem. It is only necessary to take control of the buzzer away from the rat by making buzzer termination independent of the animal's shuttling response to make the animal give up the shuttling response. Katsev went one step further; he restored control of the buzzer to the rats after the extinction phase; that is, he turned on the buzzer and kept it on until the animals crossed the barrier. The animals again acquired the shuttling habit without further shocks. Now one is tempted to say that the buzzer is aversive due to its previous association with the shock, and buzzer termination is the reward. However, Berman and Katsev (1972) went on to show that an even more effective way to extinguish the shuttling response is to sound the buzzer and prevent the animal from making the response by inserting a barrier between the two compartments. Now the buzzer sounds and the animals are forced to remain in the compartment but are not shocked. Thus the rat has been given explicit information that the buzzer–shock rule has been abolished, so it gives up the shuttling response and makes little effort to terminate the buzzer.

As a number of people have pointed out, shock avoidance is difficult to extinguish because the animal has no information that shock has been turned off. Smith (unpublished) has set up the same sort of conditions for appetitive responses. In essence, he trains animals that if they press the lever in the presence of a signal they will receive food at the time when the signal terminates. When this habit is well established, he removes the response contingency and delivers food at the end of the signal regardless of whether the rat presses the bar or not. These conditions set up an appetitive habit which is as difficult to extinguish as the shock avoidance habit, and it is difficult to extinguish for the same reason, namely, because the animal receives no information concerning the transition from the acquisition game to the extinction game.

The real lesson of the shuttlebox game is that, as B.F. Ritchie has pointed out, there is an "incurable vagueness" in reinforcement theories which allows *post hoc* specification of reinforcement. Each attempt to specify the reinforcer in physicalistic terms fails because the *animal tends to extract information from the context of the entire acquisition situation* and to conform to whatever conditions the experimenter sets up. In the final analysis, *reinforcement theory will have to give way to information theory* which is flexible enough to handle the complexity of the changing flux of acquisition and extinction conditions.

V. REWARD IS RELATIVE

For years, conditioning theorists have attempted to pin down rewards in concrete terms of two coordinates, namely, the length of the deprivation period and the magnitude of the physical reward. However, some recalcitrant experimenters have continued to produce paradoxical data which force us to abandon these absolutist axioms in favor of less restrictive relativistic views of animals' coping behavior.

For example, let us now take the problem that Crespi (1942) designed for his rats and consider the problem his rats pose for conditioning theorists today. He trained rats to run down an alley for a food reward. Some rats always received a "small" reward, and some always received a "large" reward. After extended practice, running speeds stabilized for both groups. Crespi then switched reward conditions for half of each group, and, as a control, he maintained the remaining rats under their original reward conditions. The rats switched from small to large ran faster than those controls maintained under large reward. It was as if they were "elated," Crespi said. The animals switched from large to small ran slower than those controls maintained under small reward, as if they were "deflated."

Crespi's results indicated that animals can and do relate rewards in one condition to rewards received under other conditions. This view did not find much favor with conditioning theorists, for several reasons. First, since the days of John B. Watson, behaviorists have insisted on "objectivity" modeled after classical physics. They have insisted on defining their basic categories of stimulus, response, and reinforcement in physicalistic, quantitative terms. To do otherwise was to court charges of "mentalism" and "subjectivity."

Second, in keeping with the classical physics model, behavior was conceptualized as individual unitary responses causally related to immediately antecedent stimuli, as when the trajectory of a baseball is related to the impact with the bat, regardless of what happened on previous pitches. So to the question "What is learned?" these theorists reply "Responses to stimuli." They have long held that a reward can only directly effect a response if it follows the response closely in time—approximately 0.5 sec later and certainly less than 10 sec later. This view persists to the present day (Perkins, 1968).

A contrasting view was presented by theorists influenced by gestalt psychology, notably Karl S. Lashley (1951) and Edward C. Tolman (1949). The latter proposed that animals learned "means–ends relationships" or "what leads to what." Tolman's animals reacted to events in a context and learned about outcomes and environments. Recent data obtained with complex multiple schedules in the Skinner box support this molar view of behavior. A multiple schedule is an arrangement of two or more successively presented schedule components, each signaled by a different stimulus signifying the response requirements and the reward sequences of its given component. Each component is cycled independently of the animal's responses. Ferster and Skinner (1957) have published an encyclopedia on such schedules and argue that the components are usually quite independent of each other.

Studies by Reynolds (1961) challenge this notion of independence and indicate the operation of a "Crespi-like effect" between components. He trained his pigeons on a multiple schedule with two components. At first, both components were identical, except for their signals, S_1 and S_2. Each was a variable-interval 3-min schedule where food was delivered after a varying interval whose average duration was 3 min. Such a schedule typically leads to a high and steady rate of pecking. With each component of the multiple schedule providing the same frequency of reward, he obtained equivalent response rates from both components.

In the next stage, he continued the variable-interval schedule in one component (S_1) but changed the other (S_2) to extinction so that there were no rewards for key pecking. The response rate declined in the S_2 extinction component as expected. However, Reynolds noted an *increased* response rate in the other unchanged S_1 component, even though the rate and total number of rewards in S_1 *remained constant*. Reynolds referred to these response rate changes as "behavioral contrast" (Fig. 2).

There are earlier reports of contrast-like effects in the literature. What seems to be a classical conditioning analogue of behavioral contrast was discussed by Pavlov (1927). In his experiment, it was found that the magnitude of salivation was greatest if the positive stimulus ($S+$) was presented immediately after the negative stimulus ($S-$). Other early examples of contrast-like phenomena were reported by Crespi (1942) and Solomon (1943) in discrete trial situations. Contrast-like effects have been obtained with humans by O'Brien (1968), with rats by Lawson *et al.* (1968) as well as with pigeons, and with aversive stimuli as well as with food rewards. Last, the

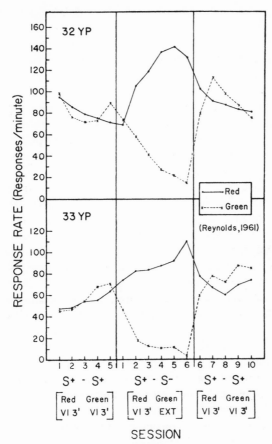

Fig. 2. Behavioral contrast obtained in two pigeons by Reynolds (1961) in a multiple schedule with two components signaled by red and green, respectively. In phase 1 (left), both components are identical. In phase 2 (center), reward is removed from the green component and pecking increases in red, although the rewards in red remain constant. In phase 3 (right), reward is reinstated in green so both components are identical. [By permission from Terrace in Honig (1966).]

contrast effect does not require extinction in one component. One component need only be changed to produce a lower rate of reward, and response rate will increase in the other component. Furthermore, if shock is added to one component, reducing response rate without affecting rate or total rewards delivered, the response rate will increase in the other component as if shock-free rewards are more "valuable."

These behavioral contrast studies pose great difficulties for the tradition-al view that responses are bound to stimuli by immediate reward and that associative strength generalizes to other stimuli in proportion to their similarity to the conditioned stimulus. An attempt to salvage the traditional view attributes the increase in responding to "frustration" generated by the failure of reward in the previous component (Scull *et al.*, 1970). It is reasoned that "behavioral contrast depends only on the immediately preceding trial situation [and] . . . reinforcement density or other factors related to a comparison of $S+$ and $S-$ would seem to be inappropriate as would, in-deed, the use of the word 'contrast' to identify the phenomenon." The issue is clearly drawn.

It may be plausible to describe the animal as "frustrated," but these theorists have previously used the frustration construct to explain response decrements (Dunham, 1968). Now it explains response increments. They cannot have it both ways. They must come to grips with the data which demonstrate that the very same schedule component of reward may be "attractive" or "unattractive" depending on the context of the multiple schedule in which it is presented (Clarke, 1972).

Behavioral contrast can operate over a 24-hr separation of the two components. Bloomfield (1967) presented daily 1-hr sessions of rewards signaled by one of two alternative visual stimuli until comparable response rates developed to each. Next, he extinguished S_1 and found that responses to S_2 increased even though the components were separated by 24 hr. If contrast operates over 24 hr, then it can hardly be attributed to the "im-mediately preceding" trials only; all preceding trials must have a contextual effect. Reynolds suggested the following relativistic conception of the conditions for contrast. The frequency of reinforcement in the presence of a given stimulus, *relative to "the frequency during all of the stimuli that suc-cessively control an animal's behavior,"* determines how the animal will react to that reward. This sounds much like Helson's (1964) adaptation level theory, which is usually applied to perceptual judgments.

The word "all" brings us squarely to the question of the limits or bound-ary conditions. Premack (1969) ran rats in a multiple schedule and, 16 hr later, ran the same rats on single schedule in a different environment. A decrease of reward in one component of the multiple schedule produced contrast, not in the other component, but in the single schedule 16 hr later in the other environmental setting. Premack asks whether any two cases of reward will interact, or whether there are limits in time and kind beyond which two experiments will not affect one another. Until we determine these limits, there will be serious methodological difficulties. Premack points out that the limits in time and kind are surprisingly broad.

We cannot suggest an answer to Premack's question; we merely point out that the same four pellets may "elate" a rat previously trained on one pellet or "deflate" a rat previously trained on eight. The notion of absolute magnitude of reward operating on the immediately preceding trials must give way to relative value of reward embedded in a broad context of the animal's recent history.

VI. REWARD IS REVERSIBLE

Behavioral contrast complicated the behavioral notion of reward, but not fatally, for in principle one can always scale rewards in relativistic terms. The empirical law of effect still survived. In an important and often cited paper, Meehl (1950) defended this law on the grounds that reinforcers had transsituational generality; that is, what reinforces behavior in one situation will do so in all others. If food will support bar pressing in a hungry rat, it will also maintain maze running and wheel turning in the rat under the same conditions of deprivation. Lists of stimuli could be drawn up, analyzed for common elements, and categorized as rewarding, neutral, and punishing. Such lists of palpable substances and stimuli would provide another referent for reward, thus saving theorists from response-inferred circularity of the following form: a reinforcer is an event that strengthens (or increases the probability of) a response on which it is contingent. When terms are substituted, this statement becomes the following: a strengthener is an event which strengthens a response. Only another referent independent of the behavioral endpoint prevents this tautology.

David Premack developed yet another system of reward relativity or reward reversibility, which poses overwhelming difficulties for the empirical law of effect. In a number of interesting and ingenious tests, he showed that the reinforcement relation between two events could be reversed. In one of his earliest studies (Premack, 1959), children were left in a room to engage in a number of activities as freely as they wished. Two of the available activities were eating small candies and playing a pinball machine. The amount of time each child spent in each of the activities was recorded. Some children spent more of their time eating candy than playing the pinball game; Premack called them the "eaters." Others spent most of their time with the pinball machine, and he called them "players." The "eaters" then had to play pinball in order to eat candy, and their pinball activity increased. The "players" had to eat candy in order to play pinball, and their eating activity increased. In a similar experiment, he demonstrated that water-deprived rats would run in an exercise wheel to obtain water and that exercise-deprived rats would drink water in order to run in the wheel (Premack, 1965).

Premack maintains that his paradigm can be stated as a general principle applicable to all animals in all situations, since animals engage in numerous activities each of which can be assigned a value. One index of this value is the proportion of time spent in a given activity, which is how Premack assigns response probability (p). If, for example, p values are determined for five activities and they are ordered in the hierarchy $a>b>c>d>e$, then a will reinforce any other activity when a is made contingent on that activity, but e will not reinforce any other activity. An arrangement where e follows and is contingent on a conforms to the punishment paradigm leading to a decrease of the probability of a.

Premack states the requirements that have to be met for his proposition to apply: (1) a situation without constraints during which p values can be assigned to each of the available activities and (2) a contingency situation in which the subject must conform to the response requirements in order to be allowed to perform a certain activity. In addition, Premack's principle has an indifference postulate: it is not how p values are generated but the values themselves that determine what will reward what and to what extent. For example, there are a number of ways to bring about a high p for drinking in rats. One can deprive the rats of water or lace the water with saccharin, but it is a matter of indifference just how the high p for drinking is produced. It will have the same incremental effect for a lower p activity such as lever pressing, when a contingency relationship is set up, i.e., when the lever must be pressed before the fluid is delivered for drinking. The value p is not necessarily fixed and immutable. If by some manipulation we altered the order of our five activities so that they came out to be $a>b>e>c>d$, then e could be used as a reward for c and d and as a punisher for a and b. Premack tested this proposition with monkeys in a free-response period where five manipulanda were available. When the favorite high-p activity was made contingent on a low-p activity, the value of the low-p activity was increased and thus it could be used as a reward for other activities which were previously preferred (Premack, 1963).

Premack used this finding as the basis for an attack on the transsituational assumption, a corollary of which is the division of stimuli and/or responses into three classes with fixed membership: rewarding, neutral, and punishing. Certainly, food has functioned as a reward for a variety of activities such as bar pressing, but only because we have fixed it so by keeping the animal from food while providing him with a bar for most of the time. Justifiably, Premack says this is no more the "natural" state of affairs than the reverse case where the so-called primary rewards (food and water) are continuously present while so-called neutral manipulanda are not. It is just as rational to deprive an animal of running and then force him to drink in order to run as it is to deprive him of water and then force him to run in order to drink. Premack has established the reversibility of rewarding activities without question.

Premack's view has received a great deal of attention lately. He has tested his system within a number of settings, with a number of species, and with remarkable success. Those committed to the "empty organism" find his indifference principle attractive, a principle which reflects Premack's own operant background. Different kinds of rewarding activities may be subserved by different neural systems, but we need not concern ourselves with their particular structure and function, we merely need to compute their p values. We need not concern ourselves with the environmental conditions attracting or repelling our subjects, we need only compute the p values of approach activity to each environmental feature. Finally, this view does justice to the animal's ability to weigh up, to relate, and to be responsive to events more distant in time than 0.5 sec. It can handle the behavioral contrast data because it deals with each activity in a broader context, all activities over hours.

There are some practical shortcomings to Premack's system. For one, p values for eating, drinking, and copulating are not stable but dependent in part on internal cycles. The attractiveness of a female waxes and wanes with estrus. All this can make the computation of the p values an enormous time-consuming chore if we ignore what we know about biological determinants of behavior, but there is a much more serious question. Can we assume that an animal's behavior in a free-access situation will be a reliable guide to his actions when constraints are imposed by making one activity contingent on the other? If we stress an animal with deprivation, or fear, or pain, will the hierarchy of p values shift in an orderly way or will the pattern change in ways that can only be foreseen if we understand the animal's individual history and the evolution of his species?

VII. REWARD IS DISPLACEABLE

Premack's data are undeniably and impressively consistent. Nevertheless, some recent work by John Falk and his co-workers requires, at the very least, some qualification of Premack's principle. These studies provide evidence of dynamic interactions between motivational systems that confound attempts to get at reward value in the manner advocated by Premack. Falk demonstrated that the amount of "extraneous" activity, and therefore its empirical probability value, changes dramatically when the deprived animal is working for an appropriate reward which is delivered intermittently, as if doing nothing while waiting is aversive.

An early study will suffice as an introduction to Falk's work. Falk (1969) deprived rats of food until they were at 80% of their *ad lib* weight. They were then trained to press a bar for a food reward. The animals could obtain food on the average of once every minute, a variable-interval 1-min schedule. Water was always available in both the home cage and the experimental chamber. The typical finding was that after receiving the food pellet, the animals would eat it and then drink about 0.5 ml of water. During the daily (3.17-hr) session, the animals drank an average of 92 ml of water. This was more than three times the pre-experimental water intake for an entire 24-hr period. In the home cages, the animals took less than 1 ml of water. There are two findings of note. First, the animals reduced and restricted their water intake between experimental sessions. Second, this strikingly high water intake, about one-half of their body weight in 3.17 hr, was much more than that which can be induced by other means such as water deprivation, heat stress, or osmotic-loading techniques.

Falk has carried out an extensive series of experiments which have allowed him to dismiss explanations based on feeding-induced dehydration. Hydration of the animals just (15 min) before the start of the daily session resulted in no more than a slight attenuation of the phenomenal drinking which Falk has labeled "psychogenic polydipsia." He also discovered that animals would press a bar as many as 50 times for water while responding on a variable-interval (1-min) schedule for food. The generality and importance of the polydipsia-like phenomenon have been demonstrated with other animals and arrangements. Falk mentions one experiment which showed that monkeys given *noncontingent* food once every 15–20 min ate large amounts of wood shavings after each food pellet. This has been termed "schedule-induced pica." Note also that the "dry-mouth theory" of polydipsia scarcely accounts for the pica discovery. Lester (1961) has been able to use the Falk arrangement to get rats to drink alcohol so as to cause long periods of inebriation.

"Superstitious conditioning" has been offered as a possible explanation. Such conditioning presumably occurs when an animal's emitted response is accidentally followed by reinforcement. Skinner (1948) has described stereotypical responses in pigeons on an intermittent schedule where reinforcement is independent of the birds' responding. But this stereotypical behavior is probably not due to accidental conditioning; rather, it may be an effect similar to Falk's. It is more likely that the pigeon is first temporally conditioned to the intermittent schedule and then begins to display stereotypical behavioral patterns during the waiting period. A man does not pace the maternity ward corridors because his wife produced a baby the last time he paced there, or because pacing brings the baby faster; he does it because he has to wait there until the baby comes. Doing nothing while waiting may be as aversive for pigeons as it is for man.

The polydipsic phenomenon has attracted a good deal of attention of late. Staddon and Simmelhag (1971) consider it to be one of two basic categories of behavior: "interim activities" (such adjunctive behavior as polydipsia, pica, and schedule-induced aggression) and "terminal responses" (those consummatory behaviors that occur just before and at the time of reward delivery). Segal *et al.* (1965) and Hinde (1966) provide similar formulations.

Falk's effect creates major difficulties for the Premack principle. Let us imagine an animal maintained in its home cage with free access to water but no food for 20 hr. We then place this animal in a box with food, water, a lever, and a bar available for an hour. Such an animal would distribute his time as follows: eating most of the time (a), drinking a little (b), pressing the lever a little (c), and pressing the bar a little (d). A Premack formulation of probabilities would be $a > b = c = d$. Now if we made eating (a) contingent on lever pressing (c) on an intermittent schedule and drinking (b) dependent on bar pressing (d), then Premack would predict that lever pressing would increase ($a > c$) but that bar pressing would not ($b = d$). Falk's data indicate that both lever pressing and bar pressing would increase markedly, an effect unforeseen by Premack. It is obvious that a "static" formulation like Premack's is of little *practical* help in predicting such interactions and is unable in *principle* to account for these data.

A more intelligible approach has been suggested by Staddon and Simmelhag (1971): "Evolution is notoriously opportunistic in the sense that adaptation is achieved by whatever structural or functional means happen to be available . . . we suggest that the means for insuring that the animal will not linger in the vicinity of food (or other reinforcers) at times when it is not available may be provided by the facilitation of drives other than the blocked one." At times of nonreward, they postulate the simultaneous suppression of (say) the food state and the facilitation of other states associated with other rewards. In this respect, their view resembles that put forward by Konorski (1967).

Hinde (1966) writes: "when mutual incompatibility prevents the appearance of those types of behavior which would otherwise have the highest priority, patterns which would otherwise have been suppressed are permitted to appear." The critical requirement for the emergence of such "adjunctive" behavior is that the animal be prevented from leaving the situation, with its resulting block of a strong drive, i.e., escape. The displacement activity studied by ethologists arises out of similar situations where one activity is blocked. Strict conditioning hypotheses seem to fall short of adequacy in dealing with reward, while ignoring the particulars of the situation, the nature of the beast, and the evolutionary forces at work. Premack limits his attention to probability values and in doing so ignores these vital considerations, vitiating his otherwise very elegant and powerful system. *We have to be concerned with the organization of behavior and the nervous system.* This was once the focus of the neobehaviorist systems of Karl Lashley (1951) and Donald Hebb (1949). It must be readmitted as a guiding theme in the study of animal behavior.

VIII. ARBITRARY OPERANTS OR DIRECTED RESPONDENTS

The specific ways in which the pigeon makes contact with its Skinner box environment have been examined in detail with videotape recordings, high-speed closeup photographs, and continuous recordings of contact force on the key by Herbert Jenkins and Bruce Moore (1972). Their data are both intrinsically interesting and theoretically important because they indicate that the pigeon's key-pecking behavior, which appeared to be an instrumentally conditioned operant, may also be viewed as a classical conditioned response. Does the pigeon really peck at the key because pecking is instrumentally reinforced or because the key (CS) is paired with food (US) and pecking (CR) is the appropriate consummatory response (UR)?

A variant of Skinner's paradigm requires illumination of the key as a signal indicating that if the key is depressed it will deliver a food reward. Under these conditions, the pigeon's behavior can be shaped by rewarding successive approximations to key pecks until the pigeon actually pecks at the lighted key and delivers the food by its own response. Soon after, the pigeon will quickly achieve a high stable key-pecking rate to the lighted key and a low rate to the unlighted key. Brown and Jenkins (1968) made a simple alteration in this experimental setup. They removed the control of food delivery from the peck at the lighted key and presented the key light and food in temporal sequence regardless of the pigeon's response. In effect, they changed the Skinner box into a Pavlov chamber. The pigeon then shaped its own pecking at the lighted key in a surprisingly lawful way, apparently pecking at the lighted key as if it were food. The phenomenon was named "autoshaping."

Jenkins and Moore (1972) took the next necessary step when they demonstrated that the topography of the autoshaped response is related to the consummatory pattern. Food-deprived birds reinforced with grain pecked at the key rapidly with a crisp, forceful series of pecks as if they were pecking grain. Water-deprived birds reinforced with water depressed the key less rapidly and less forcefully, all the while opening and closing their beaks rhythmically and licking with their tongues, as if they were drinking by "pumping" water, as pigeons do. Birds trained to peck at one signal for food when hungry and another signal for water when thirsty delivered "grain pecks" and "water pumps" appropriately. The form of the response depended on the reinforcer rather than the dominant drive state; thirsty birds responded to the food signal with grain pecks, not water pumps, and the converse was also true.

The autoshaped response poses these problems: for the Skinnerian system, the autoshaped response appears to be a Pavlovian evoked reaction, not an arbitrary emitted act; for Mowrer's two-factor theory, the skeletal muscle (voluntary) response of pecking is classically conditioned as if it were a smooth muscle (involuntary) response; for information theory, the pigeon acquires this response as if it had information concerning the relation of the signal to the reinforcer, but acquires a response which controls neither the signal nor the reinforcer. It should be recalled that in the shuttlebox problem, when control of the signal was taken away from the rat, the shuttling response was extinguished.

For Skinnerians, the challenge is clear: *they must demonstrate that a given reward can be used to increase the probability of a response which is not part of the consummatory pattern to that reward.* Howard Rachlin (1969) responded by conditioning pigeons to peck to escape shock. However, the issue is not resolved, because Rachlin's birds batted the key with their wings and pecked at it with their beaks. Wing bats and pecks are part of the pigeon's defense patterns. Food rewards do not ordinarily increase wing batting. A dissociation experiment where wing bats and pecks are orthogonally opposed to shock and food would probably demonstrate that learning does not occur equally under all four conditions.

The Skinner box is a strange world for a pigeon in search of grain to eat. In its natural niche, the bird searches visually for grain, identifies the grain, pecks at it, and swallows the reinforcer all in one pattern meaningfully directed to grain on the ground. Herbert Jenkins says that the artificial niche of the Skinner box pulls apart the pigeon's integrated grain-pecking pattern, separating the occasion to peck, the target to peck, and the reinforcer to peck. The key light comes on, signaling the occasion to eat before grain arrives. The bird orients to the light and delivers its peck to the lighted key, but the bird finds it difficult if not impossible to watch a light on one side of the box which signals it to peck at a key on the opposite side. If given two targets on the same key, it will peck at a smooth disc and avoid a pointy star. Its peck is made with the beak opening precisely wide enough to grasp the disc (Jenkins and Sainsbury, 1969; Moore, 1972) (see Fig. 3).

Let us accept the notion that much of behavior, like anatomical structure, has been shaped to some consummatory survival purpose by natural selection in the species' history. Within these biological constraints, behavior has some plasticity, but we would do well to ask, as Keller Breland was wont to ask, "What is this bird trying to do?" When he shaped the pigeon's

Fig. 3. The pigeon, when rewarded with food, demonstrates a precise (grain peck) consummatory response to the disc but avoids the star, although any peck delivered within the large circle has the same outcome. When water is presented, another (water pump) response is manifested. The reward need not be contingent on the peck for the pecking to precede its delivery. [Reconstructed from Moore (1972), Jenkins and Moore (1972), and Jenkins and Sainsbury (1969).]

neck-stretching act with food, Breland surmised that the bird was trying to fly away but the ceiling of the Skinner box restricted its flight pattern. When the ceiling was removed, the neck stretching was followed by flight. This anecdote, which was related to us by Hardy Wilcoxon (1971), indicates that acts which are part of one biological chunk may be brought into the service of another purpose if the artificial environment restricts the original sequence. This anecdote also raises the question of which consummatory pattern is being reinforced, food getting or flying away? Experimental analyses in terms of biologically meaningful behavioral chunks may elucidate the problem of "instinctual drift" raised by the Brelands (1961, 1966) in which instrumentally conditioned behavior of well-trained animals ultimately tends to break down in the direction of species-specific behavior. Pigs give up dropping their tokens in the slot and begin rooting them, raccoons wash their tokens, and other animals try to eat theirs. Bruce Moore cites apparatus specifications which prevent biting and maximize lever excursions, thus making the rat's behavior more Skinnerian and less Pavlovian. A more useful approach would be to analyze the behavior of animals into the plastic appetitive component and the fixed consummatory response and the antecedent stimulus conditions into sign stimuli and releasing stimuli, respectively, as ethologists do.

IX. THE COPING ORGANISM AND THE ARTIFICIAL NICHE

By virtue of its long evolutionary history, each species fits its natural niche like a key in a lock. While the antecedents of evolution are lost in history, the results of natural selection are evident in the natural niche and in the structure of the organism. Field studies of the species and niches and neurological studies of integrative circuits connecting receptor inputs to effector outputs are basic to an understanding of what is happening to an animal coping with the "artificial niche" of the experiment. This artificial niche serves to test the animal's plasticity, to stretch it, to mold it into new patterns. The natural niche provides the zero point from which the extent of plasticity is measured. *Ultimately, any hypothesis concerning this plasticity must be verified in the neurophysiological structure of the animal.*

The rat in a field of X-rays is in a strange artificial environment. The radiant energy floods his internal fluid milieu as well as his external world, producing chemical changes everywhere, but the structure of the animal's receptor systems imposes specificity on the diffuse physical stimulus. Chemical changes at the nasal mucosa are transduced into a "novel odor," eliciting arousal and orienting responses in the rat. Chemical changes at the human retina are transduced into a "greenish glow." Chemical breakdown products circulating in the internal milieu cause rats to become listless and ill (Garcia and Ewin, 1968b; Garcia and Koelling, 1971; Garcia et al., 1967).

It is this radiation illness which poses a special artificial problem for the rat, since in his natural niche he has never coped with a "place" (i.e., a locus in space) which made him sick or, more properly, made his ancestors sick. In his natural niche, the rat is an omnivorous exploiter of foodstuffs, making him the ubiquitous companion of man all over the earth. Apparent food sources can turn out to be nutritious or poisonous when tested, and the rat possesses an expert testing system for food (Rozin and Kalat, 1971). He eats a small amount of the novel substance on the first meal, and if its effects are nutritious he eats more next time. If it is poisonous, he avoids eating it again. Since the effects of food are delayed, he must have a mechanism which combines the flavor of food in the mouth with visceral effects over long flavor–illness intervals (Revusky and Garcia, 1970). Furthermore, when he next encounters the flavor, he must reject it, so the most effective mechanism is one which reduces the palatability of the tainted food: the temporal information yielded by the ingestion sequence is of little value. In other words, the animal acts as if the tainted food now "tastes bad," not as if he learned that "this flavor will make me ill an hour from now." Thus his very adaptation to his natural niche makes him maladapted to the artificial niche of the radiation field, because he responds more readily to radiation sickness by dietary selection than by motor movements which would take him out of the radiation field.

In contrast, consider the defensive reactions by which the rat copes with threats in the external world. When he hears a noise, as that of a predator, he orients in the appropriate direction; the predator closes in and attacks his cutaneous surface, causing pain and escape behavior. The spatiotemporal information of this attack sequence is important, since it gives the rat a vector direction on the predator which is useful in escape maneuvers. In contrast to flavor–illness, with noise–pain pairing the rat reveals that he has acquired temporal information from the conditioning sequence, for if the shock is delayed he will adjust his defensive responses to the onset of the shock—not to the onset of the buzzer. The rat is relatively better adapted to cope with buzzer–shock except that shock delivers pain disembodied from stimulus features naturally accompanying an attack. If another rat is placed in the shock box, then the shocked rat is apt to give up his conditioned avoidance and attack the other rat in the natural "consummatory" manner of response to peripheral pain.

The double dissociation experimental design (see Fig. 4) which opposes two cues, flavor *vs.* noise, against two consequences, pain *vs.* nausea, under a fourfold set of conditions is the best way to illustrate the bias of the rat to form some associations rather than others. Several of these experiments have been conducted, but the recent one by Domjan and Wilson (1972) is the most elegant. They presented the flavor to water-satiated rats, flushing the oral cavity with sweet (saccharin) water via an oral cannula. Under these conditions, the rats did not drink; thus the sweet cue could be imposed on the

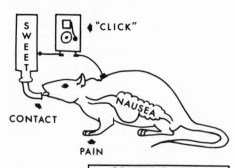

		CONSEQUENCES	
		NAUSEA	PAIN
CUES	SWEET	ACQUIRES AVERSION	DOES NOT
	"CLICK"	DOES NOT	LEARNS DEFENSE

Fig. 4. The double dissociation design where two cues are pitted against two reinforcers. Conditioning is not obtained freely in all four cells, indicating the biased associative mechanism of the rat.

rat independent of approach and choice behavior, similar to the imposition of the noise cue via an electric buzzer. They imposed illness treatment by lithium chloride injection and the pain treatment by implanted dorsal skin electrode immediately following a 35-sec presentation of the CS. After three conditioning trials, they gave the animals two tests: one was a choice between drinking noise-water or tapwater presented simultaneously, the other a choice between sweet-water or tapwater. They conducted the experiment in two ways: noise and sweet presented in combination to two groups in conjunction with illness or pain, and noise and sweet independently presented to four groups. The results were virtually identical for both experiments (see Fig. 5), indicating that the relevant cue was not overshadowing the irrelevant cue. The pain animals avoided noise-water but not sweet-water, while the illness animals avoided sweet-water but not noise-water. They revealed an associative bias more appropriate to their natural niche than to the artificial one created in the laboratory.

These dissociation data appear to contradict the Pavlovian hypothesis that any stimulus can become a conditional stimulus. Pavlov (1928) explicitly stated, "any natural phenomenon chosen at will may be converted into a conditioned stimulus. . . . Any visual stimulus, any desire sound, any odor, and stimulation of any part of the skin whether by mechanical means or by the application of heat or cold . . . ." Pavlov (1930) was also very specific concerning the conditioned response to the signal. He wrote, "If this phenomenon precedes the act of eating, once or several times, it will later provoke a food reaction; it will become, so to speak, a surrogate for food . . . the animal moves towards it and may even take it into its mouth if the object is tangible." This issue can be stated pragmatically. Can flavor be used to signal shock? Can place cues be used to signal illness? The answer to both questions is yes. However, the topography of the conditioned response is of interest.

Let us examine what was learned when thirsty rats were placed in a gray shuttlebox with an electric grid floor and with sweet-water at one end and saltwater at the other end (Garcia *et al.*, 1970). One flavor was shocked within 2 sec of the rat's first lick if he did not leave the compartment, while the other fluid was safe and the rat was allowed to drink for 2 min. The two fluids were switched unsystematically so that the flavor of the fluid in the bottle was the only cue to the shock.

Ultimately, the rat's learned to use flavor to avoid shock. They approached the waterspout cautiously, stretched their necks, and sampled the fluid. If it was the shock fluid, they turned and went to the opposite compartment and repeated their sampling performance. If the fluid was safe, they stayed and licked for 2 min. Taste in this case acts precisely as does a noise signal. The animal's defensive reactions to shock (US) are now evoked by taste (CS) in a Pavlovian manner. But the flavor of the shock fluid does

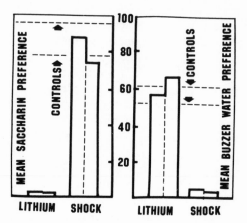

Fig. 5. Sweet-water and noise-water preference of rats conditioned with a sweet flavor and noise during drinking was paired with the nausea of lithium illness and the pain of electric shock in the double dissociation design (see Fig. 4). The left side of the bar was obtained by two groups (illness and pain) receiving both noise and flavor, while the right side was obtained with four independent groups. [From Domjan and Wilson (1972).]

not become distasteful or aversive; the same rats accepted it readily when it was presented in the home cage (see Fig. 6).

When rats were given a 3-sec drink of a very sweet (7 g/liter saccharin) solution to drink followed immediately by shock, they learned to taste and sniff the waterspout and refuse to drink during the 3-sec presentation. However, if the 3-sec drink was followed by a 30-sec interval before the shock was presented, the rats continued to drink relatively undisturbed even though the sweet taste must have lingered in their mouths when shock was delivered (Garcia et al., 1972). As before, the taste (CS) when paired with shock (US) operated as does a noise signal. The onset of the taste was the important feature of the CS, and duration added little to the effectiveness of that CS. The CS–US interval measured in seconds from onset to onset is a critical parameter, and when that interval was extended to 30 sec the defensive reactions to shock did not disturb the 3-sec drink of the shock flavor. The shock flavor, again, was not aversive.

Illness induced by X-rays has been used instead of shock in the shuttle-box apparatus, but the schedule and procedure differed because of the differences in the nature of the reinforcing stimulus. Two distinctive compartments differing in brightness and floor structure were used. In one experiment, rats were penned in one side for 12.5 min and 25 r was delivered in the final 2.5 min. The animals were removed immediately after exposure. On the following day, the animals were penned in the opposite compartment and exposed to 0 r (sham exposure). This treatment was repeated eight times with 2–4 days between X-ray trials. On the test day, the animals were placed in the shuttlebox without exposure and the barrier between the two compartments was removed. The animals were given a free choice of residence for 2 hr. They did not show the "fear" of the X-ray compartment which is so characteristic of shocked animals; rather, they explored both compartments but "chose" to spend most of the time in the sham compartment (Garcia *et al.*, 1961).

Since radiation sickness does not occur until an hour or two after exposure, this avoidance is rather remarkable, since it indicates that the rat can make a place–illness association over a CS–US interval of an hour or more. In some unpublished research (Garcia *et al.*, 1959), this experiment was repeated except that after exposure in one compartment the animals were immediately sham-exposed in the opposite compartment. No conditioning occurred, presumably because the delayed effects of X-rays followed both sets of place cues. Place avoidance due to illness is not a robust effect when compared to flavor–illness aversions. Such place avoidance requires approximately ten times the exposure and conditioning trials and even then extinguishes much more rapidly than a conditioned flavor aversion.

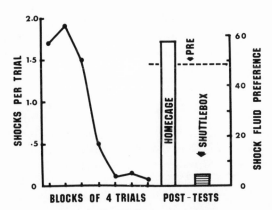

Fig. 6. The graph (left) illustrates mean shock per trial as rats learn to use a flavor cue to avoid shock in a shuttlebox. The bars (right) indicate the rats' postconditioning preference for the shock fluid in the shuttlebox where flavor–shock pairings occurred and for the same fluid in the home cage where they had never been shocked. [By permission from Garcia *et al.* (1970).]

The conditioned aversive behavior of the rat is not in keeping with a strict Pavlovian stimulus substitution hypothesis. For example, the aversive reactions (CR) to saccharin (CS) do not resemble the behavior (UR) of a rat to lithium illness (US), which is to say that *the conditioned response does not resemble the unconditioned response*. Lithium illness produces an inactive flaccid animal which lies down most of the time, but the animal reacts to the aversive saccharin flavor by jerking his head back, grooming, and rubbing his chin on the floor. He is active and vigorous and in no way appears listless and ill. Furthermore, in the X-ray shuttlebox the rat ultimately learns to move out of the X-ray compartment. In fact, other experiments (Garcia and Ervin, 1968a) have demonstrated that hungry and thirsty rats will eat and drink in the radiation compartment; thus their conditioned reactions do not resemble the behavior of a sick animal. Rather, the motor movements out of the radiation compartment reflect an appropriate reaction to place cues associated with illness. The rat integrates all the information and responds accordingly, revealing enough plasticity as a species to evolve, perhaps, an efficient motor avoidance of "hot spots" in a radiation-contaminated world of the future.

X. NEEDS, RECEPTORS, AND NEURAL CIRCUITS

Drive and drive stimuli are considered inadequate or superfluous concepts in operant behaviorism (Skinner, 1950), yet when these behaviorists set out to conduct an experiment they deprive their animals of food and/or water, keeping them at 80% of normal weight. They habituate animals to schedule and apparatus, since laboratory lore has taught them that laboratory animals are apt to spend their time exploring irrelevant features of their experimental space instead of coping with lever and $S\Delta$. When the animals are ready to eat or drink, they reward appropriate responses with food and/or water. When animals are satiated, the session is necessarily over because animals will begin exploring again and then go to sleep. Operant behaviorists behave like drive theorists—but they do not discuss drives.

The problem with "drive" is that it is conceptualized in a rather odd way as if it were a general factor to be specified by an informal factor analysis. For example, Miller (1959) insists that we seek commonality in several sets of behavioral tests before we specify hunger, but consider the morass which we would enter if we set out to specify all the conditions under which animals will and will not eat and then perform a genuine factor analysis.

Alternatively, others have assumed drive to be a general state of tension under which drive stimuli goad the animal into responding and learning. This conception was "destroyed" by experiments which showed that, for example, hungry rats would learn to run to a bottle containing saccharin-water with no food value to reduce hunger, thirsty rats would lick at streams of cool air which contained no water to reduce their thirst, and sexy male rats would learn to run to be near females even when copulation was impossible. If anything, the conditioned responses in all these cases resulted in increased tension, yet these conditioned responses were difficult if not impossible to extinguish given that sweetness, streaming air, and a receptive female, respectively, were provided even if consumation was blocked, increasing drive rather than reducing it.

Drive can best be conceptualized as an empirical fact. When we deprive an adult rat of water for a day, we drastically alter his internal environment, changing, for example, the tonicity of his body fluids. Internal receptors sensitive to osmotic pressures report to central integrative mechanisms, and the rat attempts to cope with the problem. If we mimic water by allowing a cool stream of air from a spout to play on his peripheral mouth receptors, the rat responds in the appropriate consummatory fashion. An ethologist would say that given high drive then even this impoverished stimulus releases the consummatory pattern.

We can force an animal into a state of thirst by injecting him with hypertonic saline, or we can "fool" him into a state of thirst by washing the surface of his osmoreceptors with minute amounts of hypertonic saline. We can mimic satiation by blocking the report of internal osmoreceptors or washing their surface with minute amounts of hypotonic fluid. In all these cases, we are manipulating the information concerning drive and reward given to the coping animal. *The appropriate way to specify drive is to point to the internal conditions, to the internal monitoring receptors, and to the peripheral releasing stimuli for the consummatory pattern.*

The heuristic value of a neurological hypotheses is exemplified in the differential role of odor and flavor in eating behavior. Odor seems to be as basic to food intake regulation as taste. Common sense proclaims that flavor of food is 90% odor: a head cold blocks our sense of smell, rendering our food tasteless. The olfactory receptors are chemical analyzers strategically placed for sniffing of food before it is taken into the mouth. Both man and rat utilize the olfactory sense to test foodstuffs, but the neurological projections of gustation and olfaction differ markedly.

Garcia and Ervin (1968b) distinguished gustatory–visceral (e.g., taste–illness) conditioning from telereceptive–cutaneous (e.g., noise–pain) in the rat on neurological grounds. The taste afferents project directly to the nucleus solitarius in the brain stem as do the internal afferents from the viscera. In addition, this structure is intimately related to the area postrema approximately 1 mm away, an area which contains receptors monitoring the internal milieu. The nucleus solitarius could serve as an "and-gate" integrating taste and visceral information before transmitting it forward to higher centers. On the other hand, auditory afferents project via several relays to the posterior thalamus, and cutaneous afferents converge to the same area. Auditory and cutaneous stimuli produce responses from the same units in that area, suggesting the existence of an and-gate integrating auditory and cutaneous information.

The olfactory system poses a problem, since it clearly does not project directly to the solitarius; rather, it has rich projections to the limbic system, particularly the amygdala. This neural complex has been dubbed the "nose brain." In our laboratories, Hankins (1972) paired peppermint odor with illness under conditions where the odor emanated from the locus of the waterspout. Thirsty rats displayed neophobia to the odor, approaching the waterspout and carefully and sporadically testing and drinking, indicating that odor was a strong salient stimulus. However, an odor–illness aversion was established only when illness followed odor immediately and repeatedly. Four odor-immediate illness trials were required to produce an aversion comparable in degree to a single saccharin–illness trial where the illness treatment was delayed for 30 min. When the odor–illness interval was extended to 10 min the aversion was greatly attenuated, and when the interval was extended to 30 min no conditioning was apparent. In contrast, taste–illness aversions are readily established with illness delayed several hours or more. These results indicated that odor cues alone are relatively ineffective in food–illness aversions.

Hankins then rendered other rats anosmic and tested them in the usual taste–illness paradigm. Rats made anosmic by topical application of zinc sulfate were slightly, but not significantly, superior to intact animals in acquisition of taste–illness aversions, indicating that odor does not play a necessary role in food aversions. Other experiments with nonodorous saccharin revealed that the mucosally anosmic rats were slightly, but not significantly, superior to intact controls in learning the aversion. In fact, anosmic animals were also slightly superior to intact animals in noise–shock conditioned suppression, as if deprivation of olfactory stimulation enhances learning via other modalities.

Gustation plays the unique role in adjusting palatability according to visceral feedback, thus behaviorally controlling the internal environment, as Richter (1943) taught us some time ago. Green and Garcia (1971) demonstrated that the same illness can be used to either enhance or diminish the palatability of a given flavor. They employed an apomorphine injection to produce a sudden acute illness of an hour or two followed by a rapid recuperation. When a flavor was presented prior to illness, it became aversive as if it were poison. When the same flavor was presented after illness but prior to recuperation, its value was enhanced as if it were medicine. Thus we do not have to postulate a separate receptor system for each essential nutrient (e.g., each vitamin) and its flavor; all that is required is the capacity to discriminate flavor and to sense general malaise and recovery in order to modulate diet appropriately.

The olfactory system operates as a telereceptor, perhaps more comparable to vision than to gustation. With immediate reinforcement and repeated trials, odors may be associated with illness, but we cannot be certain that this weak effect is not mediated via odorous molecules going into solution on the moist surface of the tongue. The rat can use odor to avoid tasting poisoned food, after taste–illness pairings have made the flavor aversive, but it is clear that the animal learns that odor cues will lead to an aversive taste, as if by peripheral means (rather than by central integration) gustation guides olfaction so that food which tastes aversive eventually comes to smell aversive.

Peripheral guidance is an important integrative function within the development of an infant animal, the recovery of a brain-damaged animal, the coordination of a split-brain animal, and the reciprocal relation of an infant to parent. Phillip Teitelbaum (1971) has recently reviewed the striking parallel between the development of feeding in infants and the recovery of feeding in rats with lateral hypothalamic damage. Both infants and brain-damaged animals respond reflexively in consummatory fashion to releasing stimuli. Like an infant, the hypothalamic animal eats when milk or wet palatable food is placed in his mouth but refuses both dry food and water as if he requires a super-releasing stimulus for consummatory behavior and as if his appetitive behavior has been completely abolished. The hypothalamic rat at first cannot regulate food intake to caloric need nor water intake to dehydration. Gradually under the nursing care of the experimenter he will eat dry food, but he will drink water only when suffering from a dry mouth due to eating dry food. He will not drink in response to dehydration. The sequence of recovery is always the same, as the hypothalamic animal is guided by peripheral means until he learns feeding patterns necessary for survival.

The infant displays a similar development. At first, the mother must guide the infant, presenting the hairless nipple to the nuzzling infant to release the sucking response. Before weaning, the caloric and hydration controls of the infant are poor and are under maternal guidance; that is, the maternal behavior patterns pace and meter the requirements of the infant. Gradually the infant develops appropriate control in the same sequence as the hypothalamic adult does, ultimately accepting dry food and water. Bennet Galef and his associates (Galef and Clark, 1971, 1972) have suggested two ways by which parental guidance works in the rat. First, the diet of the mother imparts a characteristic flavor to mother's milk and through gustatory–visceral conditioning the nursing pups acquire a preferential taste for the same diet as the mother. Second, the weanling pups are attracted to the external feeding site by the physical presence of the mother. Gradually a rat pup acquires his own appetitive tactics through telereceptive conditioning and achieves consummation independent of the mother.

Peripherally guided integration is often perceived as an "artifact" by those who searched for central transmission of the memory engram in split-brain animals. I. Steele Russell (1971) reviewed this problem. One hemisphere of the brain can be rendered temporarily decorticate or closed by application of potassium chloride, and the animal can be trained with only one hemisphere operating or open. When the untrained one is opened by recovery and both hemispheres are operating together for weeks with midline commissures intact, the acquired information will not spontaneously be transmitted from the trained hemisphere to the naive one. At least one more acquisition trial is needed with both hemispheres open, in which the trained hemisphere "teaches" the naive one. Coordinated bilateral receptor and response mechanisms operating in tandem make the trained hemisphere a superb teacher, so that much less training is required now than in the original training. While this process does not appear as elegant as transmission of an engram via the midline commissures, it is sufficient to coordinate even a split-brain human, who does not appear to miss the corpus callosum when bilateral peripheral receptor systems are allowed to operate in tandem (Gazzaniga, 1970).

Differential central integration has been demonstrated with the dissociation experiment by Brenda McGowan et al. (1969, 1972), who compared the effects of limbic lesions on noise–pain and taste–illness conditioning. Lesions of the lateral septum or the ventral hippocampus disrupted the adaptive acquisition of fear to an external noise signaling a cutaneous painful insult but enhanced the adaptive acquisition of an aversion to a flavor preceding an internal malaise. These lesions produce a rat which is finicky about food, hyperactive, and angry to external stimuli yet very slow to interrupt drinking in order to defend against external threats. Medial septal rats are more sensitive than intact controls to internal contingencies of flavor–illness, but they are also able to cope with noise–pain contingencies. Amygdaloid lesions produce animals which are unable either to utilize external signals to defend against pain or to reject flavors leading to illness.

It is much easier to make central lesions which disrupt external (noise–shock) control and spare internal (flavor–illness) control than to make lesions which have the converse effect. In fact, many lesions which disrupt external conditioning actually enhance internal conditioning as if internal control is the more basic function and as if intact "higher systems" may actually inhibit internal integrative functions. On the other hand, amygdaloid lesions have a devastating effect on both internal and external integration. The amygdaloid animal seems responsive enough to both sets of cues and consequences, but he does not seem to be plastic enough to associate either set effectively in order to form new fears or new disgust reactions.

XI. SUMMING UP: THE ORGANISM–INFORMATION APPROACH

In this discussion, we have confined ourselves, in the main, to laboratory studies and looked at "reward" because we believe, as Robert Bolles (1967) has written, that the *empirical law of effect* has a particular appeal for psychologists, not only because it is crucial to control of the animal in the laboratory, but also because "there is always the hope that it may be derived as a consequence of some still more basic and more powerful theoretical proposition. The search for the basis of reinforcement becomes all the more pressing as we clear away the motivational debris and find just how central a position the empirical law of reinforcement has in the explanation of behavior."

The problem with the empirical law of effect is that it does not remain empirical very long. For example, as long as we deal with a food-deprived animal and reward him with food, we can empirically and precisely specify the reward in grams per kilogram of body weight, independent of the animal's behavior. But the same animal, when not deprived, will learn to cope with a laboratory situation for "nothing." Now, the opportunity to manipulate, to explore, or to merely observe is labeled a reward, reflecting the assumption that if learning has occurred there must have been some reward, even though it cannot be empirically specified.

We have examined the two major positions on reinforcement, namely, the Pavlovian stimulus substitution through temporal contiguity hypothesis and the Skinnerian response probability increase through reinforcement hypothesis. These two positions have so dominated our attention that we have lost sight of other possibilities which have existed for years. Bolles (1967) writes, "Hobhouse (1901) attributed learning to the confirmation and the lack of inhibition of whatever behavior the individual undertook. Any response which carried through all the existing motor tendencies without leading to their disruption was learned, according to Hobhouse. A similar position was advanced by Holmes (1911) who suggested that learned responses were simply parts of some large activity in progress, and that the part would be learned to the extent that it is congruous or compatible with the total activity. Peterson (1916) proposed that any response which was undertaken would be learned if it could be carried out to completion."

The common theme in this alternative is that behavior is a *patterned sequence* beginning with *arousal and orientation*, passing through a *coping appetitive phase*, and terminating in a *consummatory phase*. If coping behavior is blocked before consummation, or if consummation is delayed, then the segments of the original sequence may be *displaced*, with adjunctive behaviors directed at features of the environment which have *demand or incentive* characteristics. For example, the hungry rat waiting for his intermittent food reward drinks water because it is there. The hungry pigeon walks in circles because it must wait within the confines of the Skinner box.

Many of the contradictions are clarified when we specify which segment of this extended behavioral sequence we are collecting data in. In the arousal and orienting phase, we employ the habituation paradigm of "no reward." We merely present a novel stimulus and record the magnitude of some index of the orienting response. With repeated trials, this magnitude diminishes, indicating that the animal "remembers" even if these trials are spaced hours apart (i.e., he has learned something from previous trials). If we now change the stimulus, the magnitude of the orienting response is increased, indicating that the animal has "discriminated" the stimulus change. If we continue to change the stimulus features, the animal continues to observe as if we are rewarding him with information.

When habituation occurs, the reinforcement theorist has to say that the animal learns to not respond because "nothing of consequence" follows the stimulus in a consistent fashion. This reveals again his persistent faith that contiguous events within each S–S or S–R or R–US trial can explain all that happens in learning. The following fact is ignored. The animal must be able to remember what went on in the previous trials on the previous days, thus indicating the capacity to associate events over hours.

Sensing the futility of dealing with the habituation paradigm, the reinforcement theorist dismisses habituation as a nonassociative process. He habituates his animals thoroughly and deprives them so that he gathers his data in the appetitive phase. In the appetitive phase, the animal is searching, guided by sign stimuli received via his telereceptors. Sign stimuli so perceived are neither attractive nor aversive, nor do they usually acquire such properties; they are merely informative, possessing some probabilistic relation to consummation. Sign stimuli are perceived in a context, for a sudden silence is as informative as a sudden noise.

During the appetitive phase, Tolman's (1949) description is most apt. The animal's behavior seems purposeful and insightful as he learns means–ends relationships. The responses toward sign stimuli are flexible. The thirsty ant or the thirsty rat will learn to turn right or to turn left in order to obtain water. The animal reads the complex environmental field and acts, not with random probes, but as Krechevsky (1932) postulated, with a hypothesis inherent in his evolved structure. His receptors act like filters, accepting only a narrow segment of a given energy spectrum. Integrative structures operate as does an electronic window, responding only to patterned sequences. The sick rat acts as if "it must have been something I ate," because he has been wired that way. Within that constraint, the rat has enough plasticity to learn which flavor led to the illness. In his natural niche, *plasticity within constraint* is more efficient than a random net capable of associating any combination of stimuli on the basis of intensity, recency, frequency, and effect.

Reinforcement, when stated abstractly as reward or punishment, loses the informative properties which direct the animal's hypotheses. The particular nature of illness forces the retrieval of gustatory information stored in the neural memory mechanism hours before. The particular nature of cutaneous pain forces the animal to retrieve more recently stored memories from telereceptive domains. When returned to the same situation on the next day, or the next week, he retrieves experiential information stored from earlier trials and modified by subsequent tests.

During the consummatory phase, the nature of the reward imposes a much closer constraint on the topography of behavior. The animal acts toward those sign stimuli which lie in close spatiotemporal proximity to reward in a consummatory way, as Jenkens and Moore have demonstrated (see Fig. 5). By our account, if a well-trained pigeon is placed on a fixed-interval schedule with response-contingent reinforcement it should deliver tentative "test pecks" at the key minutes before reward delivery. The test peck should be topographically similar for either food or water, but as the reward draws near in time the form of "test peck" should change into either a "grain peck" or a "water pump," depending on the nature of the reward.

If we must organize our conception of behavior around a unitary theme, let it be information, not reinforcement. The organism can be viewed as an information-seeking entity operating in a complex environmental field. When its homeostatic processes are at equilibrium, when it is not in need, the organism is not necessarily quiescent. *It is apt to be curious and seek information from its surroundings because natural selection has favored the animals which continuously incorporated information. When it has finished, it may then be quiescent, but in time it will again explore restlessly the world that changes over time. Viewed this way, curiosity, play, and exploration are not "needless" phenomena but the basic motive of the organism.*

When the animal is deprived, he is informed of the problem with which he must cope by internal receptors, as when the osmoreceptors report hypertonicity. Because this information is the result of specific sensors operating via labeled lines, the animal can be fooled by cannulae or electrodes that alter conditions at the sensing head or tap the labeled line. The animal's appetitive behavior ends and consummation begins when peripheral receptors contact the substance needed to restore his homeostatic need, as when a cool fluid plays on the mouth receptors. Here again, the animal can be fooled, as when a stream of cool dry air mimics the sensations of water because of specialized sensors and labeled lines.

Neal Miller (1963) once wrote that traditional conditioning theories were adequate "in predicting stupid behavior but much less convincing in predicting intelligent behavior." Nowhere does this statement apply better than to reactions to sign stimuli in the appetitive portion of the extended behavioral sequence. It is here that the conditioning hypothesis can be juxtaposed against the information hypothesis.

When a signal such as a tone has preceded a reward such as food during conditioning, the rat will continue to press longer during extinction if the tone is presented. Since he is now pressing for the tone without food, the conditioning theorist assumes that the tone has acquired conditioned reward properties. Similarly, if a neutral tone is followed by a punishing shock it will later suppress behavior as if it is conditioned punishment. An information theorist would say that the tone in both cases has gained information value, and since information is useful it is not punishing. Lockard (1963) taught rats to enter one of two arms of a T-maze. In one arm they received a shock (primary punishment), and in the other arm they received a tone preceding the shock (conditioned punishment plus primary punishment). A conditioning theorist would have to predict that shock alone would be preferred to the conditioned punishment plus shock. An information theorist would predict a preference for signaled shock, because the signal divides the interval into safe and shock periods, while unsignaled shock produces a constant state of ambiguity. Lockard's rats chose signaled shock.

Weiss (1972) compared the effects of signaled shock *vs.* unsignaled shock on ulceration produced in rats implanted with a tail electrode. The animals receiving unsignaled shock displayed four times the ulceration, as if the informative signal protected the animals from the stress of unexpected shock (see Fig. 7).

Parallel results were achieved with signaled rewards. Schuster (1969) gave pigeons a choice between two keys which delivered rewards at the same rate when pecked. On one key a light always preceded the reward, while on the other side the light was presented many more times. The pigeons preferred the key where the light signaled reward, not the side which presented more lights and the same amount of food. Still more damaging for conditioning notions are studies which show that birds will work to turn on a light which may signal the unavailability of food. Hendry's (1969) birds pecked at a white key for intermittent food rewards. Next to the food key was a white information key which, when pecked, would turn either (1) red, signaling that food was available at the food key, or (2) green, signaling

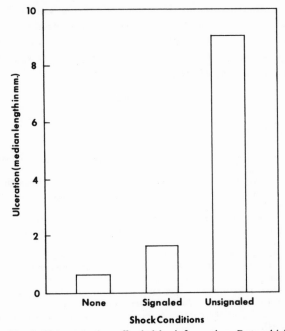

Fig. 7. The protection afforded by information. Rats which are given an auditory signal prior to unavoidable shock suffer much less ulceration than those which receive un-signaled shock delivered randomly.

that food was not available at the food key. The birds could work at the food key only and receive all the available reward; nevertheless, the birds learned to peck at the key which would inform them of where they were on the intermittent schedule. From a conditioning point of view, the birds did not get a primary reward from the information key and received conditioned punishment as well as conditioned reward, so, at best, that key should have been neutral. Clearly, the information hypothesis is more flexible and covers more situations than the reinforcement hypothesis.

Finally, let us offer a word of caution against the use of information theory to explain all behavior. Adjunctive or "displaced" behavior places a burden on both the information and the reinforcement hypotheses, since its antecedents are completely outside the acquisition variables and can be predicted only from a thorough understanding of species-specific hierarchical behavior patterns and an understanding of the demand characteristics of the environment incidental to acquisition. The phobic patient does not merely fear those things which lie along an orderly gradient from that which causes him pain and anxiety; he fears high places, open spaces, strange people, hairy animals, creeping bugs, and slithering snakes. So do other primates. Rationally, we deem these fears inappropriate, because these things have never actually caused him pain nor do they inform him about the source of his pain. Intuitively, we recognize them as appropriate because we, too, are human (Seligman, 1971).

XII. REFERENCES

Berman, J. S., and Katsev, R. D. (1972). Factors involved in the rapid elimination of avoidance behavior. *Behav. Res. Therap.* (in press).

Bitterman, M. E. (1965). Phyletic differences in learning. *Am. Psychologist* **20**:396–410.

Bloomfield, T. M. (1967). Some temporal properties of behavioral contrast. *J. Exptl. Anal. Behav.* **10**:159–164.

Bolles, R. (1967). *Theories of Maturation,* Harper and Row, New York.

Breland, K., and Breland, M. (1961). The misbehavior of organisms. *Am. Psychologist* **16**:661–664.

Breland, K., and Breland, M. (1966). *Animal Behavior,* Macmillan, New York.

Brown, P. L., and Jenkins, H. M. (1968). Auto-shaping of the pigeon's key peck. *J. Exptl. Anal. Behav.* **11**:1–8.

Butler, R. A. (1953). Discrimination learning by rhesus monkeys to visual-exploration motivation. *J. Compa. Physiol. Psychol.* **46**:95–98.

Clarke, J. C. (1972) Behavioral contrast: Response suppression, relative rate of reward, or value? (in press).

Crespi, L. (1942). Quantitative variation of incentive and performance in the white rat. *Am. J. Psychol.* **55**:167–517.

Domjan, M., and Wilson, N. E. (1972). Specificity of cue to consequence in aversion learning in the rat. *Psychon. Sci.* **26**:143–145.

Dunham, D. J. (1968). Contrasted conditions of reinforcement: A selective critique. *Psychol. Bull.* **69**:295–315.

Falk, J. L. (1969). Conditions producing psychogenic polydipsia in animals. *Ann. N. Y. Acad. Sci.* **157**:569, 593.

Ferster, C. B., and Skinner, B. F. (1957). *Schedules of Reinforcement,* Appleton-Century-Crofts, New York.

Fleer, R. (1972). Some behavioral observations on the ant with special reference to habit reversal learning. Doctoral dissertation, State University of New York at Stony Brook, New York.

Galef, B. G., and Clark, M. M. (1971). Social factors in the poison avoidance and feeding behavior of wild and domesticated rat pups. *J. Comp. Physiol. Psychol.* **75**:341–357.

Galef, B. G., and Clark, M. M. (1972). Mother's milk and adult presence: Two factors determining initial dietary selection by weaning rats. *J. Comp. Physiol. Psychol.* **78**:220–225.

Garcia, J., and Ervin, F. R. (1968a). Appetites, aversions and addictions: A model for visceral memory. In Wortis, J. (ed.), *Advances in Biological Psychiatry,* Plenum Press, New York.

Garcia, J., and Ervin, F. R. (1968b). Gustatory–visceral and telereceptor–cutaneous conditioning—adaptation in internal and external milieus. *Commun. Behav. Biol.* **1**:389–415.

Garcia, J., Ervin, F. R., and Koelling, R. A. (1967). Toxicity of serum from irradiated donors. *Nature* **213**:682–683.

Garcia, J., Hawkins, W., Robinson, J., and Vogt, J. (1972). Bait-shyness: Tests of CS–US mediation. *Physiol. Behav.* (in press).

Garcia, J., Kimeldorf, D. J., and Hunt, E. L. (1959). Unpublished research, U.S. Naval Radiological Defense Laboratory.

Garcia, J., Kimeldorf, D. J., and Hunt, E. L. (1961). The use of ionizing radiation as a motivating stimulus. *Psychol. Rev.* **68**:383.

Garcia, J., and Koelling, R. A. (1971). The use of ionizing rays as a mammalian olfactory stimulus. In Beidler, L. M. (ed.), *Chemical Senses (Handbook of Sensory Physiology,* Vol. IV), Springer-Verlag, Berlin.

Garcia, J., Kovner, R., and Green, K. F. (1970). Cue properties *vs.* palatability of flavors in avoidance learning. *Psychon. Sci.* **20**:313–314.

Gazzaniga, M. S. (1970). *The Bisected Brain,* Appleton-Century-Crofts, New York.

Green, K. F., and Garcia, J. (1971). Recuperation from illness: Flavor enhancement for rats. *Science* **173**:749–751.

Hankins, W. G. (1972). Sensory control of bait-shyness. Doctoral thesis, State University of New York at Stony Brook, New York.

Hebb, D. O. (1949). *The Organization of Behavior,* Wiley, New York.

Helson, H. (1964). *Adaptation-Level Theory,* Harper and Row, New York.

Hendry, D. P. (1969). Reinforcing value of information: Fixed-ration schedules. In Hendry, D. P. (ed.), *Conditioned Reinforcement,* Dorsey Press, Homewood, Ill.

Hinde, R. A. (1966). *Animal Behavior: A Synthesis of Ethology and Comparative Psychology,* McGraw-Hill, New York.

Honig, W. K. (ed.) (1966). *Operant Behavior: Areas of Research and Application,* Appleton-Century-Crofts, New York.

Hull, C. L. (1943). *Principles of Behavior,* Appleton-Century-Crofts, New York.

Jenkins, H. M., and Moore, B. R. (1972). *The Form of the Auto-Shaped Response with Food or Water Reinforcers* (in press).

Jenkins, H. M., and Sainsbury, R. S. (1969). The development of stimulus control through differential. In Mackintosh, N. J., and Honig, W. K. (eds.), *Fundamental Issues in Associative Learning,* Dalhousie University Press, Halifax.

Katsev, R. D. (1965). The extinction of avoidance responses. Unpublished doctoral dissertation, University of California, Berkeley.

Kish, G. B. (1966). Sensory reinforcement. In Honig, W. K. (ed.), *Operant Behavior: Areas of Research and Application,* Appleton-Century-Crofts, New York.

Konorski, J. (1967). *Integrative Activity of the Brain: An Interdisciplinary Approach,* University of Chicago Press, Chicago.

Krechevsky, I. (1932). "Hypothesis" in rats. *Psychol. Rev.* **39**:516–532.

Lashley, K. S. (1951). The problem of serial order in behavior. In *Cerebral Mechanisms in Behavior: The Hixon Symposium,* Wiley, New York.

Lawson, R., Mattis, P. R., and Pear, J. J. (1968). Summation of response rates to discriminative stimuli associated with qualitatively different reinforcers. *J. Exptl. Anal. Behav.* **11**:561–568.

Lester, D. (1961). Self-maintenance of intoxication in the rat. *Quart. J. Studies on Alcohol* **22**:223–231.

Lockard, J. S. (1963). Choice of a warning signal or no warning signal in an unavoidable shock situation. *J. Comp. Physiol. Psychol.* **56**:526–530.

Lorenz, K. L. (1966). *Evolution and Modification of Behavior,* Methuen, London.

McGowan, B. K., Garcia, J., Ervin, F. R., and Schwartz, J. (1969). Effects of septal lesions on bait-shyness in the rat. *Physiol. Behav.* **4**:907–909.

McGowan, B. K., Hankins, W., and Garcia, J. (1972). Limbic lesions and control of the internal and external environment. *Behav. Biol.* **7**:841–852.

Meehl, P. E. (1950). On the circularity of the law of effect. *Psychol. Bull.* **47**:52–75.

Miller, N. E. (1959). Liberalization of basic S-R concepts: Extensions to conflict behavior, motivation, and social learning. In Koch, S. (ed.), *Psychology: A Study of Science,* Vol. 11, McGraw-Hill, New York.

Miller, N. E. (1963). Some reflections on the law of effect produce a new alternative to drive reduction. In Jones, M. R. (ed.), *Nebraska Symposium on Motivation,* University of Nebraska Press, Lincoln. Neb.

Moore, B. R. (1972). The role of directed Pavlovian reactions in simple instrumental learning in the pigeon. In Hinde, R. A., and Hinde, J. S. (eds.), *Constraints on Learning* (in press).

Mowrer, O. H. (1960). *Learning Theory and Behavior*, Wiley, New York.

Mowrer, O. H. (1963). *The New Group Therapy*, Van Nostrand, Princeton, N. J.

O'Brien, F. (1968). Sequential contrast effects with human subjects. *J. Exptl. Anal. Behav.* 11:537–542.

Pavlov, I. P. (1927). *Conditioned Reflexes,* G. V. Anrep (trans.), Oxford University Press, London.

Pavlov, I. P. (1928). *Lectures on Conditioned Reflexes,* International Publishers, New York.

Pavlov, I. P. (1930). A brief outline of the higher nervous activity. In Murchison, C. (ed.), *Psychologies of 1930,* Clark University Press, Worcester, Mass., pp. 207–220.

Perkins, C. C., Jr. (1968). An analysis of the concept of reinforcement. *Psychol. Rev.* 75: 155–172.

Premack, D. (1959). Toward empirical behavior laws. 1. Positive reinforcement. *Psychol. Rev.* 66:219–233.

Premack, D. (1963). Rate differential reinforcement in monkey manipulation. *J. Exptl. Anal. Behav.* 6:81–89.

Premack, D. (1965). Reinforcement theory. In Levine, D. (ed.), *Nebraska Symposium on Motivation,* University of Nebraska Press, Lincoln, Neb., pp. 123–180.

Premack, D. (1969). On some boundary conditions of contrast. In Tapp, J. (ed.), *Reinforcement and Behavior,* Academic Press, New York.

Rachlin, H. (1969). Auto-shaping of key pecking in pigeons with negative reinforcement. *J. Exptl. Anal. Behav.* 12:521–531.

Revusky, S., and Garcia, J. (1970). Learned associations over long delays. In Bower, G., and Spence, J. T. (eds.), Vol. 4, Academic Press, New York.

Reynolds, G. S. (1961). Behavioral contrast. *J. Exptl. Anal. Behav.* 4:107–117.

Richter, C. P. (1943). Total self-regulatory functions in animals and human beings. *Harvey Lect. Ser.* 38:63–103.

Ritchie, B. F. An incurable vagueness in psychological theories. Mimeograph, University of California.

Rozin, P., and Kalat, J. W. (1971). Specific hungers and poison avoidance as adaptive specializations of learning. *Psychol. Rev.* 78:459–486.

Russell, I. S. (1971). Neurological basis of complex learning. *Brit. Med. Bull.* 27:278–285.

Schuster, R. H. (1969). A functional analysis of conditioned reinforcement. In Hendry, D. P. (ed.), *Conditioned Reinforcement,* Dorsey Press, Homewood, Ill.

Scull, J., Davies, K., and Amsel, A. (1970). Behavioral contrast and frustration effect in multiple and mixed schedules in the rat. *J. Comp. Physiol. Psychol.* 71:478–483.

Segal, E., Oden, D. L., and Deadwyler, S. A. (1965). Determinants of polydipsia: IV. Free reinforcement schedules. *Psychon. Sci.* 3:11–12.

Seligman, M. E. P. (1970). On the generality of the laws of learning. *Psychol. Rev.* 77:406–418.

Seligman, M. E. P. (1971). Phobias and preparedness. *Behavior Therapy* 2:307–320.

Skinner, B. F. (1938). *The Behavior of Organisms,* Appleton-Century-Crofts, New York.

Skinner, B. F. (1948). "Superstition" in the pigeon. *J. Exptl. Psychol.* 38:168–172.

Skinner, B. F. (1950). Are theories of learning necessary? *Psychol. Rev.* 57:193–216.

Skinner, B. F. (1959). A case history in scientific method. In Koch, S. (ed.), *Psychology—A Study of a Science,* McGraw-Hill, New York, p. 2.

Smith, M. (1970). Unpublished research, State University of New York at Stony Brook, New York.

Solomon, R. L. (1943). Latency of response as a measure of learning in a "single-door" discrimination. *Am. J. Psychol.* 56:422–432.

Spence, K. W. (1947). The role of secondary reinforcement in delayed reward learning. *Psychol. Rev.* 54:1–8.

Spence, K. W. (1956). *Behavior Theory and Conditioning,* Yale University Press, New Haven.

Staddon, J. E. R., and Simmelhag, V. L. (1971). The "superstition" experiment: A re-examination of its implications for the principles of adaptive behavior. *Psychol. Rev.* **78:** 3–43.

Teitelbaum, P. (1971). The encephalization of hunger. In *Progress in Physiological Psychology,* Vol. 4, Academic Press, New York, pp. 319–350.

Thorndike, E. L. (1911). *Animal Intelligence,* Macmillan, New York.

Tinbergen, N. (1951). *The Study of Instinct,* Clarendon Press, Oxford.

Tinklepaugh, O. L. (1928). An experimental study of representative factors in monkeys. *J. Comp. Psychol.* **8:**197–236.

Tolman, E. C. (1949). There is more than one kind of learning. *Psychol. Rev.* **56:**144–155.

Weiss, J. M. (1972). Psychological factors in stress and disease. *Sci. Am.* **226:**104–113.

Wilcoxon, H. (1971). Personal communication, George Peabody College, Nashville, Tenn.

Chapter 2

IMITATION: A REVIEW AND CRITIQUE[1]

J. Michael Davis

Department of Psychology
Duke University
Durham, North Carolina

I. ABSTRACT

Imitation in nonprimates is reviewed along theoretical and experimental lines. Difficulties with definitions and classifications of imitative phenomena are noted. The history of the topic is considered in terms of three basic paradigms: (1) early comparative psychology, (2) S–R/reinforcement learning theory, and (3) classical ethology. The current status of work on bird vocalization ("vocal imitation" and "vocal mimicry"), "social facilitation," and "observational learning" is evaluated. It is concluded that an inductive approach is needed in the study of imitative behavior, starting with the selection of reliably observed, naturally occurring imitative phenomena, followed by appropriate experimental analyses of the various determinants of each phenomenon.

II. INTRODUCTION

Voltaire's admonition to define one's terms seems particularly appropriate when one speaks of "imitation," for the term has been applied to a variety of behavioral phenomena. It should be noted immediately, therefore

[1]Support was provided by a Public Health Service (NIMH) predoctoral fellowship to the author and by National Science Foundation and National Institute of Mental Health grants to Dr. J. E. R. Staddon.

that the word is used in this chapter in a very general sense to refer to a broad heterogeneous class of social behavioral phenomena (in nonprimates), whose defining characteristic is, roughly speaking, some sort of similarity in behavior among two or more individuals. Under this broad heading there have, of course, been distinctions made, such as W. H. Thorpe's (1963) distinction between so-called *true imitation*, by which is meant "the copying of a novel or otherwise improbable act or utterance, or some act for which there is clearly no instinctive tendency" (p. 135), and *social facilitation*, which "can be described as 'contagious behaviour,' where the performance of a more or less instinctive pattern of behaviour by one will tend to act as a releaser for the same behaviour in another or in others" (p. 133). True imitation, according to Thorpe, "apparently involves self-consciousness and something of intent to profit by another's experience" (p. 135).

This distinction is not original with Thorpe, dating back at least to Lloyd Morgan's (1896) differentiation between "instinctive" and "voluntary," or intelligent, imitation. And, as might be expected, it is not the only classification that has been made of imitative behavior. Some of the various terms which have been associated with imitation, in addition to the word itself with sundry modifiers, and social facilitation, already mentioned, include "allelomimetic behavior," "contagion," "copying," "empathic learning," "identification," "local enhancement," "matched-dependent behavior," "mimesis," "mutual mimicry," "observational learning," "secondary conditioning," "suggestion," and "vicarious learning."

One superficial impression created by this multiplicity of terms, perhaps, is that the topic of imitation can be broken down into a number of precise categories. Actually, this is far from being true. Apart from the obvious difficulties in defining terms by using equally vague concepts or processes (such as "instinct," "self-consciousness," and "volition"), there are often more subtle forms of question begging in the use of the above terms. For example, a number of studies purporting to demonstrate observational learning, or "true imitation," have frequently been dismissed as merely the result of "local enhancement" (Roberts, 1941; Thorpe, 1963) or "stimulus enhancement" (Spence, 1937). While it may be true that the probability of a correct response can be increased through a model's "directing the animal's attention to a particular object or to a particular part of the environment" (Thorpe, 1963, p. 134), this would seem to raise the equally important question of *how* it is that one organism's attention is (sometimes) directed to, say, a particular manipulandum by observing another organism or its behavior. In other words, the question of what, exactly, determines the observing animal's response in these situations remains unanswered.

In any case, it should be recognized that the fundamental concern in regard to terminology, definitions, and classification schemes has been to separate the different possible types of causal processes underlying various imitative phenomena. (See Foss, 1965, and Gilmore, 1968, for two of the more comprehensive analytical discussions along this line.) The problem, however, is that the concern for identifying the various causal processes has basically been of an "armchair" nature—speculative rather than investigative, *a priori* rather than *a posteriori*. To be sure, the former sort of activity is quite legitimate, even necessary, as part of the scientific method, but only insofar as it leads to empirical knowledge of the actual determinants of imitative behavior. Given the present lack of such knowledge, any more ado about finding "the" proper classification scheme or "the" most precise terminology would clearly be as unproductive as it has been in the past.

It follows that the use of terms such as "social facilitation" and "observational learning" in this chapter makes no implications concerning their validity or meaningfulness. They are admittedly labels for ignorance and should never be reified or represented as anything other than vague abstractions. Their employment here is simply for reasons of convenience and popularity of usage.

III. HISTORY

The history of work on imitation may be viewed in terms of three basic paradigms or characteristic approaches: (1) early comparative psychology, (2) S–R/reinforcement learning theory, and (3) classical ethology.

A. Early Comparative Psychology

During the late 1800s and early 1900s, there were a number of anecdotal reports, naturalistic observations, and even experimental studies dealing with imitation in animals. The fundamental issue for comparative psychology was the notion of the continuity of mind, and the problem of imitation was one important aspect of the study of animal intelligence (Boring, 1950). Although *instinctive* imitation (roughly equivalent to social facilitation), like instincts in general, was widely acknowledged without question (indeed, was viewed as an *answer* to questions about the cause of behavior), the existence of a higher-order *intelligent*, or inferential, imitation was frequently the focal point of the continuity-of-mind dispute.

Romanes' (1885) affirmative stand on the question of intelligent imitation was founded on a collection of anecdotal reports and naturalistic observations (including some of his predecessor, Darwin). Later this position was given some experimental support by the work of Berry (1906, 1908)

with rats and cats, Porter (1910) with various species of birds, and Kinnaman (1902) with rhesus monkeys. But the negative experimental findings
of Small (1900) with rats, Thorndike (1898, 1901) with chicks, cats, and dogs,
and Watson (1908) with monkeys made for one of those controversies typical
of the history of psychology: the issue was never really resolved; it just died
of irrelevancy. The behavioristic revolt dismissed "mind," "consciousness,"
"ideation," and similar notions, and thereby knocked many of the props out
from under the problem of inferential imitation. In addition, the growing
realization that "instincts" were rather vacuous as explanatory concepts,
and the negative reaction which followed, resulted in a tendency to ignore
even instinctive imitation.

One may tend to think of this early period as little more than a historical
relic because of the well-known problems with anecdotal reports and a
number of difficulties with experimental design, which need not be reviewed
here. But it still offers at least one lesson for contemporary researchers by
illustrating the pitfalls of framing a problem in terms of preconceived notions.
Imitative phenomena must be considered as problems worthy of study in
their own right, apart from their possible implications for current theoretical
issues, regardless of whether the issue be continuity of mind or reinforcement
as a necessary factor in learning.

B. S–R/Reinforcement Learning Theory

Interest in research on imitation was generally at a low ebb during the
early behavioristic period. It was not until Miller and Dollard wrote *Social
Learning and Imitation* (1941) that behaviorists paid much attention to the
topic. Miller and Dollard's analysis of imitation was based specifically on
Hullian learning theory, but their position is representative of the entire
"S–R/reinforcement" paradigm.[2] According to this view, the organism first
makes some random response in the presence of a stimulus. If the organism
is consistently rewarded after making that particular response in the presence
of that particular stimulus, the relationship between the stimulus and the
response is strengthened, and the probability of the occurrence of the response is increased until the response is said to be learned.

Miller and Dollard reported some experiments in which albino rats
learned to turn right or left in a T-maze, depending on the response made
by a "leader" rat with which they were individually paired. Although the
authors' studies were not reported in full, they also claimed to find that sub-

---

[2]The designation "S–R/reinforcement" is adapted from Hilgard and Bower's (1966) discussion of observational learning (pp. 534–538). Although inaccurate in some respects
(e.g., Skinner, 1937), this label nevertheless loosely describes the Hullian and Skinnerian
traditions of theoretical thought which have dominated animal learning psychology.

jects showed no initial tendency to imitate prior to being rewarded, that the subjects either imitated or nonimitated the leader according to the reward contingencies, and that the subjects generalized their behavior to other leaders, to other situations, and across drives. Their conclusion was that imitative behavior develops because one individual's response, which just happens to match that of another individual, is rewarded and comes to occur at the cue of the other's response. This "matched-dependent"[3] behavior was also demonstrated in a number of later studies, including those by Bayroff and Lard (1944), Solomon and Coles (1954), Church (1957a,b), Conners (1966), Stimbert et al. (1966), and Haruki and Tsuzuki (1967).

It is obvious that Miller and Dollard, and others who fall in the S-R/ reinforcement category, were talking about a rather narrow conception of imitative behavior. They were, of course, only demonstrating the interesting but not at all surprising fact that a rat's behavior could be controlled by environmental stimuli, which in this case happened to be the behavior of another rat. The fact that the imitator's response (turning right or left in a T-maze) was similar to the leader's was, however, simply an artifact due to the limited choice of responses allowed by the apparatus (see Bayroff and Lard, 1944). There would hardly seem to be justification, then, for generalizing from this particular situation to all (see footnote 3) cases of imitative behavior, especially those cases which apparently occur prior to the administration of response-contingent reward under conditions allowing a much greater range of behavior.

In addition to this rather artificial type of imitative behavior, there were other imitative phenomena that were considered explicable under the S-R/ reinforcement paradigm, especially through recourse to the principle of conditioned reinforcement, which refers to the reinforcing properties of a stimulus acquired through its contiguous association with primary reward. The well-known fact that many animals eat more in groups than individually was experimentally demonstrated by James and his associates with puppies (James, 1953, 1954, 1960, 1961; James and Cannon, 1955; James and Gilbert, 1955). One of James' basic procedures (e.g., James and Gilbert, 1955) was to feed one group of subjects as a group for a period of days while feeding a second group as individuals. Otherwise, the subjects were allowed constant

[3]Miller and Dollard (1941, p. 91f) considered all cases of imitation to be comprised of three "submechanisms": (1) "same" behavior, which is only coincidentally similar behavior in two individuals; (2) "matched-dependent" behavior, which is the fundamental type and is described here; and (3) "copying," which involves the copier's responding on the basis of criteria for similarity and difference between his and a model's behavior. "The essential difference between the two processes is that in matched-dependent behavior the imitator responds only to the cue from the leader, while in copying he responds also to cues of sameness and difference produced by stimulation from his own and the model's responses" (Miller and Dollard, 1941, p. 159).

social contact. Then he started feeding the two groups either singly or socially on alternating days. He found that the puppies of the first group ate more on days of social feeding and less on days of individual feeding. But subjects in the second group required about 2 weeks before they started showing this social facilitatory effect. These results were said to suggest that "facilitation is a learned behaviour specific to the feeding situation," where "the presence of other animals in the feeding situation may act as a secondary [conditioned] reinforcement of the behaviour" (James and Gilbert, 1955). This interpretation illustrates two characteristics of the S–R/reinforcement paradigm. One is the tendency to employ the notion of conditioned reinforcement indiscriminately, even where it clearly explains nothing (since, for one group, the *absence* of others might equally well have been associated with the feeding situation). Less obvious, perhaps, is the tendency to attempt to account for the development of behavior by begging the question of where the behavior originated, for James was actually only saying that socially facilitated behavior occurred because socially facilitated behavior was reinforced.

A somewhat similar view was taken by Holder (1958), who stated in the summary of her paper that "social facilitation and inhibition of response in rats are dependent upon the way in which rats as stimuli have been associated with learning processes in the past." In her case, the response involved was lever pressing, so it seems quite reasonable, again, that rats could respond to the sight of another rat as a discriminative stimulus associated with either reward (thereby producing response facilitation) or nonreward (response inhibition), just as rats would respond to an inanimate, nonsocial stimulus. But, of course, it does not follow that all cases of social facilitation and inhibition of responses in rats are necessarily dependent on this learning process, as her statement seems to imply. More importantly, though, it does not follow, as one might be led to think, that a "social" discriminative stimulus is no different from any other discriminative stimulus. Stimbert (1970) made a direct comparison of social and nonsocial discriminative stimuli in a four-choice open-field situation and found that acquisition was faster and resistance to extinction greater for a group which had a leader rat to follow than for a group which had a strip of tape on the floor leading to the correct goal box. He thus concluded that the social discriminative stimulus was more efficacious than the nonsocial stimulus (but cf. Masaaki, 1953). This finding of course raises further questions about the reasons for the efficacy of "social" stimuli (e.g., olfactory *vs.* visual cues, movement *vs.* nonmovement of the stimulus, and social *vs.* nonsocial rearing of the subjects), but at least it suggests that more than a superficial analysis using the S–R/reinforcement paradigm is needed in accounting for these social facilitatory effects.

Another imitative phenomenon which has been the object of theoretical interest for at least one S–R/reinforcement theorist (Mowrer, 1950, 1960*b*) is vocal mimicry in "talking" birds. In essence, Mowrer claimed that because a human trainer provides food and meets certain other primary needs of these birds, the sounds made by the trainer thereby acquire conditioned reinforcing properties. Thus birds that are physically capable of producing these sounds during their random vocalizing come to do so with greater frequency because the sounds provide conditioned reinforcement.

Although Mowrer's account of this phenomenon brings up the notion of conditioned reinforcement once again (indeed, he has constructed an entire theory of behavior and symbolic processes on it—Mowrer, 1960a,b), a detailed examination of this construct is beyond the scope of this chapter. Generally speaking, though, one may well wonder whether a concept which is used to explain "everything" really explains anything at all. (For more specific comments, the reader is referred to some questions raised in recent studies by, e.g., Sheldon, 1971, and Stubbs, 1971, and to relevant criticisms made by Herrnstein, 1969.) Fortunately, the issue can be avoided in this case, since most of the pertinent work has been limited to simpler situations using food as a direct reward. There have been a number of studies demonstrating that vocalization by "talking" as well as "nontalking" species can be trained as an operant response (e.g., Ginsberg, 1960, 1963; Lane, 1961). Yet, a study by Foss (1964), designed to test Mowrer's theory, found that mynahs (*Gracula religiosa*) learned to reproduce equally well both a reinforced whistle (contiguous with food) and a nonreinforced whistle. Thus reinforcement may be a sufficient condition but certainly not a necessary condition for mynahs to imitate certain sounds. Again, the conclusion seems to be that something is missing in the theoretical accounts of imitative behavior provided by the S–R/reinforcement paradigm.

This conclusion, arrived at and stated in various ways, is being reached by an increasing number of workers today. Seligman (1970), Staddon and Simmelhag (1971), Rozin and Kalat (1971), and Shettleworth (1972) have recently presented and reviewed some of the anomalies which suggest that modifications or revisions are needed in some of the traditional principles of behavior which make up the S–R/reinforcement paradigm as described here. These apparent anomalies include long-delay learning of taste avoidance (Garcia *et al.*, 1966), autoshaping (Brown and Jenkins, 1968), automaintenance (Williams and Williams, 1969), instinctive drift (Breland and Breland, 1961), species-specific defense reactions in avoidance learning (Bolles, 1970), and the development of bird song (Thorpe, 1961; Hinde, 1969; Marler, 1970; Nottebohm, 1970), among others. Whether these anomalies amount to a "crisis" sufficient for the S–R/reinforcement paradigm to undergo a "scientific revolution," in Kuhn's (1962) terms, is difficult to say at present. In any case, there seems to be a growing awareness among animal learning psychologists (cf. Lockard, 1971) that behavior (including that studied exclusively in the laboratory) must be considered as adaptive, ecologically relevant phenomena subject to natural selection if a truly adequate conception of the causes of behavior is to be attained.

C. Classical Ethology

A prominent characteristic of classical ethology is the variety of ways in which the concept of behavioral imitation is touched on. With its roots in naturalistic study, it is not surprising that ethology has provided numerous observations on various imitative phenomena. For example, there have been observations of parental tuition or, at least, involvement in the development of feeding habits in young birds (summarized by Cushing, 1944; more recently see Norton-Griffith, 1967; note also Barnett, 1968, and Ewer, 1969, regarding mammals). There have also been reports of the spread of novel food choices among groups of birds (Buxton, 1948; Fisher and Hinde, 1949; Pettersson, 1956; Stenhouse, 1962; Newton, 1967; also cf. L. Tinbergen, 1960; Murton, 1971a,b; Murton et al., 1971), as well as the rejection of unpalatable food after observing another individual's reaction (Swynnerton, 1942; Rothschild and Ford, 1968). In addition, traditions have been reported for such things as habitat selection (Miller, 1942), covey size (Errington, 1941), migration routes (Hochbaum, 1955), and noxious food avoidance (Steiniger, 1950). Moreover, various reproductive behaviors have been claimed to be socially facilitated (see, e.g., Craig, 1913; Darling, 1938; Emlen and Lorenz, 1942; Armstrong, 1951). There also have been observations on the "transfer of mood" within flocks of birds (e.g., Lorenz, 1937; Kortlandt, 1940; Nice, 1943). And, of course, the number of observations related to the fact that animals eat more in social situations has perhaps been exceeded only by the general knowledge that many birds show imitation in their vocalizations.

Another well-known characteristic of ethology is its merging of naturalistic and experimental studies. In this respect, however, classical ethology has tended to weight the distribution on the naturalistic side of the continuum, particularly with regard to imitative phenomena. From the numerous observations cited above, there have followed very few experimental studies that do much more than demonstrate one or another phenomenon under controlled conditions, thereby presumably conferring "objectivity" to its existence, but seldom systematically manipulating very many of the elements in the phenomenon's causal matrix.[4] Thus too often one finds studies with titles of the general form, "Social facilitation of _____ behavior in the _____," whose abstracts seem to add only the words "was (or was not) demonstrated."

---

[4]One notable exception is the work on habitat selection and its determinants in early experience (see, e.g., Klopfer, 1962, 1970). This work is not discussed here because it has demonstrated that the process involved is not essentially of a social nature.

This is not to say, of course, that classical ethology has not been concerned with the causal determinants of imitative behavior. But if one views behavior on three different time scales of causation—the historical or phylogenetic, the developmental or ontogenetic, and the contemporaneous (cf. Tinbergen, 1951)—then it would have to be said that classical ethology has, as far as imitative behavior is concerned, emphasized the first (in the visage of survival value or functional significance) at the expense of the latter two, even (or particularly?) if one includes some vague references to the "conditioned reflex or trial-and-error learning" basis for some imitative phenomena. Even in the former regard, though, the ethological approach has failed to make good use of one of its basic techniques—comparative study. There have been a number of demonstrations of various sorts of imitation in various species, of course, and there have even been a few studies using more than one species at a time (e.g., Turner, 1964; Alcock, 1969a). But the latter studies provide no explicit rationale for their choice of species, and the former studies use quite different experimental procedures. Thus it is difficult to see the basis for making comparisons among species, especially when the subjects come not only from different families but from different orders and classes. Certainly, knowledge of neuroanatomical correlates of imitative behavior is quite lacking (note Konorski, 1963; Hale, 1956), so there is no basis for comparison along this line.

There has been more recently, however, a trend within contemporary ethology which attempts to correlate ecological factors with behavioral and social phenomena (see, e.g., Klopfer, 1962, 1969, 1970; Crook, 1965, 1970). This suggests that one basis for comparing various species lies in their differing (or similar) ecological adaptations. For example, Klopfer (1958, 1959a, 1961) has hypothesized that differences in imitativeness of birds' food habits would be expected to correlate with the species' forms of social organization or with their basic feeding characteristics (see below, p. 60f). It must be recognized, however, that the ultimate value of this type of approach (other than its heuristic value in suggesting which species to examine first) depends on an adequate specification or measurement of both the behavioral and the ecological factors to be correlated. Unfortunately, except possibly for the topic of bird vocalization, the state of knowledge concerning most imitative phenomena (to be discussed more fully in the following section) does not allow this approach to be of much help in interpreting and relating past work.

IV. CURRENT STATUS

Thus far, the topic of imitation has been dealt with in general terms. Particular studies have been used only to illustrate characteristic views or difficulties of three historical paradigms. More recent work requires closer examination, although the mass of relevant literature still necessitates some general comments about groups of studies. The current status of this work is considered under three traditional headings which seem to be popular areas of interest under the broad rubric of imitative behavior: bird vocalization, social facilitation, and observational learning. Again, it cannot be emphasized too much that these categories, especially the latter two, are not used because they represent necessarily meaningful or valid distinctions. Even if a classification scheme could now be devised which would adequately capture the many intuitive as well as logical distinctions that can be made concerning imitation, it is doubtful that this would add very much to our understanding of imitative behavior. There simply does not exist at present the empirical foundation necessary for forming a general framework or coherent pattern of imitative phenomena. And until there are more adequate efforts to lay this foundation through experimental analysis, attempts to formulate a "better" classification system are merely speculation.

A. Bird Vocalization

The topic of bird vocalization has often been divided into two areas[5]: (a) "vocal imitation," that is, imitation of conspecifics, as in the development of bird song; and (b) "vocal mimicry," or imitation of apparently nonfunctional sounds, as in "talking" birds. Work on the former problem (see, e.g., Thorpe, 1961; Hinde, 1969; Marler, 1970; Nottebohm, 1970) has shown that in a number of oscines certain aspects of song are socially learned, with the exact parameters of the developmental process varying for different species. For example, Thorpe (see 1961) found that young chaffinches (*Fringilla coelebs*) develop structurally atypical songs when hand-reared in isolation from only a few days of age. If young chaffinches are reared in a group, but isolated from normal song, they develop similar, more elaborate (but still abnormal) songs. Most interestingly, if young chaffinches are not isolated until their first autumn, the following spring they develop an essentially normal song, thus demonstrating an early sensitive period for this social learning. With the white-crowned sparrow (*Zonotrichia leucophrys nuttalli*), Marler (see 1970), too, has found abnormality of song in individual nestling-isolates. But rearing the young of this species in isolated groups produces equally aberrant song, unlike the case with chaffinches. Furthermore, the

---

[5]It should be noted that Thorpe's (1967) position and the attitude of this chapter are quite compatible on the subject of categorization in this regard, for Thorpe (1967, p. 246) noted "that this division is of little fundamental or physiological importance, and that the classes are not mutually exclusive," but he nevertheless retained the distinction essentially because of its practical usefulness.

sensitive period for this early learning of dialect is shorter in the white-crowned sparrow (10–15 days) than in the chaffinch (3 months).

A number of other oscines have also been studied, revealing diverse patterns of song development. For example, even relatively closely related species such as the white-crowned and song sparrows (*Z. leucophrys* and *Melospiza melodia*) and the Arizona and Oregon juncos (*Junco phaeonotus* and *J. oreganus*) develop their songs quite differently in isolation (see Marler, 1970). The isolated song sparrow, although having the most complicated song of the four species, develops normal song. The isolated juncos produce somewhat abnormal songs, but not as deviant, relatively speaking, as that of the isolated white-crowned sparrow. Moreover, Arizona juncos raised in group-isolation develop normal songs, quite unlike the case with white-crowned sparrows.

These findings represent only a very small sample of the work that has been done in this area. In addition, studies by Nicolai (1956, 1959, 1964), Mulligan (1966), Lemon (1968*a,b*; Lemon and Scott, 1966; Dittus and Lemon, 1969), Immelmann (e.g., 1969), Mundinger (1970), and others have well demonstrated the varied roles of imitative learning in the development of bird song, at all stages of life. This is not even to mention other work on related problems such as sensitive periods, sensory feedback, and neuro-anatomical and physiological bases for avian vocalization. (For more detailed coverage of the work concerned with vocal imitation, the reader is referred to a recent review by Nottebohm, 1972, in addition to articles cited earlier, especially Hinde's, 1969, volume.) Suffice it to say here that no obvious pattern for explaining the findings obtained thus far has emerged. As Nottebohm (1972, p. 116) has said in regard to the evolution of vocal learning, "Vocal ontogeny is not a trait that falls along an obvious continuum," for "Several pathways could lead to vocal learning as observed in extant species. The plausibility and survival value of all intervening stages is not always immediately apparent."

Thorpe (see 1967) has attempted to clarify the situation at least in regard to vocal mimicry in "talking" birds, using some of the above work as well as studies of antiphonal singing. He noted that the chaffinch (*F. coelebs*) imitates only sounds of a certain tonal quality and that its song serves as a territorial proclamation, but that, on the other hand, a species such as the greenfinch (*Chloris chloris*) is more labile in its imitativeness and uses its song less for territorial proclamation and more for coordinating and maintaining pair-bond functions. It seems reasonable to suppose, then, that a greater degree of variability in vocalization would aid in individual recognition in the latter case, while balanced by certain constraints needed, as in the former case, to maintain species identity. Viewed in this way, the striking imitative abilities of certain tropical avian species could be a manifestation of one extreme of an ability whose function is "to provide for social recognition and cohesion rather than for territorial defence" (Thorpe, 1967, p. 251).

Thorpe reported some work on antiphonal singing to support his view. Two hand-reared bou-bou shrikes (*Laniarius aethiopicus*), a male and a female, developed a simple antiphonal song in which the male sang two notes, followed by the female's singing four rather quavering, juvenile notes. When the female died of an illness, the male sang the full duet, imitating the characteristic quavering notes, and flew about in an agitated manner for several days, apparently searching for his missing mate. Another female was then placed with the male and, after about 6 months, a new duet developed between these two. Thus Thorpe concluded that "one of the major functions of the imitative ability of such birds is to establish and strengthen the social bonding" (Thorpe, 1967, p. 253).

This interpretation, it should be noted, is consistent with the view that ecological variables may correlate with observed differences among species' abilities to imitate. Hooker and Hooker (1969), for example, suggested that duetting occurs in a number of species whose dense tropical forest habitats or nocturnal activity patterns make auditory signals more efficient than visual signals for the purpose of social bonding. However, Diamond (1972) has reported instances of duetting in species which usually remain in immediate visual contact and which do not necessarily live in dense thickets. Moreover, recent field study of the mynah (*G. religiosa*) by Bertram (1970) has indicated that mates of this species do *not* engage in duetting or matching of calls, while territorial neighbors of the same sex *do*. Thus the problem of vocal mimicry in mynahs and other "talking" birds and its relationship to antiphonal singing remains open.

That some other form of social bonding, such as "imprinting," might be involved in the extreme imitative tendencies of birds such as mynahs and parrots is of course a possibility that has been mentioned (Thorpe, 1969; Bertram, 1970). But, so far as is known, no experimental study has been conducted to clarify its role for "talking" birds. Moreover, at least in the case of the mockingbird (*Mimus polyglottos*) and other birds which in their natural state freely imitate other species, there is no obvious reason to suppose that any sort of "social bonding" is involved. But, of course, knowledge about the determinants of the renowned mimcry of the mockingbird is, if anything, even more scarce than it is for the mynah, so it is impossible (besides being inadvisable) to extrapolate from one case to the other.

In spite of the many unanswered questions concerning imitation in bird vocalizations, and notwithstanding the present lack of a coherent pattern in which to fit the diverse findings on this subject thus far, there is an unmistakable impression that real progress is being made in this field, especially when compared with work on other types of imitative phenomena. Part of this progress is obviously due to the technological advances of high-fidelity recording and audiospectrographic analysis. But it may be significant also that the basic phenomenon being studied has been so frequently and regularly observed over the years. Perhaps even more significant than its well-established foundation in natural history, however, is the fact that this work was undertaken in inductive fashion, that is, without the aim of proving or disproving any particular hypothesis or theory about the development of bird song. It is indeed remarkable to note how far one can advance in this relatively immature science, the study of behavior, without theoretical and conceptual trappings.

B. Social Facilitation

It is when one examines the vast literature on "social facilitation" that one especially realizes that this term could not be anything more than an abstraction reflecting many different causal mechanisms. There have been reports of social influences on almost every conceivable behavior, it seems. In addition to the naturalistic observations already listed on p. 50 and the almost proverbial social facilitatory effects on feeding in various classes of animals (e.g., Bayer, 1929; Harlow, 1932; Ross and Ross, 1949*a,b*), there have been reports concerning nest building in ants (Chen, 1937), rising for air in tarpons (Shlaifer and Breder, 1940), hatching in quail (see Vince, 1969), preening in flies (Connolly, 1968), circadian rhythms in mice (Kavanau, 1969), the "following response" in ducklings (Ramsay and Hess, 1954; Klopfer, 1959*b*), copulation in rats (Larsson, 1956), and runway or maze learning in cockroaches (Gates and Allee, 1933; Zajonc *et al.,* 1969), fish (Welty, 1934; Hale, 1956), parakeets (Allee and Masure, 1936), chicks (Smith, 1957), dogs (Scott and McCray, 1967), and, of course, rats (Bayroff and Lard, 1944). There have also been various studies concerning social influences on avoidance learning (e.g., Klopfer, 1957; Angermeier *et al.,* 1959), fear reduction (e.g., Davitz and Mason, 1955), discrimination learning (e.g., Howells and Vine, 1940), classical conditioning (Kriazhev, 1929; but *cf.* Brogden, 1942), operant conditioning (e.g., Skinner, 1962; Gilbert and Beaton, 1967; Oldfield-Box, 1967; Lescrenier and Wijffels, 1969), and sundry other topics.

One of the most commonly known phenomena under this heading is the social facilitation of pecking in chicks. Tolman and his associates have pursued this problem in a number of studies (Tolman, 1964, 1965*a,b,c,* 1967, 1968, 1969; Tolman and Wilson, 1965; Tolman and Wellman, 1968; Wilson, 1968). In the first-cited paper, Tolman concluded that facilitation of ingestion was mediated by the tendency of a chick to peck at the bill of a companion (or at a tapping pencil), where such pecking could in fact lead to food. Thus

separating the subject from his companion by a transparent partition prevented the subject from obtaining food when the companion was pecking,
and hence no facilitation of ingestion resulted. However, instead of attempting to analyze further the dimensions of the "pecking stimulus," Tolman
and his co-workers chose to pursue their analyses along other lines, such as
measuring the effects of the subject's level of deprivation, the companion's
level of deprivation, the proximity of the companion, the companion's activity (feeding *vs.* nonfeeding), and other miscellaneous factors such as dominance and rearing conditions, as well as parameters of the test situation.

There have been a number of conflicting results from these studies. In
one report rearing conditions were not found to be a significant factor
(Tolman, 1964), while in another they were (Wilson, 1968). Sometimes the
presence of a separating transparent partition abolished the facilitation
phenomenon (Tolman, 1964; Tolman and Wilson, 1965), but at other times
it had little or no effect (Tolman, 1965a, 1968). In one paper, Tolman rejected
the interpretation that emotionality mediates the difference between isolated
and social chicks' feeding rates (Tolman and Wilson, 1965), but then he subsequently reversed his position by confirming the emotionality hypothesis
(Tolman, 1965a). It should be noted that Tolman did account for the latter
difference as being due to the duration of testing: in short 10-min tests
emotionality is more significant than in 1-hr tests which, presumably, allow
adaptation to the situation. But other apparent inconsistencies have been
allowed to exist without explicit resolution. Thus one is left with the rather
unsatisfactory conclusion that "something" about a companion's feeding
behavior facilitates a chick's feeding behavior in addition to the effect
produced just by the companion's visible presence.

A different approach to a similar problem was taken by Hailman in his
monograph, "The ontogeny of an instinct" (1967). Part of his very thorough
investigation of the development of the pecking response of the laughing
gull chick (*Larus atricilla*) was concerned with extending and reinterpreting
the earlier work of Tinbergen and Perdeck (1950) by identifying the stimulus
parameters most effective in eliciting pecking by the chick. He found that the
most effective stimulus was a vertical rod, about 9 mm wide, moving hori-
zontally (across its axis) at about 80 beats per minute. Furthermore, the
optimal stimulus contrasted with its background, being either blue or dark
on a white background or red on a dark background. That many of these
dimensions are found in the parent's (or sibling's) bill is, of course, no
coincidence, nor is the fact that pecking at the parent's bill normally secures
food for the young chick. But to have simply described the whole affair as an
"instinctive response" would, as Hailman amply demonstrated, have ob-
scured many subtle yet quite important variables.

This analysis of Hailman's obviously does not explain social facilitation of pecking or eating in the domestic chick, but it does suggest that one can learn a great deal about the characteristics of a stimulus (whether it be a parent's beak, another chick, or a tapping pencil) which is effective in eliciting or facilitating pecking, without making excursions into factors such as emotion and deprivation level of a companion. This is not to say that one should always study immediate stimulus variables in preference to motivational variables, ontogenetic factors, questions of survival value, or other types of causal influences. There are no precise rules for determining priorities in this regard. But observational study of the particular phenomenon under consideration can often suggest which sorts of factors need only routine parametric study and which seem more likely to be "critical" variables. For example, one would probably not concentrate on social factors if a given imitative phenomenon seemed to occur reliably without regard to the companions' dominance ranking, familiarity, sex, or other such characteristics. Or in cases where stimulus variables seem fairly straightforward, one might wish to consider first the question of functional value by, for example, comparing two similar species in which the imitative phenomenon does occur in one instance and does not occur in another.

Of course, it is not enough to answer just one of these various types of questions; to different extents, all are important. So the emphasis here should not be so much on determining which questions are the most significant as on, in fact, asking and answering all questions about the occurrence of an imitative behavior. The primary concern, then, is that each social facilitatory phenomenon undergoes thorough experimental analysis in order that *all* of its determinants can be understood as completely as possible at every level.

C. Observational Learning

A number of experiments have been conducted to determine what effect observing one individual's response (usually in some sort of task situation provided by the experimenter) has on an observer when placed in the same situation. The question of *what* is learned through observation has been treated in basically two different ways by these studies. One approach emphasizes *rate of acquisition* as the dependent variable; the other approach emphasizes *choice of response* in determining whether or not something was learned through observation (cf. Bitterman, 1962; Hall, 1963). These two categories are based simply on the type of data taken by an experimenter, which allows the possibility that a particular study may be placed under both headings if both types of measures of learning are used. Usually, though, an experimenter's emphasis on one or the other can be easily detected.

There are quite a few studies which measure rate of acquisition, also known as "profit through observation." The basic procedure is to expose one subject (the "observer") to the performance of another individual, usually a conspecific (the "actor"), and then measure the savings in the amount of time (or number of trials) required by the observer to reach some specified criterion. Excluding studies using primates and the Warden and Jackson (1935) duplicate-cage apparatus, the operant chamber, or Skinner box, has been employed most frequently in this type of investigation of observational learning. There have been at least six recent studies concerned with observational learning of the bar-press response in rats (Corson, 1967; Powell, 1968; Powell *et al.,* 1968; Jacoby and Dawson, 1969; Russo, 1971; Gardner and Engel, 1971).

All of these studies included both observation and test periods either successively or simultaneously in daily sessions, which of course means that comparisons with "trial and error" and other control groups were necessary in order to separate the effects of practice during the test periods from the effects of the observation periods *per se.* The results are often less than clear-cut. For example, using *sessions* to criterion, Corson (1967) found a significant difference between his "observation group" of hooded rats and his "shaped group" (subjects which were trained to bar-press by the traditional method of being rewarded for closer and closer approximations to the correct response), but actual *time* to criterion was virtually identical for both groups. Powell (1968), as well as Powell *et al.* (1968), failed to find evidence of learning by observation in albino rats—not a very surprising finding, perhaps, given the poor visual abilities of their subjects (Lashley, 1930). Jacoby and Dawson (1969) attempted to improve on the preceding three studies by using hooded rats "in a situation which would optimize the visual process." Both their "observation group" and their "shaped group" required significantly fewer trials to criterion than did either of two other control groups, but did not differ significantly from each other unless the total number of lever presses to criterion was used as the dependent variable.

Russo (1971) similarly found shorter average time to criterion in an observation group of hooded rats, so long as they received free pellets contingent on the actor's response. The time to criterion of a second experimental group, which received no free reinforcements during observation of the actor's performance, was not significantly different from that of a control group. Russo noted, however, that, unlike the control group, both of the experimental groups centered their activity around the lever during test periods, the first experimental group being more persistent than the second.

Gardner and Engel (1971) investigated "imitational and social facilitatory aspects of observational learning in the laboratory rat" and found no differences in the rate of acquisition of bar pressing between groups paired either with actors performing the response or with actors simply receiving free food in the absence of a manipulandum. However, both experimental groups reached asymptote considerably quicker than a control group learning individually by trial and error. This enhancement of acquisition through "the mere presence of another animal in a learning situation" was attributed by the authors to "increased activity and fear reduction."

A study by Chesler (1969) using kittens also involved administering both observation and test periods daily. She found that kittens which observed their mothers pressing a lever to obtain food (which the kittens could sometimes share with the actors) acquired the response in a median of 4.5 days, while those which observed a strange female required 18.0 days, and kittens which learned by trial and error never even approached criterion. Similar differences were also noted for bringing the response under the control of a discriminative stimulus.

A related study by John et al. (1968) included one experiment which was essentially the same as the one just described. A second experiment examined acquisition of an avoidance response to electric shock signaled by a buzzer. The procedure involved first one test or "empathy" trial, then 20 trials observing a "student" cat learning the avoidance response (jumping a hurdle) followed by 20 trials observing a "teacher" cat performing the conditioned response at asymptote. All of the observers except one (who had taken ill) clearly reached criterion much more quickly than the students; indeed, two observers performed at criterion level (90 % successful avoidance) after receiving only one or two shocks.

Cats were also used in earlier studies by Herbert and Harsh (1944) and Adler (1955), and in both the experimenters found suggestive evidence of improvement in time to success using problems such as pulling a ribbon or pawing a lever. But Herbert and Harsh made no statistical analysis of their averaged data, so it is difficult to evaluate their results, and Adler acknowledged that the effect he obtained was rather weak when he said, "The finding that this advantage [of observation] is not very permanent, and that individual differences in trial and error learning tend to show up strongly in subsequent trials, points to the fact that cats do not make much use of observation learning" (1955, p. 174).

A study by Oldfield-Box (1970) using an unspecified breed of rats in a combination runway/bar-press situation revealed no consistent improvement in speed of "forming an association" until the observation chamber was expanded from a small box in the runway to a duplicate, adjacent runway. However, this was only a preliminary study, and the results were presented in a very qualitative fashion.

More recently, Lore et al. (1971) devised an experiment on observational learning employing a passive avoidance response. Domesticated rats observed other rats learning to avoid physical contact with a lighted candle. Mean time to criterion (20 consecutive minutes of avoidance) was significantly less for observers (24.1 min) than for controls (31.8) or demonstrators (31.5). Data on number of errors (nose contacts) were consistent with these results, but not as convincing.

A number of studies using rate of acquisition measures are further distinguished by their emphasis on differential responding to discriminative stimuli as an indicator of learning influenced by observation. Groesbeck and Duerfeldt (1971) employed this type of procedure to analyze some of the factors involved in an observational learning situation. Water-deprived rats observed actors performing various aspects of a Y-maze discrimination task. "Modeling" (seeing an actor knock down the cue card blocking the goal alley) was judged to be the aspect most important for the observers' later performance. The "informational content" (seeing both the cue card and a water bottle) and the "following component" (a tendency to go in the same direction as an actor, irrespective of the placement of the cue cards) were also found effective, but "vicarious reinforcement" (seeing an actor drink from the water bottle) had either no effect or a detrimental effect on the observers' averaged rate of acquisition of the discrimination. Although apparently successful in separating a number of often confounded variables, this study does not include enough conditions to allow conclusions concerning each factor's relative importance.

A similar study by Kohn and Dennis (1972) also demonstrated a facilitatory effect on discrimination learning following several observations of a model rat's performance. The authors further showed that reversal of the cues retarded subsequent learning of the discrimination task by observers. Rats which saw a model making no response (remaining in the choice chamber without approaching either of the cue panels) and rats which saw only the cue panels in the same position on each trial did not differ significantly from control subjects which learned by trial and error. However, another group which saw the cue panels changed in position on each trial in a nonregular sequence showed a facilitatory effect almost as great as that of the group which viewed a model. That attentional mechanisms and "enhanced distinctiveness of cues" (cf. Spence's, 1937, "stimulus enhancement") seem to be involved here, as the authors suggest, and in many other reports of observational learning, may be undeniable. But the exact nature of these factors and the reasons why they seem to operate in some situations but not in others remain obscure.

Discrimination learning in birds under social influence was the subject of a series of papers by Klopfer (1958, 1959a, 1961), represented here by his 1959 report. Greenfinches (*C. chloris*) were required to discriminate between regular sunflower seeds and aspirin-filled sunflower seeds (the latter being unpalatable) on the basis of a white pattern with red cross-stripes or a plain white pattern, respectively. Actors' and observers' rates of learning the discrimination were compared under various social conditions. Klopfer found no evidence of "visual imitation"; that is, the observers did not learn to avoid a stimulus by viewing the learning process of their partners. Moreover, the presence of an untrained observer resulted in slower learning of the discrimination by actors and, unless overtrained, a decrement in the actors' performance level when the untrained observer was introduced. On the other hand, the presence of a trained partner did not impede rate of learning, but neither did it result in any improvement compared to those subjects trained alone. It would seem, then, that the most potent stimulus was the sight of a conspecific feeding, regardless of the discriminative stimulus associated with the food being eaten. Klopfer concluded that "a feeding response can be established more readily than an avoidance response . . . apart from social facilitation or local enhancement-type effects."

Klopfer noted that this positive-feedback effect would be maladaptive in a species that was less conservative in basic feeding habits than the greenfinch, and therefore would not be expected to occur in opportunistic feeders. He confirmed this prediction in a later paper (1961) in which he compared the rate of learning a discrimination in pairs of greenfinches and pairs of great tits (*Parus major*) (the latter being more exploitative feeders) and found that pairs of great tits learned just as quickly as single tits.

A study by Cronhelm (1970) also used a discrimination task, but in an operant conditioning situation. Seven-day-old chicks (*Gallus gallus*) were first shaped to peck a key for access to food and then were paired off as actors and observers. The observers viewed the actors for four sessions over 2 days as the actors came to respond differentially to two colors successively presented on the response key. One color was correlated with a variable-interval schedule of reinforcement which generated relatively high rates of pecking, and another color was correlated with a fixed-interval schedule which generated relatively low response rates. Both schedules, however, were said to allow the same frequency of reinforcement. During the 2 days of the actors' training, the observers were separated by a transparent screen which prevented access to either the food or the key. Sixteen hours after the last observation session, the observers were placed in the operant chamber side of the apparatus. In their first test session, the observers (as a group) pecked differentially at the discriminative stimuli at a level equivalent to the actors' performance in their second to third sessions. Or, considering days instead of individual sessions, the observers discriminated significantly better on their first day than the actors did on their first day. Although Cronhelm stated that response *patterns* were imitated, the form of her data allows only an inference of some sort of acceleration in learning the discrimination.

Certain findings of Turner (1964) are noteworthy in regard to Cronhelm's work, for Turner found that young chicks, as a group, tended to peck preferentially at grains of the same color (orange or green) as those pecked by a surrogate hen and that they tended to peck the grain sooner than control subjects. But Turner's procedure differed from Cronhelm's in that there was no delay intervening between the chicks' observations and their opportunity to peck at the grain. In addition to chicks, Turner worked with chaffinches and sparrows, examining a number of variables in social feeding situations (age and species of the subjects, location and familiarity of the food, attractiveness of the actor while feeding *vs.* nonfeeding, etc.). The multiplicity of his experimental conditions and the diversity of his results defy simple summarization, but they do suggest in particular that observing another bird feeding on novel foods (e.g., green pastry dough) can facilitate acquisition of the response, even though the observer may not encounter the food immediately or in the same place.

The suggestion of novel food habits being learned introduces the second type of experimental procedure for studying observational learning. This approach measures "what is learned" by noting *which* response or behavior pattern an observer performs after exposure to an actor's behavior. Most of the studies in the "choice of response" category involve discrimination tasks, but they generally differ from the already mentioned studies on speed of discrimination learning in that they analyze the response itself in greater detail and do not use time-based measures.

Two recent studies by Alcock (1969*a, b*) will be considered together under this heading. In one of these studies (1969*a*), Alcock used white-throated sparrows (*Zonotrichia albicollis*), fork-tailed flycatchers (*Muscivora tyrannus*), and black-capped chickadees (*Parus atricapillus*) to determine what effect observing a conspecific taking food concealed in a three-compartment tray would have on the observer's behavior when alone. His data were scores which were meant to reflect "how close the bird came to uncovering the hidden mealworm without hesitation," so in a sense he measured latency rather than choice of response. At any rate, his rather qualitative analysis suggested that observers tended to explore the tray more than controls, but that no observer obtained the food while alone unless it had first obtained food during the observation period.

In his second study (1969*b*), using only fork-tailed flycatchers, Alcock found that obtaining immediate reward was not always necessary for a different sort of observational learning to occur. After determining that his subjects showed a generalized avoidance to both a distasteful aposematic butterfly and its imperfect mimic, he allowed the observers to view an actor eating two of the mimics and then tested them alone. Only half of the observers would then mandibulate the mimic, but this was significantly greater than the control subjects' reactions. (Using number of "pecks" as a measure did not yield significant results, but the broader measure "mandibulation" did.) Thus Alcock concluded that "fork-tailed flycatchers appear to be capable of modifying their behaviour toward a prey species as a result of watching another bird's treatment of this prey."

A report by Dawson and Foss (1965) suggested observational learning of response topography by budgerigars (*Melopsittacus undulatus*). Over a period of 8 days, each observer saw his pretrained demonstrator remove a lid from a container of seeds, once per daily trial. Then the observers were given eight daily test trials alone. Although the observers showed no improvement in time to success on their first trial (24 hr after the last demonstration trial) or in number of trials to asymptote, they did adopt the same characteristic techniques of removing the lid as their respective partners had used. Thus two birds which had been paired with demonstrators which edged the lid off with their beaks did likewise; one which had seen his demonstrator lift the lid with his beak did the same; and two birds which had observed a demonstrator use his foot to dislodge the lid adopted that technique. Moreover, the subjects maintained the same modes of response when tested once at 3 and again at 6 months later.

Finally, recent work by Galef and his associates (Galef, 1971; Galef and Clark, 1971a, b, 1972; Galef and Henderson, 1972) is worthy of note for its comprehensive analysis of a social learning phenomenon hitherto known as a case of imitation. Although the special abilities of rats for learning to avoid eating noxious substances have been the object of a great deal of recent experimental investigation (see Rozin and Kalat, 1971, for review), this behavior has seldom been considered in a social context. In a series of ten experiments reported in their first study (Galef and Clark, 1971a), it was determined that pups born into a colony of wild rats avoided food that the adults had previously been poisoned on (and hence avoided), even though the food in question was always palatable and never poisoned during the pups' exposure to it. Moreover, the pups maintained this avoidance for some time after separation from the adults. Taken together, the results of their experiments were said to suggest a three-stage process underlying this phenomenon: first, the pups follow adults to food; second, they apparently learn cues associated with that food; and third, they show a generalized neophobia to novel or unfamiliar foods. Thus the phenomenon appeared to be not so much a case of active avoidance behavior as a case of exclusive approach responses to certain foods.

Even so, the nature of the interaction between pups and adults which results in the pups' approaching only certain food sources needed further clarification. Follow-up studies by Galef (1971) and Galef and Clark (1971b) indicated that pups normally do not actually follow or "trail" the adults from the nest area to the food source. Thus "pups do not actively imitate the adults but rather tend to move into areas near the adults and begin eating whatever food they find there" (Galef, 1971). Probing one step further, Galef and Henderson (1972) found that cues associated with the lactating mother's milk, as well as the mere presence of adults at the feeding sites (Galef and Clark, 1972), are sufficient to influence the food choices of the young during weaning.

The studies reviewed in this section have represented basically one or the other of two approaches to the problem of observational learning, measuring learning either as rate of acquisition or as choice of response. While a variety of conceptual grounds for using the former method may exist, it should be noted that there are a number of instances in which social influences on learning are known to operate on a much broader time scale than minutes or hours, even in animals in which one might be willing to allow the possibility of immediate "insight." For example, in teaching the chimp, Washoe, to communicate using the hand sign language for the deaf, the Gardners found that few, if any, of her first signs were ever acquired by immediate imitation (Gardner and Gardner, 1969). Rather, she would sometimes make the signs months later for events, or associated objects, that had been highly routine and ritualized activities. Likewise, the number of months intervening between early sensitive period and production of full song in the chaffinch (Thorpe, 1961) is worth considering when one expects to find immediate effects of social interactions on behavior.

These facts suggest that "speed of learning," as affected by social interactions, may be an irrelevant or even misleading measure of observational learning. Rather than assuming that "learning" is a unitary process that can be measured by the rate of acquisition of some arbitrary response such as bar pressing, a more important and prerequisite consideration would seem to be determining *which* responses or behavior patterns are susceptible to social influences. This can be done more efficiently and informatively not by contriving experiments which demonstrate some sort of "effect of observation" (depending on how the data are measured and manipulated), but by taking a reliable natural phenomenon and analyzing its causal determinants.

V. CONCLUSIONS

Perhaps the most simple yet accurate summation of what has been said in this chapter is that an inductive approach is needed in the study of imitative behavior (note that the proper object of study is imitative behavior and not the abstraction "imitation"). There is a need, as Tinbergen has so appropriately put it in a slightly different context, "to ignore the term and to return to the phenomena (which are singled out for study because they are suspected of having a different causation than other phenomena)" (Tinbergen, 1963,

p. 413). This need to "return to the phenomena" is particularly striking in the case of observational learning, where the preliminary stage of observing and describing naturally occurring phenomena—a stage which psychology as a whole has frequently been justifiably accused of neglecting (e.g., Verplanck, 1970)—seems to have been entirely omitted in most studies. This omission no doubt accounts in part for the notable lack of progress in this area as compared with the work on bird vocalization. But, as too many studies on social facilitatory phenomena illustrate, observing and demonstrating a reliable imitative phenomenon are only the first steps toward, and not a guarantee of, success in the endeavor to understand such behavior.

Research on imitative phenomena also requires a shift away from the customary "method-oriented" approaches to the study of behavior in favor of "problem-oriented" approaches (cf. Platt, 1964). Thus, instead of allowing the traditional equipment and techniques of experimental psychology to determine the problem, the particular problem or phenomenon under investigation must determine the types of experimental methodologies that are appropriate and necessary to reach a satisfactory understanding of its determinants. Moreover, it should always be remembered that the determinants of a behavioral phenomenon are not only multiple, but of several different orders as well, and that no single level of analysis, whether it be in terms of, e.g., survival value or ontogeny or immediate stimulus variables, is sufficient in itself for a truly complete understanding of a behavioral phenomenon.

Until a vigorous inductive approach to the many problems and questions lumped under the heading of "imitation" is pursued, the term "imitation" and its subvarieties will continue to be merely labels for ignorance.

VI. ACKNOWLEDGMENTS

The influence, encouragement, and assistance of Dr. J. E. R. Staddon and Dr. Peter Klopfer in the development and preparation of this chapter are most gratefully acknowledged. Dr. Carl Erickson and Dr. James Kalat kindly read the manuscript and offered helpful criticisms.

VII. REFERENCES

Adler, H. E. (1955). Some factors of observational learning in cats. *J. Genet. Psychol.* **86:** 159–177.
Alcock, J. (1969*a*). Observational learning in three species of birds. *Ibis* **111:**308–321.
Alcock, J. (1969*b*). Observational learning by fork-tailed flycatchers (*Muscivora tyrannus*). *Anim. Behav.* **17:**652–658.
Allee, W. C., and Masure, R. H. (1936). A comparison of maze behavior in paired and isolated shell-parakeets (*Melopsittacus undulatus,* Shaw) in a two-alley problem box. *J. Comp. Psychol.* **22:**131–155.

Angermeier, W. F., Schaul, L. T., and James, W. T. (1959). Social conditioning in rats. *J. Comp. Physiol. Psychol.* **52**:370–372.

Armstrong, E. A. (1951). The nature and function of animal mimesis. *Bull. Anim. Behav.* **9**:46–58.

Barnett, S. A. (1968). The "instinct to teach." *Nature* **220**:747–749.

Bayer, E. (1929). Beiträge zur Zweikomponententheorie des Hungers. *Z. Psychol.* **112**:1–54.

Bayroff, A. G., and Lard, K. E. (1944). Experimental social behavior of animals. III. Imitational learning of white rats. *J. Comp. Physiol. Psychol.* **37**:165–171.

Berry, C. S. (1906). The imitative tendency of white rats. *J. Comp. Neurol. Psychol.* **16**:333–361.

Berry, C. S. (1908). An experimental study of imitation in cats. *J. Comp. Neurol.* **18**:1–25.

Bertram, B. (1970). The vocal behaviour of the Indian hill myna, *Gracula religiosa*. *Anim. Behav. Monogr.* **3**:79–192.

Bitterman, M. E. (1962). Techniques for the study of learning in animals: Analysis and classification. *Psychol. Bull.* **59**:81–93.

Bolles, R. C. (1970). Species-specific defense reactions and avoidance learning. *Psychol. Rev.* **77**:32–48.

Boring, E. G. (1950). *A History of Experimental Psychology,* Appleton-Century-Crofts, New York.

Breland, K., and Breland, M. (1961). The misbehavior of organisms. *Am. Psychologist* **16**:681–684.

Brogden, W. J. (1942). Imitation and social facilitation in the social conditioning of fore-limb-flexion in dogs. *Am. J. Psychol.* **55**:77–83.

Brown, P. L., and Jenkins, H. M. (1968). Auto-shaping of the pigeon's key-peck. *J. Exptl. Anal. Behav.* **11**:1–8.

Buxton, E. J. M. (1948). Tits and peanuts. *Brit. Birds* **41**:229–232.

Chen, S. C. (1937). Social modification of the activity of ants in nest-building. *Physiol. Zool.* **10**:420–437.

Chesler, P. (1969). Maternal influence in learning by observation in kittens. *Science* **166**:901–903.

Church, R. M. (1957*a*). Transmission of learned behavior between rats. *J. Abnorm. Soc. Psychol.* **54**:163–165.

Church, R. M. (1957*b*). Two procedures for the establishment of "imitative behavior." *J. Comp. Physiol. Psychol.* **50**:315–318.

Connolly, K. (1968). The social facilitation of preening behaviour in *Drosophila melanogaster*. *Anim. Behav.* **16**:385–391.

Connors, K. R. (1966). An exploratory study of operant conditioning and generalization of imitative behavior in a rat. *J. Sci. Lab. Denison Univ.* **47**:35–39.

Corson, J. A. (1967). Observational learning of a lever pressing response. *Psychon. Sci.* **7**:197–198.

Craig, W. (1913). The stimulation and the inhibition of ovulation in birds and mammals. *J. Anim. Behav.* **3**:215–221.

Cronhelm, E. (1970). Perceptual factors and observational learning in the behavioural development of young chicks. In Crook, J. H. (ed.), *Social Behaviour in Birds and Mammals: Essays on the Social Ethology of Animals and Man,* Academic Press, New York, pp. 393–439.

Crook, J. H. (1965). The adaptive significance of avian social organizations. *Symp. Zool. Soc. Lond.* **14**:181–218.

Crook, J. H. (ed.) (1970). *Social Behaviour in Birds and Mammals: Essays on the Social Ethology of Animals and Man,* Academic Press, New York.

Cushing, J. E. (1944). The relation of non-heritable food habits to evolution. *Condor* **46**:265–271.

Darling, F. F. (1938). *Bird Flocks and the Breeding Cycle,* Cambridge University Press, Cambridge, England.

Davitz, J. R., and Mason, D. J. (1955). Socially facilitated reduction of a fear response in rats. *J. Comp. Physiol. Psychol.* **48**:149–151.

Dawson, B. V., and Foss, B. M. (1965). Observational learning in budgerigars. *Anim. Behav.* **13**:470–474.

Diamond, J. M. (1972). Further examples of dual singing by southwest Pacific birds. *Auk* **89**:180–183.

Dittus, W. P. J., and Lemon, R. E. (1969). Effects of song tutoring and acoustic isolation on the song repertoire of cardinals. *Anim. Behav.* **17**:523–533.

Emlen, J. T., and Lorenz, F. W. (1942). Pairing responses of free-living valley quail to sex-hormone pellet implants. *Auk* **59**:369–378.

Errington, P. L. (1941). An eight-winter study of central Iowa bobwhites. *Wilson Bull.* **53**:85–102.

Ewer, R. F. (1969). The "instinct to teach." *Nature* **222**:698.

Fisher, J., and Hinde, R. A. (1949). The opening of milk bottles by birds. *Brit. Birds* **42**: 347–357.

Foss, B. M. (1964). Mimicry in mynas (*Gracula religiosa*); a test of Mowrer's theory. *Brit. J. Psychol.* **55**:85–88.

Foss, B. M. (1965). Imitation. In Foss, B. M. (ed.), *Determinants of Infant Behaviour*, Vol. III, Methuen, London, pp. 185–199.

Galef, B. G. (1971). Social effects in the weaning of domestic rat pups. *J. Comp. Physiol. Psychol.* **75**:358–362.

Galef, B. G., and Clark, M. M. (1971*a*). Social factors in the poison avoidance and feeding behavior of wild and domesticated rat pups. *J. Comp. Physiol. Psychol.* **75**: 341–357.

Galef, B. G., and Clark, M. M. (1971*b*). Parent–offspring interactions determine time and place of first ingestion of solid food by wild rat pups. *Psychon. Sci.* **25**:15–16.

Galef, B. G., and Clark, M. M. (1972). Mother's milk and adult presence: Two factors determining initial dietary selection by weaning rats. *J. Comp. Physiol. Psychol.* **78**: 220–225.

Galef, B. G., and Henderson, P. W. (1972). Mother's milk: A determinant of the feeding preferences of weanling rat pups. *J. Comp. Physiol. Psychol.* **78**:213–219.

Garcia, J., Ervin, F. R., and Koelling, R. A. (1966). Learning with prolonged delay of reinforcement. *Psychon. Sci.* **5**:121–122.

Gardner, E. L., and Engel, D. R. (1971). Imitational and social facilitatory aspects of observational learning in the laboratory rat. *Psychon. Sci.* **25**:5–6.

Gardner, R. A., and Gardner, B. T. (1969). Teaching sign language to a chimpanzee. *Science* **165**:664–672.

Gates, M. F., and Allee, W. C. (1933). Conditioned behavior of isolated and grouped cockroaches on a simple maze. *J. Comp. Psychol.* **15**:331–358.

Gilbert, R., and Beaton, J. (1967). Imitation and cooperation by hooded rats: A preliminary analysis. *Psychon. Sci.* **8**:43–44.

Gilmore, J. B. (1968). Toward an understanding of imitation. In Simmel, E. C., Hoppe, R. A., and Milton, G. A. (eds.), *Social Facilitation and Imitative Behavior*, Allyn and Bacon, Boston, pp. 217–238.

Ginsberg, N. (1960). Conditioned vocalization in the budgerigar. *J. Comp. Physiol. Psychol.* **53**:183–186.

Ginsberg, N. (1963). Conditioned talking in the mynah bird. *J. Comp. Physiol. Psychol.* **56**:1061–1063.

Groesbeck, R. W., and Duerfeldt, P. H. (1971). Some relevant variables in observational learning of the rat. *Psychon. Sci.* **22**:41–43.

Hailman, J. P. (1967). The ontogeny of an instinct; the pecking response in chicks of the laughing gull (*Larus atricilla*) and related species. *Behaviour* Suppl. 15.

Hale, E. B. (1956). Social facilitation and forebrain function in maze performance of green sunfish, *Sepomis cyanellus*. *Physiol. Zool.* **29**:93–107.

Hall, K. R. L. (1963). Observational learning in monkeys and apes. *Brit. J. Psychol.* **54**: 201–226.

Harlow, H. F. (1932). Social facilitation of feeding in the albino rat. *J. Genet. Psychol.* **41**:211–221.

68   J. Michael Davis

Haruki, Y., and Tsuzuki, T. (1967). Learning of imitation and learning through imitation in the white rat. *Ann. Anim. Psychol.* **17**:57-63.
Herbert, M. J., and Harsh, C. M. (1944). Observational learning by cats. *J. Comp. Psychol.* **37**:81-95.
Herrnstein, R. J. (1969). Method and theory in the study of avoidance. *Psychol. Rev.* **76**: 49-69.
Hilgard, E. R., and Bower, G. H. (1966). *Theories of Learning,* 3rd ed., Appleton-Century-Crofts, New York.
Hinde, R. A. (ed.) (1969). *Bird Vocalizations,* Cambridge University Press, Cambridge, England.
Hochbaum, H. A. (1955). *Travels and Traditions of Waterfowl,* University of Minnesota Press, Minneapolis.
Holder, E. E. (1958). Learning factors in social facilitation and social inhibition in rats. *J. Comp. Physiol. Psychol.* **51**:60-64.
Hooker, T., and Hooker, B. I. (1969). Duetting. In Hinde, R. A. (ed.), *Bird Vocalizations,* Cambridge University Press, Cambridge, England, pp. 185-205.
Howells, T. H., and Vine, D. O. (1940). The innate differential in social learning. *J. Abnorm. Soc. Psychol.* **35**:537-548.
Immelmann, K. (1969). Song development in the zebra finch and other estrildid finches. In Hinde, R. A. (ed.), *Bird Vocalizations,* Cambridge University Press, Cambridge, England, pp. 61-74.
Jacoby, K. E., and Dawson, M. E. (1969). Observation and shaping learning: A comparison using Long Evans rats. *Psychon. Sci.* **16**:257-258.
James, W. T. (1953). Social facilitation of eating behavior in puppies after satiation. *J. Comp. Physiol. Psychol.* **46**:427-428.
James, W. T. (1954). Secondary reinforced behavior in an operant situation among dogs. *J. Genet. Psychol.* **85**:129-133.
James, W. T. (1960). The development of social facilitation of eating in puppies. *J. Genet. Psychol.* **96**:123-127.
James, W. T. (1961). Relationship between dominance and food intake in individual and social eating in puppies. *Psychol. Rep.* **8**:478.
James, W. T., and Cannon, D. V. (1955). Variation in social facilitation of eating behavior in puppies. *J. Genet. Psychol.* **87**:225-228.
James, W. T., and Gilbert, T. F. (1955). The effect of social facilitation on food intake of puppies fed separately and together for the first 90 days of life. *Brit. J. Anim. Behav.* **3**:131-133.
John, E. R., Chesler, P., Bartlett, F., and Victor, I. (1968). Observation learning in cats. *Science* **159**:1489-1491.
Kavanau, J. L. (1969). Behavior of captive white-footed mice. In Willems, E. P., and Raush, H. L. (eds.), *Naturalistic Viewpoints in Psychological Research,* Holt, Rinehart and Winston, New York, pp. 221-270.
Kinnaman, A. J. (1902). Mental life of two *Macacus rhesus* monkeys in captivity. *Am. J. Psychol.* **13**:98-148, 173-218.
Klopfer, P. H. (1957). An experiment on empathic learning in ducks. *Am. Naturalist* **91**: 61-63.
Klopfer, P. H. (1958). Influence of social interactions on learning rates in birds. *Science* **128**:903.
Klopfer, P. H. (1959a). Social interactions in discrimination learning with special reference to feeding behaviour in birds. *Behaviour* **14**:282-299.
Klopfer, P. H. (1959b). An analysis of learning in young Anatidae. *Ecology* **40**:90-102.
Klopfer, P. H. (1961). Observational learning in birds: The establishment of behavioural modes. *Behaviour* **17**:71-80.
Klopfer, P. H. (1962). *Behavioral Aspects of Ecology,* Prentice-Hall, Englewood Cliffs, N. J.
Klopfer, P. H. (1969). *Habitats and Territories,* Basic Books, New York.
Klopfer, P. H. (1970). *Behavioral Ecology,* Dickenson, Belmont, Calif.

Kohn, B., and Dennis, M. (1972). Observation and discrimination learning in the rat: Specific and nonspecific effects. *J. Comp. Physiol. Psychol.* **78**:292–296.

Konorski, J. (1963). Analiza patofizjologiczna Róznych Rudzajoiv Zaburzen Mowy I Próba ich Klasyfikacji. *Rozprawy Wydziatre Nauk Medycznych (Warsaw)* R. 6 **2**:11–32. Cited by Thorpe, W.H. (1967). Vocal imitation and antiphonal song and its implications. In Snow, D. W. (ed.), *Proceedings of the XIV International Ornithological Congress,* Blackwell, Oxford, England, pp. 245–263.

Kortlandt, A. (1940). Eine Übersicht der angeborenen Verhaltungsweisen des Mittel-Europäischen Kormorans (*Phalacrocorax carbo sinesis,* Shaw and Nodd), ihre Funktion, ontogenetische Entwicklung und phylogenetische Herkunft. *Arch. Néerl. Zool.* **4**:401–442. Cited by Armstrong, E. A. (1965). *Bird Display and Behaviour. An Introduction to the Study of Bird Psychology,* rev. ed., Dover, New York, p. 181.

Kriazhev, V. I. (1929). The objective investigation of the higher nervous activity in a collective experiment. *Vysshaya Nervnaya Deyatel'nost* **1**:291. (*Psychol. Abst.* **8**:2532, 1934.)

Kuhn, T. S. (1962). *The Structure of Scientific Revolutions,* University of Chicago Press, Chicago.

Lane, H. (1961). Operant control of vocalizing in the chicken. *J. Exptl. Anal. Behav.* **4**:171–177.

Larsson, K. (1956). *Conditioning and Sexual Behavior in the Male Albino Rat,* Almqvist and Wiksell, Stockholm.

Lashley, K. S. (1930). The mechanism of vision: III. The comparative visual acuity of pigmented and albino rats. *J. Genet. Psychol.* **37**:481–484.

Lemon, R. E. (1968*a*). Coordinated singing by black-crested titmice. *Can. J. Zool.* **46**:1163–1167.

Lemon, R. E. (1968*b*). The relation between organization and function of song in cardinals. *Behaviour* **32**:158–178.

Lemon, R. E., and Scott, D. M. (1966). On the development of song in young cardinals. *Can. J. Zool.* **44**:413–428.

Lescrenier, M. C., and Wijffels, H. (1969). Effects of observation, audience and social facilitation on the rat (*Rattus norvegicus*) in an instrumental learning situation. *J. Psychol. Norm. Pathol.* **66**:205–227.

Lloyd Morgan, C. (1896). *Habit and Instinct,* Edward Arnold, London.

Lockard, R. B. (1971). Reflections on the fall of comparative psychology: Is there a message for us all? *Am. Psychologist* **26**:168–179.

Lore, R., Blanc, A., and Suedfeld, P. (1971). Empathic learning of a passive-avoidance response in domesticated *Rattus norvegicus. Anim. Behav.* **19**:112–114.

Lorenz, K. (1937). The companion in the bird's world. *Auk* **54**:245–273.

Marler, P. (1970). A comparative approach to vocal learning: Song development in white-crowned sparrows. *J. Comp. Physiol. Psychol.* **71**:1–25 (Monogr. Suppl. No. 2).

Masaaki, Y. (1953). Social learning in the white rat. *Jap. J. Psychol.* **24**:13–20.

Miller, A. H. (1942). Habitat selection among higher vertebrates and its relation to intraspecific variation. *Am. Naturalist* **76**:25–35.

Miller, N. E., and Dollard, J. (1941). *Social Learning and Imitation,* McGraw-Hill, New York.

Mowrer, O. H. (1950). On the psychology of "talking birds": A contribution to language and personality theory. In Mowrer, O. H. (ed.), *Learning Theory and Personality Dynamics. Selected Papers,* Ronald Press, New York, pp. 688–726.

Mowrer, O. H. (1960*a*). *Learning Theory and Behavior,* John Wiley and Sons, New York.

Mowrer, O. H. (1960*b*). *Learning Theory and the Symbolic Processes,* John Wiley and Sons, New York.

Mulligan, J. A. (1966). Singing behavior and its development in the song sparrow *Melospiza melodia. Univ. Calif. Publ. Zool.* **81**:1–76.

Mundinger, P. C. (1970). Vocal imitation and individual recognition of finch calls. *Science* **168**:480–482.

Murton, R. K. (1971a). The significance of a specific search image in the feeding behaviour of the wood-pigeon. *Behaviour* **40**:10–42.
Murton, R. K. (1971b). Why do some bird species feed in flocks? *Ibis* **113**:534–536.
Murton, R. K., Isaacson, A. J., and Westwood, N. J. (1971). The significance of gregarious feeding behaviour and adrenal stress in a population of wood-pigeons *Columba palumbus. J. Zool. (Lond.)* **165**:53–84.
Newton, I. (1967). Evolution and ecology of some British finches. *Ibis* **109**:33–99.
Nice, M. M. (1943). Studies in the life history of the song sparrow, II. The behavior of the song sparrow and other passerines. *Trans. Linn. Soc. N.Y.* **6**:1–238.
Nicolai, J. (1956). Zur Biologie und Ethologie des Gimpels (*Pyrrhula pyrrhula,* L.). *Z. Tierpsychol.* **13**:93–132.
Nicolai, J. (1959). Familientradition in der Gesanzsentwichlung des Gimpels (*Pyrrhula pyrrhula,* L.). *J. Ornithol.* **100**:39–46.
Nicolai, J. (1964). Der Brutparasitismus der Viduinae als ethologisches Problem. Prägungsphänomene als Faktoren der Rassen- und Artbildung. *Z. Tierpsychol.* **21**:129–204.
Norton-Griffith, M. (1967). Some ecological aspects of the feeding behaviour of the oyster catcher, *Haematopus ostrealegus,* on the edible mussel, *Mytilus edulis. Ibis* **109**:412–425.
Nottebohm, F. (1970). Ontogeny of bird song. *Science* **167**:950–956.
Nottebohm, F. (1972). The origins of vocal learning. *Am. Naturalist* **106**:116–140.
Oldfield-Box, H. (1967). Social organization of rats in a "social problem" situation. *Nature* **213**:533–534.
Oldfield-Box, H. (1970). Comments on two preliminary studies of "observation" learning in the rat. *J. Genet. Psychol.* **116**:45–51.
Pettersson, M. (1956). Diffusion of a new habit among greenfinches. *Nature (Lond.)* **177**:709–710.
Platt, J. R. (1964). Strong inference. *Science* **146**:347–353.
Porter, J. P. (1910). Intelligence and imitation in birds: A criterion of imitation. *Am. J. Psychol.* **21**:1–71.
Powell, R. W. (1968). Observational learning *vs.* shaping: A replication. *Psychon. Sci.* **10**:263–264.
Powell, R. W., Saunders, D., and Thompson, W. (1968). Shaping, autoshaping, and observational learning with rats. *Psychon. Sci.* **13**:167–168.
Ramsay, A. O., and Hess, E. H. (1954). A laboratory approach to the study of imprinting. *Wilson Bull.* **66**:196–206.
Roberts, D. (1941). Imitation and suggestion in animals. *Bull. Anim. Behav.* **1**:11–19.
Romanes, G. J. (1885). *Mental Evolution in Animals,* Kegan Paul, French and Co., London.
Ross, S., and Ross, J. G. (1949a). Social facilitation of feeding behavior in dogs: I. Group and solitary feeding. *J. Genet. Psychol.* **74**:97–108.
Ross, S., and Ross, J. G. (1949b). Social facilitation of feeding behavior in dogs: II. Feeding after satiation. *J. Genet. Psychol.* **74**:293–304.
Rothschild, M., and Ford, B. (1968). Warning signals from a starling, *Sturnus vulgaris,* observing a bird rejecting unpalatable prey. *Ibis* **110**:104–105.
Rozin, P., and Kalat, J. (1971). Specific hungers and poison avoidance as adaptive specializations of learning. *Psychol. Rev.* **78**:459–486.
Russo, J. D. (1971). Observational learning in hooded rats. *Psychon. Sci.* **24**:37–38.
Scott, J. P., and McCray, C. (1967). Allelomimetic behavior in dogs: Negative effects of competition on social facilitation. *J. Comp. Physiol. Psychol.* **63**:316–319.
Seligman, M. E. P. (1970). On the generality of the laws of learning. *Psychol. Rev.* **77**:406–418.
Sheldon, M. H. (1971). Stimulus functions in some chained fixed-ratio schedules of reinforcement. *J. Exptl. Anal. Behav.* **15**:311–317.
Shettleworth, S. J. (1972). Constraints on learning. In Lehrman, D. S., Hinde, R., and Shaw, E. (eds.), *Advances in the Study of Behavior,* Vol. 4, Academic Press, New York.
Shlaifer, A., and Breder, C. M. (1940). Social and respiratory behavior of small tarpon. *Zoologica (N.Y.)* **25**:493–512.

Skinner, B. F. (1937). Two types of conditioned reflex: A reply to Konorski and Miller. *J. Gen. Psychol.* **16**:272–279.

Skinner, B. F. (1962). Two "synthetic social relations." *J. Exptl. Anal. Behav.* **5**:531–533.

Small, W. S. (1900). An experimental study of the mental processes of the rat. *Am. J. Psychol.* **11**:133–165.

Smith, W. (1957). Social "learning" in domestic chicks. *Behaviour* **11**:40–55.

Solomon, R. L., and Coles, M. A. (1954). A case of failure of generalization of imitation across drives and across situations. *J. Abnorm. Soc. Psychol.* **49**:7–13.

Spence, K. W. (1937). Experimental studies of learning and the higher mental processes in infra-human primates. *Psychol. Bull.* **34**:806–850.

Staddon, J. E. R., and Simmelhag, V. L. (1971). The "superstition" experiment: A re-examination of its implications for the principles of adaptive behavior. *Psychol. Rev.* **78**:3–40.

Steiniger, F. (1950). Beiträge zur Soziologie und sonstigen Biologie der Wanderratte. *Z. Tierpsychol.* **7**:356–379.

Stenhouse, D. (1962). A new habit of the redpoll, *Carduelis flammea,* in New Zealand. *Ibis* **104**:250–252.

Stimbert, V. E. (1970). A comparison of learning based on social or nonsocial discriminative stimuli. *Psychon. Sci.* **20**:185–186.

Stimbert, V. E., Schaeffer, R. W., and Grimsley, D. L. (1966). Acquisition of an imitative response in rats. *Psychon. Sci.* **5**:339–340.

Stubbs, D. A. (1971). Second-order schedules and the problem of conditioned reinforcement. *J. Exptl. Anal. Behav.* **16**:289–313.

Swynnerton, C. F. M. (1942). Observations and experiments in Africa by the late C. F. M. Swynnerton on wild birds eating butterflies and the preferences shown. *Proc. Linn. Soc. Lond.* **1**:10–46.

Thorndike, E. L. (1898). Animal intelligence; an experimental study of the associative processes in animals. *Psychol. Rev. Monogr. Suppl.* **2**(4):1–109 (Whole No. 8).

Thorndike, E. L. (1901). Mental life of the monkeys. *Psychol. Rev. Monogr. Suppl.* **3**(5): 1–57 (Whole No. 15).

Thorpe, W. H. (1961). *Bird Song. The Biology of Vocal Communication and Expression in Birds,* Cambridge University Press, Cambridge, England.

Thorpe, W. H. (1963). *Learning and Instinct in Animals,* 2nd ed., Harvard University Press, Cambridge, Mass.

Thorpe, W. H. (1967). Vocal imitation and antiphonal song and its implications. In Snow, D. W. (ed.), *Proceedings of the XIV International Ornithological Congress,* Blackwell, Oxford, England, pp. 245–263.

Thorpe, W. H. (1969). The significance of vocal imitation in animals, with special reference to birds. *Acta Biol. Exptl.* **29**:251–269.

Tinbergen, L. (1960). The natural control of insects in pinewoods. 1. Factors influencing the intensity of predation by songbirds. *Arch. Néerl. Zool.* **13**:265–336.

Tinbergen, N. (1951). *The Study of Instinct,* Oxford University Press, London.

Tinbergen, N. (1963). On aims and methods of ethology. *Z. Tierpsychol.* **20**:410–433.

Tinbergen, N., and Perdeck, A. C. (1950). On the stimulus situation releasing the begging response in the newly hatching herring gull chick (*Larus a. argentatus,* Pont.). *Behaviour* **3**:1–38.

Tolman, C. W. (1964). Social facilitation of feeding behaviour in the domestic chick. *Anim. Behav.* **12**:245–251.

Tolman, C. W. (1965*a*). Emotional behaviour and social facilitation of feeding behaviour in domestic chicks. *Anim. Behav.* **13**:493–496.

Tolman, C. W. (1965*b*). Feeding behaviour of domestic chicks in the presence of their own mirror images. *Can. Psychol.* **6**:227 (abst.).

Tolman, C. W. (1965*c*). Social dominance and feeding behaviour in domestic cockerels. *Psychol. Rep.* **17**:890.

Tolman, C. W. (1967). The feeding behaviour of domestic chicks as a function of rate of pecking by a surrogate companion. *Behaviour* **29**:57–62.

Tolman, C. W. (1968). The varieties of social stimulation in the feeding behaviour of domestic chicks. *Behaviour* **30**:275–286.

Tolman, C. W. (1969). Social feeding in domestic chicks: Effects of food deprivation of nonfeeding companions. *Psychon. Sci.* **15**:234.

Tolman, C. W., and Wellman, A. W. (1968). Social feeding in domestic chicks: A test of the disinhibition hypothesis. *Psychon. Sci.* **11**:35–36.

Tolman, C. W., and Wilson, G. F. (1965). Social feeding in domestic chicks. *Anim. Behav.* **13**:134–142.

Turner, E. R. A. (1964). Social feeding in birds. *Behaviour* **24**:1–46.

Verplanck, W. S. (1970). An "overstatement" on psychological research: What is a dissertation? *Psychol. Rec.* **20**:119–122.

Vince, M. A. (1969). Embryonic communication, respiration and the synchronization of hatching. In Hinde, R. A. (ed.), *Bird Vocalizations,* Cambridge University Press, Cambridge, England, pp. 233–260.

Warden, C. J., and Jackson, T. A. (1935). Imitative behavior in the rhesus monkey. *J. Genet. Psychol.* **46**:103–125.

Watson, J. B. (1908). Imitation in monkeys. *Psychol. Bull.* **5**:169–178.

Welty, J. C. (1934). Experiments in group behavior of fishes. *Physiol. Zool.* **7**:85–128.

Williams, D. R., and Williams, H. (1969). Auto-maintenance in the pigeon: Sustained pecking despite contingent non-reinforcement. *J. Exptl. Anal. Behav.* **12**:511–520.

Wilson, G. F. (1968). Early experience and facilitation of feeding in domestic chicks. *J. Comp. Physiol. Psychol.* **66**:800–802.

Zajonc, R. B., Heingartner, A., and Herman, E. M. (1969). Social enhancement and impairment of performance in the cockroach. *J. Personal. Soc. Psychol.* **13**:83–92.

Chapter 3

BEHAVIORAL ASPECTS OF PREDATION[1]

John R. Krebs

Institute of Animal Resource Ecology[2]
University of British Columbia
Vancouver, British Columbia

I. ABSTRACT

In this paper, I discuss four hypotheses concerning mechanisms which might be of general importance in the behavior of predators: hunting by searching image, hunting by expectation, area-restricted search, and "niche" hunting. In addition, I review briefly two approaches which have been taken in studying predator behavior at a more general level: experimental component analysis and optimal foraging theory. I make no attempt to review comprehensively the literature on behavior of predators.

The concept of hunting by searching image has been used in a great variety of senses, and to be of value it should be used in a more restricted manner. Searching-image formation in the strict sense of "learning to see" has been demonstrated in several laboratory experiments and perhaps in some seminatural field experiments with birds. The real problem is that it is difficult to distinguish in operational terms between "learning to see" and some other phenomena which result in the predator's preferring one prey type over another. This is especially true in the field; thus the value of the concept of "learning to see" in field studies must be questioned. Gibb's hypothesis of hunting by expectation still needs to be tested in a critical way. Although the hypothesis

[1]Financial support was provided by grant N.R.C. 67–6295.
[2]Present address: Department of Zoology, University College of North Wales, Bangor, Caernarvonshire, U.K.

was put forward in the context of birds hunting for insect larvae in pine cones, essentially the same problem confronts any predator hunting for clumped prey, namely, when to leave one clump and go on to the next one. Some alternative hypotheses are discussed. Area-restricted search after capturing a prey item seems to be a fairly widespread phenomenon. At least some predators can learn to alter their searching path to improve their exploitation of particular dispersions of prey. The hypothesis of "niche" hunting is supported by some laboratory studies and indirectly by several pieces of field evidence. Experimental component analysis is an attempt to investigate and synthesize all the behavioral factors underlying the response of a predator to changes in prey density. This approach has been more concerned with determining the existence and effect of particular behavioral processes than with elucidating the mechanisms underlying each process. So far it has only been used in studying the behavior of predators in simple laboratory situations. Optimal foraging models have provided useful hypotheses for looking at particular mechanisms. More general models seem to make either very "weak" predictions about behavioral mechanisms (ones which could be produced by many other hypotheses) or ones which are not easily testable.

II. INTRODUCTION

In this paper, I shall be examining some hypotheses and approaches used by workers studying the behavioral mechanisms by which predators search for, detect, and capture their prey. I use the term "predator" in a wide sense, and not just to refer to animals traditionally described as "predatory" (see also de Ruiter, 1967). I will not attempt to deal with all aspects of predator behavior or with antipredator adaptations of prey; nor will I deal with more ecological aspects of predation such as the role of predators in regulating prey populations or influencing species diversity. Hopefully, however, a knowledge of predator behavior is helpful in understanding how and why predators respond to changes in prey density, and why predators choose to eat particular types of prey and not others. Other general discussions of predation can be found in the papers of de Ruiter (1967), Murdoch (1972), Salt (1967), Schoener (1971), and Royama (1971).

The paper falls into three parts. The first discusses the work of Holling (1965, 1966), which is an attempt to analyze predator behavior in a general way; the second examines four hypotheses concerning the behavior of predators which have been widely accepted in the ethological and ecological literature; the third looks briefly at some optimization models of predator behavior.

III. EXPERIMENTAL COMPONENT ANALYSIS

The most comprehensive attempt so far to study the factors influencing the number of prey captured by a predator is that of Holling (1959a,b, 1965, 1966). The general rationale of his approach is that a complex behavioral phenomenon such as predation can be broken down into simple components. Each component can be analyzed experimentally and the relationships within the component described mathematically. These mathematical relationships can then be synthesized into a model which hopefully describes what goes on in the real world. The validity of the model can then be tested by experiments which examine its predictions.

Throughout his work, Holling has mainly concerned himself with the way in which a predator responds (in terms of numbers of prey eaten) when it is confronted with different densities of prey (the so-called functional response of a predator to prey density). The idea was to start off by considering only how an individual predator responds to prey density and to ignore complicating factors such as hunger of the predator and social interactions between predators.

A. Analysis of the Functional Response

Holling's first study (Holling, 1959a) was of a simple field situation, in which three species of rodents were feeding mainly on the cocoons of the European sawfly (*Neodiprion sertifer*) (other prey made up only a small proportion of the predators' diet). When Holling plotted the number of cocoons eaten against their density, he obtained the type of curve shown in Fig. 1. The predators did not increase their rate of attack in a linear fashion with increasing prey density, but at first responded only slightly to the prey increase, then responded rapidly, and finally did not respond to further increases in prey density. It should be borne in mind that this picture is the result of looking at the numbers eaten in several different areas and not the result of describing the changes in behavior of particular individuals. Holling did, however, look at the changes in behavior of individual predators in the laboratory. In these experiments, he presented caged *Peromyscus* with different densities of sawfly cocoons buried in the soil at the bottom of the cage. There was also a continuous supply of an alternative food (dog biscuits), but the mice preferred cocoons when they were available. These experiments essentially confirmed the field results. A further interesting point is that two out of the three mice tested never ate a diet completely composed of cocoons at high densities even though excess cocoons were present.

Holling (1959b) subsequently examined the nature of the functional response curve more closely using an artificial predator–prey system—a blindfolded secretary searching for sandpaper discs pinned to a tabletop.

The discs were laid out at different densities, and the predator was allowed to search for 1 min at each density by tapping her fingers on the desk. The predator removed each disc as she found it. The functional response curve obtained from this experiment differed from that found with the mice (Fig. 2).

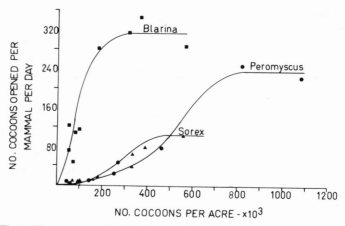

Fig. 1. The response of a vertebrate predator to changes in the density of one of many potential prey types. The data were collected in the field and refer to the predation by three species of small mammals on cocoons of *Neodiprion sertifer*. All three species of mammal show an S-shaped response curve—the so-called type-3 functional response. [After Holling (1959a).]

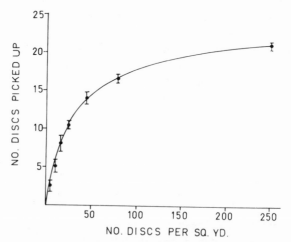

Fig. 2. The results of an experiment in which a "predator" (a blindfolded secretary) hunted for "prey" (sandpaper discs) at different densities. The curve is a so-called type-2 functional response. [After Holling (1959b).]

Holling found that the curve could be described accurately by the equation

$$y = \frac{T_t a x}{1 + a b x}$$

where T_t is the total time predator and prey are exposed to one another; b is the handling time (time taken to pursue and catch one prey and be ready to start searching again); a is a constant, the attack rate; and x is the prey density. This is the so-called disc equation (for obvious reasons). Basically the equation tells us that, all other things being equal, the reason that predators take a smaller and smaller proportion of the prey as density goes up is because as the rate of feeding goes up the predator spends a greater and greater proportion of its time handling the prey. The extreme case of this, that a chick confronted with 5000 grains cannot eat any faster than one confronted with 500 grains, is more or less trivial. However, more interesting is the fact that when several prey types are available the predator rarely takes 100% of one type (see below).

Since Holling (1959b) found that data from several studies of insect parasitoids fitted the disc equation, he named this type of functional response the "invertebrate" or type-2 response. The type-1 response shown by, for example, filter feeders is a linear rise to a plateau (Holling, 1965). The S-shaped response shown by the mice was labeled the "vertebrate" or type-3 response. It is important to note that Holling did not intend to imply that all vertebrates would always respond to changes in prey density in one way and all invertebrates always respond in another way. A predator would normally show a type-3 response only if there was more than one prey type available, since the type-3 response results from the predator's suddenly concentrating its attack on one type. If the predator failed to react to the prey at all below a certain density because, for example, the prey was a completely new type, a predator could show a type-3 response with only one prey species. If only one prey type were available, a vertebrate predator would show a type-2 response, as in Holling's disc experiment (also Ivlev, 1961).

Holling further pursued his study by examining (1) the effects of food deprivation on the type-2 response and (2) the nature of the factors underlying the S-shaped type-3 response. The effects of food deprivation were investigated using the mantid (*Hierodula crassa*) preying on houseflies (Holling, 1963, 1966). This study showed that the predator's "reactive distance," the distance from which the mantid first responded to the fly, increased with increasing hunger (and so therefore did the amount of time spent pursuing the prey), while the "digestive pause" after each capture (part of the handling time) decreased. Subsequent studies of fish have shown, however, that these findings are not applicable to all predators (Beukema, 1968;

Ware, 1971), perhaps because "hunger" in vertebrates is more complex than in invertebrates (Hinde, 1970). Thus at least in the mantid, the shape of the type-2 response results not only from an increase in the proportion of time spent handling prey at high densities but also from the fact that the predator searches less efficiently as prey density goes up and it becomes less hungry.

The S-shaped curve which Holling obtained from his studies of mice is similar to that found by Tinbergen (1960) in great tits. Tinbergen plotted "percentage in diet" against density of prey, and Holling plotted "number eaten" against density, but this does not affect the basic nature of the curve. Both authors interpret the sudden acceleration in the response curve as being a result of learning by the predator. Tinbergen thought of the learning primarily in terms of the predator's learning to "break the crypsis" of the prey, while Holling was much less specific about what the predator learns. Holling did, however, suggest a specific way in which learning could influence the predator's behavior (Holling, 1965). He argued that learning alters the threshold at which the predator will respond to the prey. As a result of one capture, the predator "learns a lesson" which, if the prey is palatable, decreases the attack threshold. Immediately the predator starts to "forget" the lesson, and, if sufficient time elapses before the next capture, the predator's attack threshold will return to the prelearning level. If, however, the predator encounters several prey in rapid succession, the "lessons" will have a cumulative effect and produce a longer-term and greater change in the attack threshold. Thus the extent to which the predator learns about a particular prey is a result of the interaction between learning and forgetting, and only when the prey density is high enough will the predator learn more rapidly than it forgets. At this point, the attack rate will suddenly increase. Holling was not specific about the time scale of the learning-and-forgetting process, but in his *Peromyscus* the scale seemed to be a matter of hours rather than minutes. Holling suggests that the type of learning he postulates differs from Tinbergen's idea in that the process is cumulative rather than a sudden threshold change. However, as we shall see later (Dawkins, 1971a), learning to break the crypsis of the prey may be a cumulative process of learning and forgetting, similar to that suggested by Holling.

Earlier it was mentioned that, in Holling's experiments with *Peromyscus*, not only did the functional response flatten off at high densities but also the response was such that the mice never ate a diet purely composed of their favorite food—even though it was highly abundant. Tinbergen observed a similar phenomenon in great tits; they never ate a completely monotonous diet even when one prey species was extremely abundant. Tinbergen explained this in terms of the predator's controlling its diet so as to maintain variety (presumably a mechanism similar to that of "specific hungers";

Rogers, 1967; Hughes and Wood-Gush, 1971). Kear (1962) showed that finches are capable of maintaining a varied diet and compensate after being forced to eat a monotonous diet. Holling, however, interprets the result in a different way. He argues that one can regard the two types of food, favored and less favored, in terms of different thresholds of response. The predator has to be hungrier (and therefore more tolerant of unpalatable food) in order to attack the second type of prey. Holling suggests that once the predator has eaten one of its favorite food items, it will rest until its hunger exceeds the threshold to start looking for one of the favorite prey. Assuming that it does not find food immediately, its hunger will go on rising and soon pass the threshold for the next most favored prey; if it now encounters one of these, it will eat it. Thus if the thresholds of the two prey are not too different from one another, and the second prey are rather common, the predator will automatically maintain a varied diet. It seems highly unlikely that this explanation could apply to Tinbergen's data, partly because the great tits were primarily hunting for food for their young, so their own hunger level should not influence the food they choose, partly because the birds do not eat enough at one time to have a "digestive pause," and partly because Holling's assumptions about the relative abundance of the favored and alternative prey do not seem to hold up. However, it is possible that Holling's explanation may account for the maintenance of a varied diet in some other cases. It should be borne in mind that Holling's and Tinbergen's explanations are not really alternatives, in that Holling refers to the proximate mechanism and Tinbergen refers to the ultimate survival value of maintaining a varied diet. As we shall see later, Royama (1970) offers yet a third explanation of why great tits should have a varied diet.

So far we have looked at what Holling calls three "basic" components of predation (common to all predator–prey interactions)—attack rate, handling time, and total time predator and prey are exposed to one another—and two "subsidiary components" (which are not universal)—learning and hunger. Holling has also examined one other subsidiary component—interference (Griffiths and Holling, 1969). (Dill, 1972, has examined another subsidiary component, prey avoidance learning.) Interference occurs when predators interact with one another in such a way as mutually to decrease their hunting efficiency (in other words, they compete with one another). From a laboratory study of the ichneumonid parasite *Pleophilus basizonicus* attacking sawfly cocoons, Holling concluded that, at realistic parasite densities, interference was a negligible factor in influencing attack rate. This applied both to direct interference (adults attacking one another) and to avoidance of hosts which had already been attacked. Although several studies have shown that interference *can* occur (e.g., Burnett, 1953, 1958;

Wylie, 1965; Messenger and Force, 1963). Griffiths and Holling dismiss these results on the grounds that they were obtained at unrealistically high parasite densities.

Hassell (1971a,b) has argued, however, that interference between searching parasites is of importance and points to weaknesses in the design of Griffiths and Holling's experiments. Hassell, working with the ichneumonid *Nemeritis canescens* attacking larvae of the almond moth *Ephestia cantella,* found that even at parasite densities similar to those used by Holling (which are roughly similar to those found in nature), interference reduced the searching efficiency of parasites. Interference took the form of "aggressive" encounters between adult parasites, the frequency of which increased with increasing parasite density. As a consequence, each individual spent a smaller proportion of its time searching for hosts at high (parasite) densities. The important difference in design of Hassell's and Griffiths and Holling's experiments was that Hassell offered the parasites a simultaneous choice of host densities, which more closely resembles the natural situation. In this situation, the parasites spent a disproportionately large amount of their time searching in the areas of high prey density. This would of course result in an increase in the effective parasite density in local areas and magnify the effects of competition between the parasites.

B. Conclusions

Holling's approach to the study of predation is clearly unique and of interest in that it is an attempt to integrate all aspects of predator behavior. Obviously, some of the details will vary from species to species (such as the effect of hunger; see above), but this does not detract from the validity of the general approach. More serious criticism might be leveled at the oversimplification involved in Holling's approach. The question of how a predator locates an area of high prey abundance is not dealt with at all. It is not clear whether a predator confronted with many different prey species all fluctuating in density more or less independently of one another and all with different spatial distributions will show the same sort of functional response to prey density as a predator hunting in a simple environment for one, regularly distributed prey species at a time.

Holling's work has been concerned with the behavior of predators only insofar as behavioral changes influence the number of prey eaten. Thus the "components" of predator behavior he discusses are features which influence the rate of prey capture. Ethologists, however, have examined predator behavior at a more detailed level, being concerned with the mechanisms underlying Holling's "components"—especially the component of learning. Some of these more detailed analyses of mechanisms are discussed in the next four sections.

IV. SEARCHING IMAGES

Since the term "hunting by searching image" has been widely adopted in the literature in recent years, I will examine in some detail the development of, and evidence for, this idea.

The first person to use the term "searching image" in referring to the hunting behavior of predators was L. Tinbergen (1960). Over some years, he studied the abundance of several pine forest insects, most of which were highly cryptic, and palatable to birds. At the same time, he recorded the types of insects brought to their young by various birds which were nesting in his study area (the main species was *Parus major*). Tinbergen found that the proportion of different prey species in the diet of the young birds did not relate in any simple way to the relative abundance of prey in the area. The parents favored some species and disliked others. Tinbergen defined this preference of the tits more precisely, in terms of a "risk index" specific to each prey species as follows:

$$N_a = R_a D_a t$$

where N_a is the number of prey species a taken by the tits, R_a is the risk of species a, D_a is the density of a in the area, and, t is the hunting time. The "risk" of a particular species varied according to factors such as its size and palatability. In general, large (palatable) prey had the highest risk, followed by small palatable prey. Warningly colored prey (e.g., various *Diprion* species) had the lowest risk.

Most important from our present point of view, the risk of a particular species changed with time, according to changes in its abundance (at least for the two species for which the most data were collected—*Panolis* and *Acantholyda*). Mook *et al.* (1960) found a similar effect in *Bupalus*. If the risk had remained constant with changing density, the proportion of a particular species in the diet at different densities could be predicted as follows (assuming that the predator encounters prey at random):

$$\text{Percent } a \text{ in diet} = \frac{R_a D_a}{R_a D_a + R_o D_o} \times 100$$

where R_a is the risk of species a, D_a is the density of species a, R_o is the risk of other species in the diet, and D_o is the density of other species (R_o and D_o were obtained by averaging all other species). Tinbergen found that the actual proportion of *Acantholyda* in the diet at different densities deviated from the theoretical expectation, following the curve shown in Fig. 3. At low and high densities, the prey formed a smaller proportion of the diet than expected; in between, the curve showed a positive acceleration. The curve is similar to the sigmoid type-3 functional response discussed by Holling

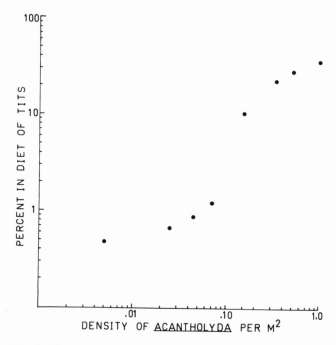

Fig. 3. The response of titmice to changes in the density of one species of prey, the larvae of *Acantholyda*. The curve is similar in shape to a type-3 functional response of Holling, but in this case the *y* axis is "percent in diet" instead of "number eaten." [After L. Tinbergen (1960).] The data for this graph were taken from Tinbergen (1960, Table X, p. 328). The points shown in the graph were obtained by dividing Tinbergen's data into eight groups according to prey density and taking an average for each group. The groups were chosen such that each contained approximately the same number of nests. It should be noted that the original data refer to different nests in different years and were very variable.

(although the two are algebraically the same only if one assumes that the total number of prey eaten is constant). Tinbergen suggested the following explanation of this curve: When the highly cryptic prey first appeared as available food for the tits, which in this instance was as a result of molting into an instar large enough for the tits to eat, the birds had difficulty in detecting the prey. As the prey density increased with time (more larvae molted into the appropriate instar), the birds came across the prey by chance more and more often. As a result of these chance encounters, the birds learned to recognize the prey against the background and the rate of predation suddenly increased. Tinbergen termed this process, primarily a perceptual phenomenon, "adoption of a specific searching image." The adaptive value of this behavior is presumably that by focusing its attention on one or perhaps two species of prey at a time, the bird can find them more efficiently than if it hunts for all species at once. Tinbergen argued that the bird will only adopt a searching image when the encounter rate, which is related to prey density, rises above a certain level and that the bird will abandon its searching image as soon as prey density drops (perhaps as a result of forgetting). Thus the bird always concentrates on the most abundant prey (all other things such as palatability being equal). As we saw earlier, Tinbergen explained the dropoff in predation rate at high prey densities in terms of the tits' need for a varied diet.

A. Evidence for Searching-Image Formation

Since Tinbergen's paper, the term "searching image" has been widely used in the literature in a variety of senses (see reviews by Croze, 1970; Dawkins, 1971a) ranging from the specific sense of "learning to see" as Tinbergen originally defined it (e.g., Dawkins, 1971a,b; Murton, 1971; Royama, 1970; Alcock, 1972) to referring to more or less any type of learning which results in a predator's preferring[3] one type of prey over another (Brown, 1969; Ware, 1971). Dawkins (1971a) has proposed that if the term "searching image" is to have any value in describing a behavioral mechanism involved in predation, it should be used in a clearly circumscribed and specific way. (Even then it describes, rather than explains, what we observe.) The types of learning which Dawkins argues must be excluded before we can accept evidence of searching-image formation in the sense of Tinbergen are as follows:

[3]Since much of the ensuing discussion involves the term "preference," it is necessary to discuss briefly the use of this term. Irwin (1958), in discussing the response of an animal to two stimuli, distinguishes between *differential response, discrimination, bias,* and *preference.* *Differential response* simply means that an animal responds to one stimulus more strongly (in some conveniently measurable way). If the differential response can be shown to be due to the animal's associating a particular stimulus with a particular outcome (e.g., triangle with food reward), one can refer to *"discrimination."* If the differential response is independent of the outcome (animal tends to choose left more than right even if the reward is equally often at left and right), it is said to be due to a *bias.* Finally, if the differential response is independent of the situation and depends purely on the outcome (e.g., animal learns to choose raisin over peanut whether raisin is hidden behind circle or square), it is said to be due to a *preference.* As Irwin himself points out, these categories are certainly not mutually exclusive; for example, one cannot show a preference without showing a discrimination. Many other authors use the term "preference" in a much less precise manner, roughly equivalent to Irwin's "differential response." In this chapter, I use "preference" in the broad sense. Searching-image formation results in a preference, and the problem is to distinguish it from all the other reasons for having a preference.

1. The predator's learning to go to a particular area to find food (Royama, 1970; Alcock, 1973; Simons and Alcock, 1971; Smith and Dawkins, 1971; Croze, 1970).
2. The predator's altering its search path after finding a prey item or learning to improve its search path so as to increase its probability of encountering prey (Dixon, 1959; Smith, 1971; Croze, 1970; Beukema, 1968).
3. The predator's learning to handle prey more effectively with increasing experience (Norton Griffiths, 1967; Orians, 1969; Recher and Recher, 1969).
4. The predator's preferring a particular prey species or avoiding a species. For example, avoidance may result from unpalatability or mimicry (Brower, 1963) or simply from novelty or rarity (Allen and Clarke, 1968; Coppinger, 1969). Preference for a particular food type may result from relatively long-term exposure, or exposure during a "sensitive period" early on in life (Rabinowitch, 1968, 1969; Capretta, 1969; Murdoch, 1969; Landenberger, 1968; Mahmoud and Lavenda, 1969). In other cases, short-term exposure (Alcock, 1973; Bryan, 1971) and oddity (Mueller, 1971) have led to a preference for a particular type of prey. In all these instances, the preference of the predator is independent of its ability to see the different types of prey, and therefore these should not be counted as cases of searching-image formation.

To these cases suggested by Dawkins we can add two other types of learning which should not be taken as evidence for "searching image."

1. Learning of specialized hunting techniques by particular individuals. For example, Tinbergen (1949) mentions an individual great tit which learned to hunt under leaves and as a result took a large number of arboreal grasshoppers. Neish (1970) found that individual *Ambystoma gracile* specialized on either benthic or water-column feeding and that this difference was probably learned. Norton Griffiths (1967) has also reported on the development in individuals of specialized feeding techniques.
2. Learning to look in a particular type of place. Krebs *et al.* (1972) found that if individual great tits found a mealworm in one of four possible types of hiding place, they would increase their searching effort in that type of place during the next few minutes.

Once we have eliminated these possibilities, we are essentially left with cases in which the predator learns to detect the prey, which is presumably cryptic and therefore difficult to detect in the first place. We will now look at some cases which do seem to show this type of learning taking place, that is, instances of true searching-image formation.

De Ruiter (1952) found that individual captive jays (*Garrulus garrulus*) and chaffinches (*Fringilla coelebs*) took a long time to find stick caterpillars (Geometridae) which were on the floor of their cage mixed with twigs from the natural resting trees. Once a bird had accidentally found one caterpillar (for example, by stepping on it), it very rapidly found others. This seems to show that at first the birds could not distinguish between twigs and caterpillars but that after one experience most of them learned to recognize the prey. De Ruiter found that the caterpillars were only protected if they were on the "right" background of twigs (the right tree species from the right place). This shows that the birds were not simply failing to realize that the caterpillars were in fact potential food.

Beukema (1968) uses the term "searching image" in a broad sense, to include any behavior of the predator that alters the risk, in Tinbergen's sense, of a particular prey species. In his laboratory work with sticklebacks he did, however, demonstrate what appears at first to be searching image in the strict sense. Individual sticklebacks were accustomed to hunting for *Tubifex* in a large tank subdivided into 18 "cells," and Beukema studied their response to the introduction of a new prey species (either *Drosophila* larvae or bits of *Enchytraeus*). At first, the sticklebacks ignored the new prey, but after something like 50–200 encounters the fish first started to eat the new prey. This initial learning is not searching-image formation; however, even after the fish had learned to eat the new prey, their reactive distance to them increased over the next 60 or so experiences (spread over a period of several weeks). This result could be interpreted as showing that the sticklebacks were "learning to see" the new prey.

An alternative interpretation of these results, however, is that the fish were not learning to detect the prey but forming a stronger and stronger preference for a prey type as a result of becoming more familiar with it.

The work of Ware (1971) helps to resolve this problem. His results are exactly parallel to those of Beukema. When rainbow trout (*Salmo gairdneri*) were first confronted with a novel, palatable food, they ignored it. After several days of exposure, they learned to accept the food, and during the next 6 days (about 40 trials), their reactive distance to the new prey increased.

Ware carried out his experiments with two main types of prey; both were standard-sized pieces of chicken liver, one "white" (blanched) liver and the other black (dyed) liver. The background of the test tank was black; thus the black prey were cryptic and the white prey conspicuous. There was evidence that this distinction held for the trout as well as for humans, since the fish reacted to the white prey from a greater distance. In one series of tests, naive fish were trained on black prey and in another series on white prey. Training consisted of repeatedly exposing the fish to the prey at 48-hr intervals. The results in both test series were similar; after an initial lag, the fish started to eat the prey, and the reactive distance then increased over several days. The fact that similar results were obtained with both cryptic and conspicuous prey shows that the change in reactive distance was not a function of the ability of the fish to see the prey. The white prey were easily visible from the start. The results show simply that the trout built up a preference for a familiar prey type and responded to it more strongly. Thus neither Beukema nor Ware has demonstrated searching image in the strict sense.

The work of Croze (1970) is the only detailed attempt so far to investigate the mechanism of searching-image formation in the field. The crucial experiments were carried out with hand-raised and wild crows hunting for artificial prey. The prey were mussel shells (*Mytilus edulis*) painted various colors (red, black, and yellow were the basic colors) and laid out on a shingle beach. Although the shells are referred to as "red," "black," and "yellow," they were muted versions of these colors and were highly cryptic to the human eye, resembling the pebbles on the beach in shape, size, and color. The shells were placed over bits of meat, which acted as a reward to stimulate the crows to hunt. In his experiments with hand-raised birds, Croze laid out several groups of three shells (one of each color) on the beach. He showed the birds a shell of one color (by pointing to it so that the bird would turn it over and find the meat) and then allowed a crow to walk around the beach looking for prey. The results showed that if the crow had had two or three "training" experiences, it would selectively detect and prey on shells of the training color. In other words, the crows were forming a searching image on the basis of two or three experiences. It seems reasonable to conclude that the crows were actually learning to see, since the prey were cryptic and other types of learning were excluded. This experiment of Croze did have a weakness in its design in that within a series of tests the prey population was apparently not replenished. Thus if the crow hunted, for example, for red shells in the first run of a series and was then shown a black shell at the beginning of the second run, the prey population was now depleted with respect to red shells. This does not, however, invalidate the general conclusion that the crows were forming searching images.

The crows were not completely specific in their searching image and would sometimes turn over stones similar in color to the shell for which they had formed a searching image. Croze investigated the specificity searching image in more detail using wild crows. In these experiments, he trained wild crows to hunt for red shells and then offered populations consisting of pairs of shells, one red and one a variant color, shape, or texture. The pairs, which were unrewarded, were spaced out on the shingle beach. In general, these results showed that the searching image for red shell was specific but was generalized to somewhat similar objects.

Den Boer (1971) studied predation by captive tits (*Parus* spp.) on two color morphs, yellow and green, of the larvae of *Bupalus piniarus*. Normally the tits ate fewer of the green morph than the yellow when the two were presented in equal numbers on green pine twigs (the natural resting place). However, the two types were taken in equal numbers when presented on a brown background; thus the birds did not have, for example, any taste preference for the green ones. In an experiment with a coal tit, Den Boer found that if he gave the bird prior experience with green larvae on green twigs, it took more green than yellow when presented with equal numbers of both on a green background. Thus the birds apparently normally have difficulty in detecting the green larvae, and as a result of experience learn to see them.

A similar effect has been described by Popham (1941, 1942), who studied predation by rudd (*Scardinius eryopthalmus*) on color morphs of the corixid bug *Sigara distincta*. Popham found that when he offered equal numbers of two brown morphs, one of which matched the background (cryptic) and one of which did not (conspicuous), the rudd took more of the conspicuous form. If a population of predominantly cryptic morphs was offered, the rudd took proportionately more cryptic morphs than they did in the 50:50 population. Popham's results are less convincing than those of Den Boer, because whenever a bug was eaten during an experiment it was replaced with a similar morph. Presumably, the bug which had just been put into the tank was more conspicuous (because of movement) than the others, and therefore more likely to be eaten. Thus once the rudd had started eating a particular morph, the experimental procedure would encourage them to carry on eating that type and give the appearance of having formed a searching image.

The field experiments of Murton (1971) indicate that pigeons (*Columba palumbus*) form searching images very rapidly, as did the birds in Croze's experiments. Murton's experiments involved wild pigeons searching for several types of more or less cryptic grain on fields of clover or stubble. The main seeds used were maize (yellow), maple peas (mottled brown), green peas (pale green), and tic beans (dark brown). These seeds were all treated with a stupefying drug which acted on the birds in about 15–30 min. The idea, then, was to lay out mixed populations of seeds at various densities, allow a flock of wild pigeons to feed on the area, collect the birds after they had drugged themselves, and examine their crop contents. This method does not give the exact sequence in which the seeds were eaten but tells the proportions of the different types eaten in relation to their availability.

When Murton laid out roughly equal densities of maple peas and tic beans, he found that of 133 birds feeding in the area, in the 15–30 min of the experiment, 72% had specialized on either one seed type or the other. Slightly fewer birds specialized on tic beans, since these were a less favored food of the whole. Murton argues that this rapid specialization of the birds is a result of searching-image formation; the birds learned about one type of seed in the first few encounters and subsequently concentrated on that type for the next 20 min. The individuals who failed to form a searching image presumably did not meet a long enough run of one type of seed to form one. Further experiments demonstrated essentially similar points—the pigeons always concentrated on a particular type of seed. The relative numbers concentrating on each type of seed depended on relative density and the birds' basic preferences. Since each test involved different individuals, it is conceivable that Murton's results simply demonstrate that different individual pigeons prefer different types of seed (see Brown, 1969). This explanation would require, however, that different flocks of birds (in different trials) were made up of individuals with different preferences, which seems unlikely.

Murton also noted an additional important point, that the risks of the various seeds varied from one test to another independently of changes in density. In some tests the birds would, for example, on average prefer maple peas over green peas, and then in a subsequent trial more birds would specialize on green peas than on maple. Murton explained this phenomenon by suggesting that the birds in the flock copy one another, so that the overall preference of the flock might be governed by the choice made by the first few individuals to start feeding. By chance, these first few individuals would choose different seeds on different trials.

In all the cases discussed so far, the evidence that the predator learns to see is indirect, although, apart from the results of Beukema and Ware, the most reasonable interpretation of the data is that this is happening (but see Shettleworth, 1972a). The most direct evidence of learning to see by a predator comes from the elegant laboratory work of Dawkins (1971a,b). She worked with domestic chicks searching for colored rice grains on various backgrounds. By matching the color of the rice and the background, the prey could be made highly cryptic and since the prey were standardized except for color and crypsis, factors such as palatability, handling time, and search pattern could be ignored.

When chicks were presented with either conspicuous or cryptic rice grains (e.g., green grain on orange, or orange on an orange background), they would immediately start to eat the conspicuous grains, but only ate the cryptic prey after an initial lag of several minutes. Once the chicks had started to find the cryptic grains, they rapidly increased their attack rate on them (Fig. 4). This experiment suggested that the chicks were initially unable to detect the cryptic grains but after a few minutes they had learned to see them. In the first tests the conspicuous and cryptic grains were of different colors, but in some later tests the grains were the same color and only the background varied. Thus we are not dealing purely with a preference for one

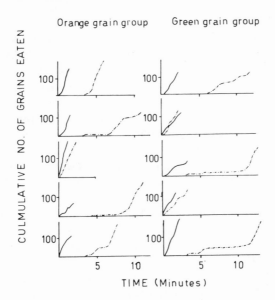

Fig. 4. Evidence of "learning to see" in ten domestic chicks. The graphs show, for each chick, the number of conspicuous prey eaten (solid line) and number of cryptic prey eaten (broken line) during a test period. The prey were orange or green rice grains on an orange or green background. In most cases, the chicks first took the conspicuous prey (which contrasted in color with the background) and only after some delay ate the cryptic prey (matching the background). [After Dawkins (1971a).]

color. When the chicks were learning to see, they were not simply learning to peer at the ground, and Dawkins argues that the change must have been at a more central level. Further experiments showed that the chicks would forget how to detect cryptic grains within about 24 hr. This forgetting could also be shown on a much shorter time scale (a few minutes). There was a correlation between the ability to detect a cryptic grain (as measured by peck latency) and the time since finding the previous cryptic grain (Dawkins, 1970). For the prey animals, this would place a premium on living as far away as possible from the next individual of the same type, as Tinbergen *et al.* (1967) and Croze (1970) have shown. The rate of forgetting is influenced by the other items eaten in the time between finding two cryptic prey. Chicks which had just been eating conspicuous orange grains (orange on a green background) were more likely to forget how to detect cryptic grains, even if the cryptic grains were also orange (on an orange background).

The types of changes which Dawkins has demonstrated in these experiments are clearly related to those labeled as examples of selective attention by psychologists (Hinde, 1970; Triesman, 1964, 1969; Broadbent, 1965; but see Shettleworth, 1972a). In particular, the Sutherland and Mackintosh (1971) theory of discrimination learning involves mechanisms closely analogous to those of searching-image formation. These authors suggest that when an animal learns a discrimination problem, it first learns to attend to particular cues associated with the stimuli and then learns what response to make. Many types of evidence support this idea; for example, rats trained to perform one discrimination involving brightness cues will subsequently do better than naive controls in other tasks involving brightness discrimination. This suggests that rats can become attuned to brightness cues. Dawkins pursued the idea that the chicks were selectively attending to particular types of cue in a manner similar to rats learning a simple discrimination. She examined the possibility that chicks hunting for conspicuous grains attended to color cues, while those looking for cryptic grains (i.e., hunting by searching image), being unable to attend to color ones—the grains were the same color as the background—switched their attention to some other cues such as shape and size. Various experiments supported this general idea. In general, chicks which had recently been pecking at conspicuous grains attended to color cues, while those looking for cryptic grains did not (Dawkins, 1971b). Thus searching image formation seemed to involve a switch of attention by the chicks.

B. Switching and Searching-Image Formation

Murdoch (1969) has discussed the concept of "switching" by predators. "Switching" refers to any situation in which a predator changes its behavior to exert disproportionately heavy predation on the prey of greatest relative abundance. Murdoch defines switching mathematically in terms of a null hypothesis: switching does *not* occur when

$$\frac{P_1}{P_2} = \frac{cN_1}{N_2}$$

where P_1 and P_2 are the proportions of two prey types eaten, N_1 and N_2 are the numbers, and c is a constant. Thus Murdoch's formulation is the same as that of Tinbergen (1960) except that Tinbergen uses the term "relative risk" to describe the proportionality constant:

$$\frac{N_a}{N_o} = \frac{R_a D_a}{R_o D_o}$$

where N_a/N_o is the relative occurrence in the diet and D_a/D_o is the relative numbers available. Both models assume that encounters between predator and prey are random.

In terms of behavioral mechanisms underlying switching, any of the types of learning discussed in this paper, including searching-image formation, could produce the effect. Murdoch himself worked with the predatory whelk (*Acanthina spirata*) and found that long-term "training" (4–9 weeks) could produce a preference for either *Balanus glandula* or *Mytilus edulis* as prey. Before training, *Acanthina* showed no particular preference for either prey type. In experiments where the predator did have a strong preference at the beginning, no switching occurred as a result of training.

C. Polymorphism and Searching Images

It has frequently been asserted that selection pressure from predators could maintain or favor the evolution of genetic polymorphisms in prey populations (Ford, 1964; Curio, 1966; Croze, 1970; Clarke, 1962, 1969). One way in which predators could maintain a polymorphism is by taking a disproportionately large number of the commoner form. This type of selection, a form of frequency-dependent selection, has been termed "apostatic" selection (Clarke, 1962). Obviously, searching-image formation is one behavioral mechanism which could underly apostatic selection, since the predator may form a searching image for the most abundant prey type (which could by morph of a species) and prey on it until it becomes rare.

However, it seems that apostatic selection will work even if a predator does not form a searching image, simply because predators like to eat food with which they are familiar and tend to avoid rare food types (see references cited earlier). Allen and Clarke (1968) studied predation by various species of wild passerines on artificial prey (pastry "larvae") of two colors, brown and green. The experiments were carried out on two types of background, brown soil and green grass. The results, although not conclusive, suggested that irrespective of background resemblance, the rarer of two morphs in a prey population suffered proportionately less predation than the commoner. These results have since been confirmed by Allen (1972).[4] Further, if the birds were familiarized over a period of several days with one color and then given a population of half green and half brown, they took more of the familiar color. This result is not, however, an example of searching-image formation but of avoidance of unfamiliar food (since the avoidance of rare types did not seem to be dependent on crypsis). Similar results have been obtained by

[4]Allen (1972) also found that if the artificial prey were presented at an extremely high density (300 in a 19-cm-diameter circle), birds chose the rarer of two morphs. Perhaps in this rather artificial situation, the birds treated the common type as a "background" against which the rare types stood out.

Manly *et al.* (1972). Although I have argued that there is a distinction between "learning to see" and "forming a preference for familiar food," the difference between the two is obviously not clear-cut and depends on exactly how "learning to see" is defined in operational terms.

Several studies have shown that predators will avoid completely unfamiliar food even if it is highly palatable and easily visible (Coppinger, 1969, 1970; Ware, 1971; Rabinowitch, 1968; Shettleworth, 1972*b*; Morrell and Turner, 1970; Reighard, 1908). In these instances, the predator's avoidance of the unfamiliar food is not apparently based on an inability to detect it and thus does not involve searching-image formation. These results are not directly comparable to those of Allen and Clarke, since these authors were not studying the avoidance of completely novel food.

Croze (1970) describes an experiment which does seem to relate searching-image formation to the maintenance of polymorphism. He studied predation by wild crows on either monomorphic or polymorphic populations of mussel shells; the morphs were color variants. The populations were of equal total numbers; therefore, the density of each morph was lower in the polymorphic than in the monomorphic population. The results showed that crows took fewer shells from the polymorphic than monomorphic populations and also that each morph suffered proportionately less mortality when it was in a polymorphic population. Croze suggests that this is because the birds failed to form a searching image when they were hunting for the polymorphic population (because the density of each morph was too low) and thus hunted less efficiently. The fact that the crows did not take runs of one particular morph supports this idea. It is possible that the well-known polymorphism of the snail (*Cepea nemoralis*) could serve to foil a predatory searching image (Clarke, 1962).

Thus hunting by searching image could act to maintain a polymorphism in a prey population, but apostatic selection can arise simply because predators prefer familiar food to rare food.

D. Conclusions

In summary, it seems that the question is not whether the phenomenon of searching-image formation exists but rather the extent to which predators hunting for their natural prey form and use searching images. A variety of psychological studies have demonstrated, at the behavioral level, that animals including man have the ability to attend selectively to certain cues in the environment or to improve their ability to detect cryptic objects by knowing what to look for (Sperling *et al.*, 1971; Engel, 1971), and the various ethological studies described above have demonstrated similar phenomena in birds looking for food.

There do not seem to be any convincing instances of searching-image formation in vertebrate predators other than birds; this may simply be because no critical experiments have been designed to test it. However, all studies which have demonstrated the existence of searching-image formation in our restricted sense have done so in a more or less artificial situation, and the gap remaining seems to be in showing that the same mechanisms operate in the real world. Unfortunately, it is not at all clear as to how this can be done. The type of observation made by Tinbergen (1960) is inadequate to tell us in detail about the predators' behavior, and yet to make direct observations and know about availability of prey, etc., is often not practicable. In particular, it often seems difficult to separate out "learning to see" and "preference for familiar food types." To do this, one needs to know the predators' previous history in terms of feeding, and the degree of crypsis of the prey in addition to having a clear operational definition of "learning to see." Murton's (1971) study seems the closest to reaching a compromise in this situation. In any case, it is highly likely that even if predators in the wild do use searching images in hunting for their prey, many other factors such as learning where to search for food (see below), time required to handle the prey, and its palatability and nutrient value will play an important role in determining which particular prey item a predator will eat.

As was mentioned earlier, the presumed survival value of forming a searching image is that the predator is more effective in finding prey when it hunts for one thing at a time, "knowing what to look for" rather than searching for many different things at once. Croze's (1970) polymorphism experiment does indeed suggest that the birds were more efficient predators when they hunted for one prey at a time. Murton (1971), however, found that pigeons which concentrated on one seed type in his experiments did not do any better (in terms of numbers of seeds eaten) than birds which failed to form searching images. Thus the evidence concerning the survival value of searching-image formation is equivocal.

V. HUNTING BY EXPECTATION

The idea of "hunting by expectation" was first suggested by Gibb (1962a) on the basis of his detailed studies of predation by blue and great tits on the larvae of *Ernarmonia conicolana* in areas of pine forest (Gibb, 1958, 1962a,b, 1966). The idea has subsequently been discussed by other authors (Emlen, 1973; Orians, 1971; Simons and Alcock, 1971; Smith and Dawkins, 1971; Tullock, 1971).

A. Field Evidence

The larvae of the eucosimid moth *Ernarmonia* feed on pine seeds between July and September. The fully fed larva then eats its way to the surface of the cone and lives in a small chamber just under the surface until it pupates in March. Usually, not more than one or two, but occasionally up to eight, larvae occupy each cone. The tits search for the larvae either by tapping the cone or perhaps by looking for slight marks on the surface. Since the tits leave a mark where they have extracted a larva, Gibb was able to estimate the number of larvae available to the tits (by examining the cones for larvae) and the number eaten. As regards the availability of *Ernarmonia,* Gibb took two measures: the *intensity,* that is, the number of larvae per 100 cones from an area, and the *density* per square meter (member of cones per square meter times intensity per cone). The study area was an extensive pine plantation which Gibb divided into a number of "compartments," each of about 6 ha. The measurements of intensity were made for each compartment, and in some cases for small 15 m^2 subplots *within* compartments. The *intensity* and *density* of larvae were not always correlated, since some areas had a poor cone crop and heavy infestation of larvae and the tits seemed to be influenced by both factors in their predation. When Gibb examined the difference in percentage predation between areas of different intensity, he found the results shown in Fig. 5. In one year predation was clearly intensity dependent; in the next year it was not, although in this year there was threshold intensity below which the tits ignored *Ernarmonia.* Gibb argues that the intensity-dependent predation was produced by the tits' learning to concentrate their searching effort in areas of high intensity. The reason why predation was not intensity dependent in the second year was that the density of cones was so

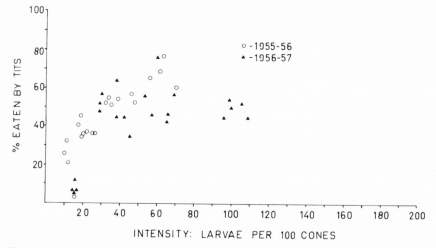

Fig. 5. Predation by titmice on the larvae of *Ernarmonia conicolana* in pine cones. In one year (1955–1956), predation was intensity dependent. In the next year, it was not, although there was a threshold below which very few larvae were taken. [After Gibb (1958).]

low that the tits were discouraged from looking for larvae even in high-intensity areas. This suggestion is borne out by the fact that within one compartment where cone density was high in the second year, predation was intensity dependent. In all cases, below a minimum intensity of about 15 larvae per 100 cones, the tits ignored *Ernarmonia*. Gibb argues that this is not because the birds were failing to detect the larvae (as Tinbergen's searching-image hypothesis would suggest) but that they examined the low-intensity cones and rejected them as being unprofitable (see also below for discussion of Royama, 1970).

The idea of hunting by expectation came from examining the pattern of predation *within* a compartment (i.e., fine-scale variations). Here again, the intensity of larvae varied from plot to plot, some trees or groups of trees being heavily infested, others hardly at all. On the whole, the tits hunted within a compartment in an intensity-dependent fashion (as mentioned above), but there were some interesting exceptions. In one area where the average intensity was 26 per 100 cones, one tree had an intensity of 256 per 100 cones. In this tree, the tits took fewer larvae from each cone than would be expected if they were searching in the usual intensity-dependent fashion. Gibb argued that the tits had learned to *expect* a certain number of larvae per cone in that area and stopped searching in a cone after it had yielded the expected number. Gibb (1962a) also looked at the data for another area and found a suggestion of the same trend—the tits seemed to form an expectation and leave a cone after finding the expected number of larvae. However, this effect applied only over a limited range of intensities; when the intensity was above 100 larvae per 100 cones, the tits seemed to stop hunting by expectation and resume their intensity-dependent predation.

Although Gibb's hypothesis was put forward to explain the specific case of tits hunting in pine cones, the problem of how long a predator should continue to hunt in a particular prey patch is a quite general one. Thus Gibb's hypothesis, if confirmed, could have quite general application.

B. Discussion

Simons and Alcock (1971) have correctly pointed out that if one plots *numbers* of larvae eaten per cone on larval intensity, the leveling off of the curve is much less marked than if one plots *percentage* eaten, as Gibb did. In other words, the evidence does not convincingly support the idea that the birds stop searching after finding a fixed number of larvae.

An alternative interpretation of Gibb's data, which retains the idea of expectation, is to postulate that the birds spend a fixed amount of *time* on each cone, rather than eating a fixed number of larvae. If, in addition, one makes the reasonable assumption that the time taken to find one larva is inversely proportional to the number of larvae on the cone it follows that

$$t = \frac{c}{L} + T_h$$

where t is the time taken to search for one larva, L is the number of larvae in the cone, T_h is the time taken to "handle" one larva, and c is a constant. If T_{max} is the maximum time the bird is willing to spend on a cone,

$$\text{Number eaten per cone} = \frac{T_{max}}{c/L + T_h}$$

$$= \frac{T_{max} L}{c + T_h L}$$

This is basically Holling's disc equation, and the number eaten will therefore follow a type-2 functional response curve (see earlier). Thus one could explain Gibb's observation that the birds take proportionately fewer larvae from abnormally high-density cones by assuming that the bird operates by spending a fixed time on each cone. Of course, in addition it is assumed that the intensity-dependent predation results from the birds' concentrating their effort in areas of high intensity.

Tullock (1971) has also discussed Gibb's data, but his discussion concerns the pattern of predation *between* compartments. Tullock does not seek to explain why the birds take fewer larvae than expected from high-intensity cones (which is what hunting by expectation is all about) but asks why the birds in general concentrate more attention on heavily infested areas. His argument is basically the same as Gibb's (and Royama's), that the bird "shops around" for the area which offers the greatest reward per unit of searching effort and spends most of its time there. This would produce the intensity-dependent predation which Gibb observed.

Gibb inferred that the birds were hunting by expectation as a result of examining cones after the tits had searched in them. This method makes the tacit assumption that each cone is gleaned by only one bird. If many birds searched in the same cone one after the other, each hunting with a different expectation, the cones would not have shown any sign of "hunting by expectation" when Gibb collected them. Although it seems possible that birds do avoid cones which have already been gleaned, data are needed to test this assumption.

One further point is worth mentioning: whether hunting by expectation should be regarded as an adaptive strategy[5] which maximizes the birds' food

---

[5]Some authors (e.g., Schoener, 1971) have proposed a distinction between a predator's *strategy* (ultimate goal) and *tactics* (techniques of achieving the goal). However, since any behavior could be viewed as a strategy or tactic, depending on the level of analysis, I have not maintained the distinction in this chapter.

intake, or whether it is purely a nonadaptive consequence of the predator's inability to adjust its searching persistence on the occasional high-intensity cones.

Charnov (1973) has developed an optimal foraging model dealing with the question of when a predator should stop hunting in a particular patch and go on to the next one. For our present purposes, a patch can be equated with a cone or a group of cones. The model, which makes the assumption that once the predator is in a patch its rate of food intake will decline with time (for example because of depletion effects), asks how the predator can maximize its rate of energy gain, or if the prey items are of equal size, maximize its rate of capture. The predator has an average rate of capture for any particular environment (e.g., a group of trees), and if the predator behaves in an optimal fashion it will leave a patch as soon as its capture rate drops below this average value. If the environment is rich, so that the average rate of capture is high, the predator will give up in a patch more readily than in a poor environment. These predictions can be tested by measuring "giving up times" (Croze 1970), which should be shorter in a rich environment than in a poor one, and should be constant over all patches within an environment. From the point of view of Gibb's data, the optimal foraging model does not predict that the birds will take a constant number of larvae from all cone types, but that the predator should take larvae in an intensity-dependent fashion, as Gibb found between compartments.

In summary, the idea of hunting by expectation in its original form is not clearly supported by the original data, although a modification of the hypothesis which suggests that the birds hunt by time expectation might fit the field evidence. Hunting by expectation is not an optimal searching strategy predicted by a simple model. The alternative hypotheses could, however, be tested by further field and/or laboratory work. Simons and Alcock (1971) have shown that caged white-crowned sparrows (*Zonotrichia leucophyrys*) can rapidly learn to focus their attention on one of four possible "cones" (blocks of wood with food hidden in six small holes) with food in it. The position of the rewarded block was varied from test to test, but the birds learned to "sample" the blocks, by looking in one or two holes, and only searched more persistently in the rewarded block. This type of learning would account for the intensity-dependent predation which Gibb observed. Smith and Dawkins (1971) have similarly shown that great tits in aviaries will learn to focus most of their hunting attention on high-density areas when offered a choice of several areas. The idea of hunting by expectation, however, has not yet been critically tested.[6]

---

[6]In a recent study of captive black-capped chickadees (*Parus atricapillus*), it was found that the birds did not hunt by expectation, but bore out the predictions of optimal foraging theory (Krebs, Ryan, and Charnov, in preparation).

VI. SEARCH PATHS

In this section, I shall consider whether or not predators search in anything approaching an optimal fashion and are able to adaptively modify their strategy according to the distribution of prey. I will not discuss the problem of how one decides when an animal is actually searching for food (Smith, 1971).

A. Random Searching

In a simulation model, Cullen (cited in Smith, 1971) has shown that in general an ongoing strategy of movement is more efficient for a predator than a completely random pattern, since it reduces the amount of re-searching of areas already exploited (see also Cody, 1971; Paloheimo, 1971a,b). The birds studied by Smith did hunt with an ongoing strategy, as did the sticklebacks studied by Beukema (1968). In spite of a lack of quantitative data, it seems likely that this result is generally applicable to predators.

B. Area-Restricted Searching

Croze (1970) and Tinbergen et al. (1967) have demonstrated experimentally that degree of predation suffered by populations of artificial prey is dependent on the distance between individual prey items. Populations of cryptic prey (in Croze's case, green pastry "larvae" on a green field; in the experiment of Tinbergen et al., camouflaged eggs on a dune area) were laid out, and wild crows were conditioned to hunt in the area for the prey. The prey were always laid out in a regular rectangular grid pattern, but the distance between prey items varied from one test to another. In general, the number of larvae surviving was greater in scattered populations than in crowded ones. Tinbergen et al. and Croze both argue that this was a result of "area-restricted searching" by the predator. After finding one prey item, the predator tended to search for a while in the immediate vicinity, thus increasing the likelihood of capture of nearby prey. Some data from natural prey populations also suggest that predator selection should lead to spacing out (Horn, 1968; Krebs, 1971).

However, Croze and Tinbergen et al. did not directly record the movement tracks of the crows during their experiments, and, as Smith (1971) has shown, their results could be equally well explained without postulating area-restricted searching. Several studies of insects have, however, directly demonstrated the occurrence of area-restricted searching (e.g., Banks, 1957;

Dixon, 1959; Kaddou, 1960; Bansch, 1966; Chandler, 1969). In most of these cases, the area-restricted searching resulted from an increase in the rate of turning of the predator after finding a prey item. The behavior is presumably adaptive in that the prey hosts of the species studied usually occur in clumps in nature. Further, since more animals in nature are clumped or random than spaced out (Southwood, 1966), a predator which is not capable of learning to modify its searching pattern to take advantage of local differences in prey spacing should choose area-restricted searching as the best "average strategy." The only detailed field study of this phenomenon is that of Smith (1971), who found that thrushes (*Turdus merulus* and *T. philomelos*) showed a small but significant area-restricted searching after capture of a prey item when they were hunting for earthworms on a meadow. The main behavioral change of the birds after finding a worm was that they increased their tendency to make several successive turns in the same direction as opposed to alternating the direction of turns, which they normally do. This would of course result in the birds' tending to stay in the area of the capture. Since the earthworms were aggregated in distribution, the thrushes were searching in an adaptive way, if one assumes that availability is reflected in the distribution. Laboratory studies by Beukema (1968) and Krebs *et al.* (1972) have also demonstrated the occurrence of area-restricted searching after a capture in sticklebacks and great tits.

C. Modification of Searching Strategy Through Learning

Beukema (1968) studied the ability of sticklebacks to improve their searching strategy as a result of experience. His apparatus consisted of a large aquarium divided into 18 hexagonal cells with interconnecting doorways. In one series of experiments, he allowed individual sticklebacks to search for a *Tubifex* worm placed in one of the 12 cells forming the outer ring of his 18-cell maze. He recorded over a series of tests the encounter efficiency E—the number of prey encountered per cell visited. Beukema calculated that if the fish had searched at random they would have encountered four prey per 100 cells visited (E of 0.04), while the optimal strategy would produce an E of 0.20. The real fish had an average E of about 0.10; their strategy was better than random, but not optimal. The main way in which the fish tracks improved on random was by having an ongoing tendency (see above). As the test series proceeded, the fish learned to decrease their rate of turning and to hunt mainly in the outer 12 cells. Both of these resulted in an increase in their encounter efficiency.

Smith (1971) examined the tracks of thrushes hunting for artificial prey distributed on a meadow in a variety of patterns. The prey were brown or green pastry "larvae" similar to those used by Croze. The dispersion patterns were random, regular (rectangular grid pattern), or clumped (clumps of five prey distributed at random). Each pattern was laid out for several days in a row. The analysis of the birds' hunting tracks showed that with the random and clumped distributions they showed area-restricted searching, while with regularly dispersed food they did not. Thus there is some indication that the birds were adaptively modifying their strategy in response to the dispersion of prey.

At higher prey densities, the birds did not show such clear area-restricted searching when hunting for clumped prey. It could be argued that as prey density increases, the proportion of the predator's time spent searching decreases; thus the advantage of having a highly adapted searching strategy also decreases, the main time-consuming operations now being capture and handling of prey. Perhaps the results were not more clear-cut because the birds were influenced by the dispersion of food they ate outside the test situation. When they were hunting for more conspicuous prey (brown "larvae"), some individuals increased their speed of movement, which again would seem to be an adaptive strategy.

VII. "NICHE" HUNTING AND PROFITABILITY

We have already noted, in the discussion of Gibb's work, that great tits are able to distinguish between areas of different prey density and concentrate their hunting effort in the most profitable areas. Croze (1970) observed that his wild crows learned to hunt in a particular area for food, and even suggested that the crows might form a one-prey–one-area association (this has been demonstrated by Dawkins, 1970, for domestic chicks).

A. Evidence

The idea of searching in profitable areas has been developed by Royama (1970) to explain his observations of prey brought to nestling great tits. Royama's data are basically similar to those of Tinbergen (1960), but his conclusions are very different. Royama found, as did Tinbergen, that the type of prey brought to the young great tits did not relate in any simple way to the overall availability of different species in the habitat. However, Royama argues that certain features of the pattern of predation are at odds with the searching-image hypothesis:

1. In the case of several species, the frequency in the diet of the birds declined dramatically before there was any decline in abundance in the habitat, which certainly would not be expected on Tinbergen's

hypothesis, as the tits should be continuously reinforced for the same searching image.

2. Some prey were very common in the area but not taken in large numbers by the birds (e.g., *Tortrix viridana*) even though they were palatable.

3. All species, even rare ones, which the tits ate were taken in "runs," whereas Tinbergen assumed random encounters between predator and prey. This point is a weak one, since Tinbergen assumed only random encounters in his null hypothesis and Dawkins (1971*b*) has shown that searching-image formation will lead to runs of a particular prey type.

4. There was some evidence that the parents were eating different species themselves from those brought to the young; this again is inconsistent with the searching-image idea.

5. Spiders were always brought when the young were at a specific age. This was not related to any sudden increase in their abundance.

Having shown that Tinbergen's idea was not consistent with his data, Royama offered an alternative explanation. He argued that one can regard the prey as occurring in a number of different "niches"[7]—these could be particular areas or particular types of place—and the bird maximizes its food intake by spending most of its time in the most profitable niche. One further assumption made by Royama is that the rate of prey capture in a niche will vary according to density in the pattern suggested by Holling's disc equation. In other words, when the density is high, the rate of food intake is limited by handling time, and the profitability of a niche reaches an asymptote even if prey density increases further. Thus the hypothesis is that the bird spends most of its effort in the most profitable niches but continues to sample different niches to check whether or not their profitability has changed. If the bird obeys this model, the time it spends in any particular niche (and therefore the predation rate on prey in that niche) will vary in a sigmoid fashion with density, as observed by Tinbergen. Further, the bird will never spend 100% of its time in a niche, because it always has to spend some time sampling other niches to see if they have become profitable. The bird may abandon a niche while the density of prey is still high, if a new, more profitable niche is found. Thus Royama's hypothesis can explain the observations of Tinbergen and the anomalies in his own data. The initial lag in responding to new prey is due to the bird's sampling the niche and deeming it unprofitable, the acceleration in response at intermediate densities is when the bird spends most of its time in the niche, and the leveling off at high densities is when the niche profitability is limited by handling time and the bird is sampling other niches.

[7]Note that "niche" is not being used in the usual ecological sense.

The weight of a particular food species will also influence profitability. When the parent is carrying food to the young, it is particularly advantageous to take large items, because of the time involved in carrying each item to the nest (the bird usually only takes one item at a time). This may explain why some common palatable small species are ignored by the birds when feeding the young. When the adult is feeding itself, the time spent carrying food does not come into consideration, so a small abundant species might be more profitable than a large rare one. This could well explain the observed diet difference between adults and nestlings. Lind (1965) and Hartwick (personal communication) have made parallel observations on the European and black oystercatcher (*Haematopus ostralegus* and *H. bachmani*), respectively. In both cases, the parents bring back food items to the young which are large in relation to those normally eaten by the parents.

B. Discussion

Evidence from several studies supports Royama's idea that predators can learn to hunt in profitable areas after sampling less profitable ones. Hassell (1971a,b) and Smith and Dawkins (1971) both found that a predator would spend a disproportionately large amount of effort in the most profitable of several areas when offered a simultaneous choice. In the experiments of Smith and Dawkins with great tits, the birds apparently did not discriminate between the various lower-density areas (although the density in each area was different).[8]

In these experiments, the birds were essentially faced with a probability learning task. Various authors (e.g., Bitterman *et al.*, 1958; Bitterman, 1965) have shown that if a rat or pigeon is trained on a spatial discrimination between two stimuli, one rewarded (for example) on 75% of occasions, the other on 25%, it will end up responding 100% of the time to the majority stimulus. This is true only if no correction is applied by the experimenter to maintain the reward rates at 75:25, and results from the fact that the animal soon starts to select the majority stimulus on more than 50% of trials and by doing so decreases its likelihood of getting any reward at all on the minority stimulus. Thus the effective reinforcement schedule becomes 75:0 (Sutherland and Mackintosh, 1971). Even if the experimenter corrects for this and keeps the reward rate at 75:25, rats and pigeons tend to select the majority stimulus on more than 75% of occasions. In this case, they do not, however,

[8]Subsequent work in a similar, but slightly more complex, laboratory setup has shown that great tits do learn to distinguish between the various lower-density areas and to some extent "rank" them in terms of profitability. Further, when the most profitable area was suddenly made unprofitable (by lowering the food density), the birds tended to switch to what had previously been the second best area (J. Smith, personal communication).

reach the optimal strategy of selecting the majority stimulus on 100% of occasions. Sutherland and Mackintosh (1971) have discussed the mechanisms underlying this "failure to maximize." They argue that it is not simply that the animal pays attention to the minority stimulus if it chances to get rewarded for selecting it, but rather that the animal sometimes attends to irrelevant cues (e.g., brightness in a spatial discrimination) and therefore makes mistakes.

Alcock (1973) found that two 30-min training sessions were enough to condition red-winged blackbirds (*Agelaius phoenicus*) to search in one of two rows of holes drilled in planks of wood. Similarly, Krebs *et al.* (1972) found that even one reward was enough to stimulate great tits to search preferentially in one of several types of potential food sites. Goss-Custard (1970*a*) has presented some field evidence which suggests that redshanks (*Tringa totanus*) concentrated their searching effort in profitable areas.

In summary, Royama's hypothesis is attractive in its simplicity, and the basic ideas seem to be supported by various other studies. It also is more amenable to testing (especially in the field) than is the searching-image hypothesis. As Royama correctly points out, his hypothesis does not exclude the possibility that predators also learn to see cryptic prey; indeed, many studies have suggested that natural predators have difficulty in locating prey which appear cryptic to humans (e.g., de Ruiter, 1952, 1955; Baker, 1970), and presumably in these instances learning to see is important. It is quite possible that the predator may learn to respond to cues associated with "niche" or those associated with the prey itself according to which is rewarded more often. Further, the predator may well learn one of these two, which then acts as a reinforcer for learning the second.

VIII. OPTIMIZATION MODELS OF PREDATOR BEHAVIOR

We have already seen how the idea of optimization has provided useful insights in the discussion of hunting by expectation, searching strategies, and niche hunting. Several authors have developed optimization models concerned with predator strategies at a more general level. In this section, I will briefly examine some of these. The literature in this field has been well reviewed by Charnov (1973). In general, optimization models ask how the predator should behave in order to maximize its fitness in the evolutionary sense. Ultimately, fitness depends of course on genotype contribution to future generations, but in the models optimization is usually considered in terms of maximizing food intake per unit effort (Schoener, 1971). In discussing optimization models, it should always be borne in mind that the way in which a predator behaves is likely to be an outcome of conflicting selection pressures and therefore a compromise, and that the optimum strategy for obtaining any particular goal will vary according to environmental changes. Further, a strategy which is optimal in the short term may not be optimal in the long term. Noting these caveats, I will look briefly at two features of predator behavior which have been discussed in terms of optimization; for more detailed discussions, see Emlen (1966, 1968, 1973), MacArthur (1972), MacArthur and Pianka (1966), Menge (1972), Orians (1971), Rapport (1971), and Schoener (1969, 1971).

A. Optimal Prey Selection

Emlen (1966, 1968), MacArthur and Pianka (1966), Schoener (1971), and Charnov (1973) have proposed models to describe the optimal choice of food items. In general, these models measure the profitability of a particular prey type purely in terms of energy gained per unit searching effort and do not take into account any such factors as palatability, searching-image formation, and avoidance of novel prey. MacArthur and Pianka (1966) (further developed by MacArthur, 1972) discuss two similar graphical models, one dealing with optimal choice of prey items, the other with optimal use of patches. Both models operate on the assumptions that the predator maximizes its average rate of prey capture, and is capable of ranking prey types or patches in terms of their profitability.

The optimal diet model proposes that the time taken to deal with one prey item is made up of two components, the search time (time traveling between prey items) and the pursuit time (time taken to capture, handle, and eat one prey). The predator ranks prey by increasing handling time, so that when one more prey type is added to the diet the average search time will decrease, because the effective density of prey has increased, and the average pursuit time will increase, because the new type of prey is of a lower rank. As long as the decrease in search time is greater than the increase in average pursuit time, it will be profitable for the predator to add extra prey types to its diet. This is illustrated graphically in Fig. 6, in which the decrease in average search time and increase in average pursuit time are plotted. The curve for change in average search time is derived from the assumption that searching time is inversely proportional to the number of prey types, and the curve for increase in average pursuit time is a guess. Where the two lines intersect, the increase in average pursuit time (P) becomes greater than the decrease in average search time (S), and the predator will lose by adding more prey types to its diet.

The predictions of this model, other than that the predator should rank prey in terms of handling time, are of a general nature. For example, if the overall density of food is shifted down, the S curve in Fig. 6 will move up, but

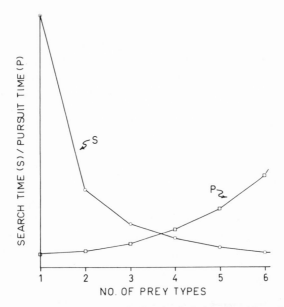

Fig. 6. A graphical model of optimal choice of number of prey types. In the particular model shown, the optimal diet contains four prey types [After MacArthur and Pianka (1966).]

the P curve will stay the same. The point of intersection of the two curves will move to the right and the optimal diet will include more items. Thus the model predicts that predators should be less specialized when food is scarce. This prediction could of course be reached from other starting hypotheses.

MacArthur (1972) has derived a more general form of the same model, based on energy gain rather than handling times. The model states that if the pursuit time per gram for a new prey type is less than or equal to the mean pursuit time per gram plus the mean search time per gram of prey items already in the diet, the predator should add the new type to its diet. Both versions of the model assume that the predator recognizes each prey type instantaneously, and that it can assess its encounter rate and energy gain from each type of prey.

Charnov (1973) has developed a model concerned with prey size selection. Assuming that the predator encounters prey at random, the model predicts that the predator will always eat prey according to their rank order of profitability and that at certain densities of large prey the predator will ignore small prey even if they are very abundant. Goss-Custard (1970b) found that redshanks will soon learn to ignore small prey even if they are much more common than large items.

Holling (1964) has examined optimal size choice in a rather different way. From a consideration of the geometry and forces of the grasping foreleg of a mantid, he predicted which size of prey the mantid would be most efficient at grasping. Behavioral experiments later confirmed that the mantid does indeed strike preferentially at prey which are optimal from the mechanical point of view. Kear (1962) has also examined choice of food items in terms of the predators' mechanical ability to deal with them. By measuring the speed with which chaffinches were able to open seeds of different sizes, she was able to rank the seeds in order of food yield per unit opening time. In choice tests, the birds did in fact prefer the two types of seed which were most "efficient," and there was some indication that interspecific differences followed the same trend. However, Kear found that there was not a simple one-to-one correspondence between efficiency and preference when all seed types were considered. Willson (1971), in a similar study of North American finches, also found that the birds did not necessarily prefer the "most efficient" seeds; this illustrates the oversimplification involved in the models discussed above.

B. Optimal Use of Patches

Royama's hypothesis suggested that great tits hunt in such a way as to utilize the most profitable "niches." The use of different niches has also been discussed by MacArthur and Pianka (1966). They refer to a niche as a "patch" and consider the question of how many patch types an optimal predator should search in. Their model, which employs the same type of reasoning as the optimal diet model, is again based on considerations of only energy and time, and patches are ranked according to their energy yield per unit time hunting. The optimal number of patches is an outcome of the interaction between reducing the time to travel between patches by adding new ones and decreasing the hunting efficiency within a patch by adding poorer-quality ones. Smith (1970) has examined the utilization of different tree species by squirrels. The squirrels utilized the most profitable species preferentially, and there was some evidence of ranking trees with one species, at least for the most favored species, lodgepole pine. J. Smith (personal communication) has evidence of great tits' ranking areas in terms of profitability. Thus the predictions of MacArthur and Pianka's model are partly borne out.

In conclusion, the predictions made by generalized optimal foraging models are on the whole too simplistic to provide much guidance in studying behavioral mechanisms of predation. In both the patch model and the optimal diet models, the predictions which are easily testable are of only a very general nature. However, discussing particular predator tactics in terms of optimization has proved fruitful.

IX. GENERAL CONCLUSIONS AND SUMMARY

I have discussed briefly two different general approaches to studying predator behavior—optimization models and experimental component analysis—and have also examined some of the more specific hypotheses concerning predator behavior. General optimization models seem at the moment to take too simplistic a view of predator behavior to be of much value to the ethologist, although in examining specific mechanisms and tactics the idea of optimization has proved useful. Experimental component analysis is of more interest, since it provides a general framework within which to view specific behavioral mechanisms and has given some insight into the way in which predators respond to prey density. It does not, however, concern itself particularly with the types of ideas about predator behavior that ethologists have studied.

The concept of searching image has been used in a variety of senses, and, as Dawkins (1971a) has argued, it should be used in a more specific way if it is to be used at all. I have used it in the specific sense of "learning to see." It is clear that birds, and perhaps other vertebrates, under artificial or semiartificial conditions do learn to see cryptic prey, but the question remains as to whether this is of importance in governing prey selection by predators in the wild. Further, it will probably be difficult to demonstrate conclusively that natural predators do form searching images, given all the other types of learning, especially a preference for familiar prey, which have to be eliminated.

Three ideas which are more amenable to investigation in the field are hunting by expectation, area-restricted searching, and niche hunting. The latter two of these are already supported by some laboratory and field data.

X. ACKNOWLEDGMENTS

I wish to thank the various members of the Institute of Animal Resource Ecology for discussion of ideas. Drs. J. Alcock, Ric Charnov and S. Shettleworth kindly allowed me to see their unpublished manuscripts. I am most grateful to Drs. J. M. Cullen, M. Dawkins, K. Krebs, W. W. Murdoch, and J. N. M. Smith for their critical comments on an earlier draft of this paper.

XI. REFERENCES

Alcock, J. (1973). Cues used in searching for food by redwinged blackbirds *Agelaius phoeniceus. Behaviour* (in press).

Allen, J. A. (1972). Evidence for stabilizing and apostatic selection by wild blackbirds. *Nature (Lond.)* **237**:348–349.

Allen, J. A., and Clarke, B. (1968). Evidence for apostatic selection by wild passerines. *Nature (Lond.)* **220**:501–502.

Baker, R. R. (1970). Bird predation as a selective pressure on the immature stages of the cabbage butterflies *Pieris rapae* and *P. brasscae*. *J. Zool. Lond.* **162**:43–59.

Banks, C. J. (1957). The behaviour of individual coccinellid larvae on plants. *Brit. J. Anim. Behav.* **5**:12–24.

Bansch, R. (1966). On prey-seeking behaviour of aphidophagous insects. In Hodek, I. (ed.), *Ecology of Aphidophagous Insects,* Academia, Prague.

Beukema, J. J. (1968). Predation by the three-spined stickleback (*Gasterosteus aculeatus* L.): The influence of hunger and experience. *Behaviour* **31**:1–126.

Bitterman, M. E. (1965). Phyletic difference in learning. *Am. Psychologist* **20**:396–410.

Bitterman, M. E., Tyler, D. W., and Elam, C. B. (1958). Some comparative psychology. *Am. J. Psychol.* **71**:94–110.

Broadbent, D. E. (1965). Information processing in the nervous system. *Science* **150**:457–462.

Brower, J. V. Z. (1963). Experimental studies and new evidence on the evolution of mimicry. *Proc. 16th Internat. Zool. Congr.* **3–4**:156.

Brown, R. G. B. (1969). Seed selection by pigeons. *Behaviour* **34**:115–131.

Bryan, J. E. (1971). Prey specialization by individual trout living in a stream and pond: Some effects of feeding history and parental stock on food choice. Ph. D. thesis, University of British Columbia.

Burnett, T. (1953). Effects of temperature and parasite density on the rate of increase of an insect parasite. *Ecology* **34**:322–329.

Burnett, T. (1958). Effect of host distribution on the reproduction of *Encarsia formosa* Gahan (Hymenoptera: Chalcidoidae). *Can. Entomol.* **90**:179–191.

Capretta, P. J. (1969). The establishment of food preferences in chicks *Gallus gallus*. *Anim. Behav.* **17**:229–331.

Chandler, A. E. F. (1969). Locomotory behaviour of first instar larvae of aphidophagous Syrphidae (Diptera) after contact with aphids. *Anim. Behav.* **17**:673–678.

Charnov, E. L. (1973). Optimal foraging theory: some theoretical explorations. Ph. D. thesis, University of Washington.

Clarke, R. B. (1962). Balanced polymorphism and the diversity of species. In Nichols, D. (ed.), *Taxonomy and Geography,* Oxford University Press, New York.

Clarke, R. B. (1969). The evidence for apostatic selection. *Heredity* **24**:347–352.

Cody, M. L. (1971). Finch flocks in the Mohave Desert. *Theoret. Popul. Biol.* **2**:142–148.

Coppinger, R. F. (1969). The effect of experience and novelty on avian feeding behaviour with reference to the evolution of warning colouration in butterflies. *Behaviour* **35**:45–60.

Coppinger, R. P. (1970). The effect of experience and novelty on avian feeding behaviour with reference to the evolution of warning coloration in butterflies. II. Reactions of naive birds to novel insects. *Am. Naturalist* **104**:323–337.

Croze, H. J. (1970). Searching images in carrion crows. *Z. Tierpsychol. Beiheft.* **5**:85 pp.

Curio, E. (1966). Die Schutzanpassungen dreier Raupen eines Schwärmers (Sphingidae) auf Galapogos. *Zool. Jahrb. Syst.* **91**:1–29.

Dawkins, M. (1970). The mechanism of hunting by "searching image" in birds. D. Phil. thesis, Oxford University.

Dawkins, M. (1971a). Perceptual changes in chicks: Another look at the "search image" concept. *Anim. Behav.* **19**:566–574.

Dawkins, M. (1971b). Shifts of "attention" in chicks during feeding. *Anim. Behav.* **19**:575–582.

Den Boer, M. H. (1971). A colour polymorphism in caterpillars of *Bupalus piniarius* (L.) (Lepidoptera: Geometridae). *Netherlands J. Zool.* **21**:61–116.

de Ruiter, L. (1952). Some experiments on the camouflage of stick caterpillars. *Behaviour* **4**:222–232.

de Ruiter, L. (1955). Countershading in caterpillars. *Arch. Néerl. Zool.* **11**:1–57.
de Ruiter, L. (1967). The feeding behaviour of vertebrates in their natural environment. In *Handbook of Physiology,* Sect. 6: *Alimentary Canal,* Vol. I: *Food and Water Intake,* American Physiological Society, Washington, D. C.
Dill, L. M. (1972). Prey avoidance learning and the functional response of predators. Ph. D. thesis, University of British Columbia.
Dixon, A. F. G. (1959). An experimental study of the searching behaviour of the predatory coccinellid beetle *Adalia decempunctata* (L). *J. Anim. Ecol.* **28**:259–281.
Emlen, J. M. (1966). The role of time and energy in food preference. *Am. Naturalist* **100**: 611–617.
Emlen, J. M. (1968). Optimal choice in animals. *Am. Naturalist* **102**:385–389.
Emlen, J. M. (1973). *Ecology: An Evolutionary Approach,* Addison Wesley, London.
Engel, F. L. (1971). Visual conspicuity, directed attention and retinal locus. *Vision Res.* **11**: 563–576.
Ford, E. B. (1964). *Ecological Genetics,* Methuen, London.
Gibb, J. A. (1958). Predation by tits and squirrels on the eucosmid *Ernarmonia conicolana* (Heyl.). *J. Anim. Ecol.* **27**:375–396.
Gibb, J. A. (1962*a*). L. Tinbergen's hypothesis of the role of specific search images. *Ibis* **104**:106–111.
Gibb, J. A. (1962*b*). Tits and their food supply in English pine woods: A problem in applied ornithology. *Festschr. Vogelschutzwarte,* Hessen, Rheinland-Pfalz und Saarland, pp. 58–66.
Gibb, J. A. (1966). Tit predation and the abundance of *Ernarmonia conicolana* (Heyl.) on Weeting Heath, Norfolk, 1962–63. *J. Anim. Ecol.* **35**:43–54.
Goss-Custard, J. D. (1970*a*). The responses of redshank (*Tringa totanus* (L)) to spatial variations in their prey density. *J. Anim. Ecol.* **39**:91–113.
Goss-Custard, J. D. (1970*b*). Factors affecting the diet and feeding rates of the redshank (*Tringa totanus*) *Symp. Brit. Ecol. Soc.* **10**:101–110.
Griffiths, K. S., and Holling, C. S. (1969). A competition submodel for parasites and predators. *Can. Entomol.* **101**:785–818.
Hassell, M. P. (1971*a*). Mutual interference between searching insect parasites. *J. Anim. Ecol.* **40**:473–486.
Hassell, M. P. (1971*b*). Parasite behaviour as a factor contributing to the stability of insect host–parasite interactions. *Proc. Advan. Stud. Inst. Dynamics Numbers Popul.,* Oosterbeck, 1970, pp. 366–379.
Hinde, R. A. (1970). *Animal Behaviour,* 2nd ed., McGraw-Hill, New York.
Holling, C. S. (1959*a*). The components of predation, as revealed by a study of small mammal predation of the European pine sawfly. *Can. Entomol.* **91**:293–332.
Holling, C. S. (1959*b*). Some characteristics of simple types of predation. *Can. Entomol.* **91**:385–398.
Holling, C. S. (1963). An experimental component analysis of population processes. *Mem. Entomol. Soc. Can.* **32**:22–32.
Holling, C. S. (1964). The analysis of complex population processes. *Can. Entomol.* **96**: 335–347.
Holling, C. S. (1965). The functional response of predators to prey density and its role in mimicry and population regulation. *Mem. Entomol. Soc. Can.* **45**:1–60.
Holling, C. S. (1966). The functional response of invertebrate predators to prey density. *Mem. Entomol. Soc. Can.* **48**:1–86.
Horn, H. S. (1968). The adaptive significance of colonial nesting in the Brewer's blackbird (*Euphagus cyanocephalus*). *Ecology* **49**:682–694.
Hughes, B. O., and Wood-Gush, D. C. M. (1971). A specific appetite for calcium in domestic chickens. *Anim. Behav.* **19**:490–499.
Irwin, F. W. (1958). An analysis of the concepts of discrimination and preference. *Am. J. Psychol.* **58**:152–163.
Ivlev, V. S. (1961). *Experimental Ecology of the Feeding of Fishes,* Yale University Press, New Haven.

Kaddou, I. (1960). The feeding behaviour of *Hippodamia quinquesignata* (Kirby) larvae. *Univ. Calif. Publ. Entomol.* **16**:181–232.

Kear, J. (1962). Food selection in finches with special reference to interspecific differences. *Proc. Zool. Soc. Lond.* **138**:163–204.

Krebs, J. R. (1971). Territory and breeding density in the great tit, *Parus major* L. *Ecology* **52**:2–22.

Krebs, J. R., MacRoberts, M. H., and Cullen, J. M. (1972). Flocking and feeding in the great tit: An experimental study. *Ibis* **114**:507–530.

Lind, H. (1965). Parental feeding in the oystercatcher (*Haematopus o. ostralegus* (L.)). *Dansk. Orn. Foren. Tidsskr.* **59**:1–31.

Landenberger, D. E. (1968). Studies on selective feeding in the starfish *Pisaster* in Southern California. *Ecology* **49**:1062–1075.

MacArthur, R. H. (1972). *Geographical Ecology,* Harper and Row, New York.

MacArthur, R. H., and Pianka, E. R. (1966). On the optimal use of a patchy environment. *Am. Naturalist* **100**:603–609.

Mahmoud, I. Y., and Lavenda, N. (1969). Establishment and eradication of food preferences in red-eared turtles. *Copeia,* pp. 298–300.

Manly, B. F. J, Miller. P., and Cook, L. M. (1972). Analysis of a selective predation experiment. *Am. Naturalist* **106**:719–736.

Menge, B. (1972). Foraging strategy of a starfish in relation to actual prey availability and environmental predictability. *Ecol. Monogr.* **42**:25–50.

Messenger, P. S., and Force, D. C. (1963). An experimental host–parasite system: *Therioaphis maculata* (Buckton), *Praon palitans* Muesebeck (Homoptera: Aphidae, Hymenoptera: Braconidae). *Ecology* **44**:532–540.

Mook, J. H., Mook, L. J., and Heikens, H. S. (1960). Further evidence for the role of "searching image" in the hunting behaviour of titmice. *Arch. Néerl. Zool.* **13**:448–465.

Morrell, G. M., and Turner, J. R. G. (1970). Experiments on mimicry. I. The response of wild birds to artificial prey. *Behaviour* **36**:116–130.

Mueller, H. C. (1971). Oddity and specific searching image more important than conspicuousness in prey selection. *Nature (Lond.)* **233**:345–346.

Murdoch, W. W. (1969). Switching in general predators: Experiments on predator specificity and stability of prey populations. *Ecol. Monogr.* **39**:335–354.

Murdoch, W. W. (1972). The functional response of predators. *Proc. 14th Internat Entomol. Congr.*

Murton, R. K. (1971). The significance of a specific search image in the feeding behaviour of the wood pigeon. *Behaviour* **40**:10–42.

Neish, I. C. (1970). A comparative analysis of the feeding behaviour of two salamander populations in Marion Lake, B. C. Ph. D. thesis, University of British Columbia.

Norton Griffiths, M. (1967). A study of the feeding behaviour of the oystercatcher *Haematopus ostrelegus*. D. Phil. thesis, Oxford University.

Orians, G. H. (1969). Age and hunting success in the brown pelican (*Pelecanus occidentalis*). *Anim. Behav.* **17**:216–319.

Orians, G. H. (1971). Ecological aspects of behaviour. In Farner, D. S., and King, J. R. (eds.), *Avian Biology,* Academic Press, New York.

Paloheimo, J. (1971*a*). On a theory of search. *Biometrika* **58**:61–75.

Paloheimo, J. (1971*b*). A stochastic theory of search: Implications for predator prey situations. *Math. Biosci.* **12**:105–132.

Popham, E. J. (1941). The variation in colour of certain species of *Arctocorisa* (Hemiptera, Corixidae) and its significance. *Proc. Zool. Soc. Lond. Ser. A* **111**:135–172.

Popham, E. J. (1942). Further experimental studies on the selective action of predators. *Proc. Zool. Soc. Lond. Ser. A* **112**:105–117.

Rabinowitch, V. E. (1968). The role of experience in the development of food preferences in gull chicks. *Anim. Behav.* **16**:425–428.

Rabinowitch, V. E. (1969). The role of experience in the development and retention of seed preferences in zebra finches. *Behaviour* **33**:222–236.

Rapport, D.J. (1971). An optimization model of food selection. *Am. Naturalist* **105**:575–587.

Recher, H. F., and Recher, J. A. (1969). Comparative foraging efficiency of adult and immature little blue herons (*Florida caerulea*). *Anim. Behav.* **17**:320–322.

Reighard, J. (1908). An experimental field study of warning coloration in coral reef fishes. *Publ. Carnegie Inst.* **103**:257–325.

Rogers, W. L. (1967). Specificity of specific hungers. *J. Comp. Physiol. Psychol.* **64**:49–58.

Royama, T. (1970). Factors governing the hunting behaviour and selection of food by the great tit (*Parus major* L.). *J. Anim. Ecol.* **39**:619–668.

Royama, T. (1971). A comparative study of models for predation and parasitism. *Res. Popul. Ecol. Kyoto Suppl.* **1**:1–91.

Salt, G. W. (1967). Predation in an experimental protozoan population (*Woodruffia–Paramecium*). *Ecol. Monogr.* **37**:113–144.

Schoener, T. W. (1969). Models of optimal size for solitary predators. *Am. Naturalist* **103**:277–313.

Schoener, T. W. (1971). Theory of feeding strategies. *Ann. Rev. Ecol. Syst.* **2**:369–404.

Shettleworth, S. J. (1972*a*). Constraints on learning. In Lehrman, D. S., Hinde, R. A., and Shaw, E. (eds.), *Advances in the Study of Behavior,* Vol. 4, Academic Press, New York.

Shettleworth, S. J. (1972*b*). The role of novelty in learned avoidance of unpalatable "prey" by domestic chicks *Gallus gallus. Anim. Behav.* **20**:29–35.

Simons, S., and Alcock, J. (1971). Learning and the foraging persistence of white-crowned sparrows *Zonotrichia leucophrys. Ibis* **113**:477–482.

Smith, C. C. (1970). The coevolution of pine squirrels (*Tamiasciurus*) and conifers. *Ecol. Monogr.* **40**:349–371.

Smith, J. N. M. (1971). The searching behaviour and prey recognition of certain vertebrate predators. D. Phil. thesis, Oxford University.

Smith, J. N. M., and Dawkins, C. R. (1971). The hunting behaviour of individual great tits in relation to spatial variations in their food density. *Anim. Behav.* **19**:695–706.

Southwood, T. R. E. (1966). *Ecological Methods,* Methuen, London.

Sperling, G., Budiansky, J., Spivak, J. G., and Johnson, M. C. (1971). Extremely rapid visual search: The maximum rate of scanning letters for the presence of a numeral. *Science* **174**:307–311.

Sutherland, N. S. (1964) The learning of discriminations by animals. *Endeavour* **23**:148–152.

Sutherland, N. S., and Mackintosh, N. J. (1971). *Mechanisms of Animal Discrimination Learning,* Academic Press, New York.

Tinbergen, L. (1949). De invloed van roofieren op de aantalssterkte van hun prooidieran. *Vakblad. Biol.* **28**:265–343.

Tinbergen, L. (1960). The natural control of insects in pine woods. I. Factors influencing the intensity of predation by songbirds. *Arch. Néerl. Zool.* **13**:265–343.

Tinbergen, N., Impkoven, M., and Franck, D. (1967). An experiment on spacing out as a defence against predation. *Behaviour* **28**:307–321.

Triesman, A. (1964). Selective attention in man. *Brit. Med. Bull.* **20**:12–16.

Triesman, A. (1969). Strategies and models of selective attention. *Psychol. Rev.* **76**:382–399.

Tullock, G. (1971). The coal tit as a careful shopper. *Am. Naturalist* **105**:77–79.

Ware, D. M. (1971). Predation by rainbow trout (*Salmo gairdneri*). The effect of experience. *J. Fish. Res. Bd. Can.* **28**:1847–1852.

Willson, M. F. (1971). Seed selection in some North American finches. *Condor* **73**:415–429.

Wylie, H. G. (1965). Some factors that reduce the reproductive rate of *Nasonia vitripermis* (Walk.) at high population densities. *Can. Entomol.* **97**:970–977.

Chapter 4

ORIENTATION OF BIRDS BY GEOMAGNETIC FIELD

S. J. Freedman

Department of Zoology
Duke University
Durham, North Carolina

Plunged into mists by the immaculate wings of swans —George Seferis

I. ABSTRACT

Recent evidence strongly indicates that homing pigeons and migrating birds use but do not require optical clues to find their way about. We review the possible nonvisual alternatives and fix our attention on information contained in the geomagnetic field as being the most likely one on present evidence. This possibility is examined in a simple manner from the point of view of seeing whether or not any serious difficulties arise in the reception system, in the secular variations of the geophysical field, or in a "navigation" strategy. As an illustration, a strategy for golden plovers is presented. We conclude that the geophysical field is a reasonable source of information for guiding birds in their wanderings.

II. INTRODUCTION

In this chapter, we attempt to assess whether or not birds can use information contained in the earth's magnetic field alone as an instrument for guiding their wanderings. It is not suggested or expected that the magnetic field is the only physical aid used (Keeton, 1969). It is well known that both migrating birds and homing pigeons can orient to celestial objects (Alder, 1963; Emlen, 1969; Hamilton, 1962; Hoffmann, 1959; Kramer, 1953, 1955, 1959; Matthews, 1953; Sauer and Sauer, 1960; Schmidt-Koenig, 1961a,b, 1963a,b; Walraff, 1960). It is not as yet known how the birds can use this orientation ability in finding their way about, if, indeed, they use it at all. We shall not here discuss this aspect of the problem, primarily because it appears to us that orientation with respect to celestial objects is not a necessary condition for birds to find their way. This view is supported by recent experiments of several types.

It appears that radar is capable of tracking single migrating birds as well as flocks of birds. This radar evidence strongly indicates that migrating birds fly in opaque cloud cover (Bellrose and Graber, 1963; Bellrose, 1967; Bruder-er, 1971; Eastwood, 1967; Lack, 1963). Keeton (1969) has been able to induce homing pigeons to return to the home loft under what appeared to be opaque cloud cover; Schlichte and Schmidt-Koenig (1972) have had pigeons home wearing frosted contact lenses. If the main thrust of these types of experiments is true, and considerable work still remains to substantiate it, the problem which is now presented is to determine what kind of information the birds can use for finding their way about without view of the sky, without view of the ground, and most probably without view of other birds. We are dealing with a system which enables birds to find their way about on the surface of the earth without the use of optical clues. It seems of considerable importance at this point to assess briefly the various types of geophysical fields available with a view toward suggesting which of them might most reasonably be used by the birds. The nonoptical clues which present themselves as candidates are of four kinds: Coriolis forces (Ising, 1946), low-frequency sound waves (Griffin, 1969), inertial navigational means (Barlow, 1964), and the geomagnetic field (Barnothy, 1964; Yeagley, 1947, 1951). We will not here enter into a detailed discussion of all four of these possible clues. Instead, we will take the point of view that the evidence points most strongly to one of them, the geomagnetic field. The Coriolis force in essence requires a view of the ground. The forces involved are small in any case, most probably too small to be used by the birds. The possible information which low-frequency sound waves can provide seems to be of two types: the landmark type and the directional bearing type. The wave length of low-frequency sound waves would seem to be so large as to make this field incapable of providing sufficiently accurate directional information. Land-mark-type information could consist of a distribution of relative nodes and antinodes in the sound field. But it is clear that many migrating birds fly over vast expanses of ocean, which is characterized acoustically as being without discontinuities in the acoustic impedance, and therefore the resulting physical field should be acoustically uninteresting and incapable of providing sufficient information. An inertial navigation system is of course theoretically possible, as illustrated by the success of our space program, but at the present time there is *no* substantiating evidence that the birds use inertial forces. Indeed, the serious problem of determining position by double integration over the small accelerations presented in a turn of large radius raises serious doubts about whether such a system could be accurate enough to guide the birds.

The situation is not so grim with respect to the earth's magnetic field. This geophysical field has often been suspected of being involved in migration and homing and just as often rejected. Recently, however, several new pieces of evidence have come to light which tend to support the hypothesis that the birds in some manner use information from this field. The experiments of Merkel and Wiltschko (Merkel *et al.* 1965; Wiltschko and Merkel, 1966; Wiltschko *et al.,* 1971) in which wild-caught European robins (*Erithacus rubecula*) in migration restlessness, housed in an optically uninteresting enclosure, showed marginally statistically significant orientation in their activity which could be shifted by an artificially applied magnetic field, and the more recent similar work of Wiltschko and Wiltschko (1972), are perhaps the best evidence obtained in the laboratory to date. It is important to note that positive results were obtained only for magnetic fields which closely mimicked the earth's magnetic field in both magnitude and inclination. This observation may well explain the failure of previous attempts to obtain orientation with respect to an artificially applied field. The magnetic sensory system seems to contain a window. Work by Keeton (1971) as well as recent unpublished work by Walcott indicates that the homing of carrier pigeons can be affected by artificially applied fields: small magnets in Keeton's work, Helmholz coils in Walcott's work. Experiments of these types have been tried before and have proved generally unreliable. However, in Keeton's work some attempt was made to have the field at the birds' heads roughly that of the earth. In Walcott's work, the field applied to the pigeons' heads was more uniform than had been tried in previous experiments. Admittedly, in these experiments the earth's magnetic field was not accurately modeled, but the results are highly suggestive of the facts that (1) when birds are denied optical clues their behavior is affected by the earth's magnetic field and (2) the better the model of the disturbing field as compared with the earth's magnetic field, the more significant the experimental results. In recent unpublished experiments by Keeton and J. Alexander, in addition, experienced homing pigeons were clock-shifted and divided into two groups. The first group was allowed a view of the sun but not allowed exercise flights in the vicinity of the loft. The second group was allowed to fly in the vicinity of the loft. When these two groups of pigeons were released some distance from the loft, the first group had vanishing directions oriented in the time-shifted direction. The second group under certain conditions seemed to ignore completely the fact that their clock had been shifted and had their vanishing directions grouped significantly in the homeward direction. Again, it is as if pigeons react to discrepant optical information by ignoring optical clues completely.

It seems worthwhile at this point to investigate purely speculatively whether there are grounds to believe that any major barrier, geophysical or physiological, exists to the birds' using geomagnetic information. In short, is the geomagnetic field sufficiently detailed to allow any and all birds to get from here to there? Are there reasonable physiological mechanisms which might be presumed to exist for information perception and processing? On the basis of current information, we think the answer is yes: the magnetic field of the earth is a perfectly reasonable means which is always available to the birds, good weather or bad, for guiding them in their wanderings.

III. THE GEOMAGNETIC FIELD

In this section and in more detail in Appendix I, we wish to point out what seem to be the salient features of the earth's magnetic field. The field is characterized by both a spatial and a temporal variation in both direction and magnitude. Within an accuracy of about 20%, the spatial distribution of the magnetic field over the earth's surface can be attributed to a dipole directed approximately from geographical north to geographical south and located at the center of the earth. Both the position and the magnitude of this dipole change secularly with time. In addition to this large-scale feature, there are local variations of field rivaling a topographic map in complexity. The spatial variation of these so-called anomalies is very small in magnitude except in the regions of large magnetic deposits. They usually amount to, perhaps, a few parts in 10,000 on a scale of miles. It is tempting to assume that the birds use these anomalies as landmarks to guide their wanderings but is not esthetically satisfying at this point in view of the small magnitude of the effects. Let us see, then, just how far we can get using the gross variations of the field. There are clearly many questions which need to be answered. (1) What are the possible receptor mechanisms in birds? It should be noted at the outset that *no* organ for magnetic reception has ever been found in any animal. We must postulate the existence of a magnetic "organ." (2) What kind of strategy do the birds use in finding their way about? This question has been discussed previously by many others. The most usual classification includes the following:

    1. Piloting, i.e., recognition and use of landmarks.

2. Compass orientation, i.e., the ability to maintain a geographical direction without reference to landmarks.
3. Reverse azimuth strategy, i.e., keeping track by some means of direction and distance during transportation and then just reversing the procedure on the way home.
4. True navigation, i.e., the ability to determine more or less exactly one's position on the earth's surface, to know the position one wishes to reach, and to plot a course between the two.

It appears to us that this division is both arbitrary and uneconomical, and, as will be discussed below, the most reasonable strategy used by migrating birds is a kind of hybrid system consisting of a compass, an odometer, and an inherited set of instructions which would correspond in usual terms to an auto map which showed only the route to be taken, not all the surrounding roads. For pigeons, for reasons discussed below, we think it is most economical at this point to assume that they use essentially the reverse azimuth game.

IV. MAGNETIC FIELD PERCEPTION

We postulate that migrating birds and homing pigeons can orient to a static magnetic field. This ability has recently been shown to exist in the case of European robins (Wiltschko and Wiltschko, 1972). We postulate also that birds can detect and integrate the electric fields associated with changing magnetic fields. It is a well-known physical law that along any line in an object moving through lines of magnetic flux an electromotive force is induced which is proportional to the speed of motion and the strength of the field. The speed of motion here refers to ground speed, and the field is the geomagnetic field. Radar studies have shown that birds are capable of maintaining a fairly constant ground speed even in the presence of strong winds. Without optical clues, one is at a loss to explain the relative constancy of the ground speed without other geophysical clues. The earth's magnetic field provides such a clue. Birds flying through the air will have a voltage induced across any part of their bodies proportional to their ground speed, provided only that they are not flying parallel to the earth's field. Although the ground speed is remarkably constant, it does vary by as much as perhaps 20%. A navigational system with an inherent 20% error would probably annihilate the race in at most a few migrations. It does not seem reasonable that migrating birds use the induced voltage directly. It is easy to show, however (see Appendix II), that a system in which the birds *integrate* the induced voltage over the time of migration or equivalently over the migration route is independent of their speed and depends only on the migration route. That is, all birds flying the same route will have the same integrated signal. We believe that this is an important result. There is no need for the birds to fly at a precisely constant air speed. There is no need for them to make a non-stop journey. They can stop for any reasonable amount of time and make small excursions from the migration route for foraging, for water, for finding shelter, etc., and still at the end of their journey have the same integrated signal. It is unfortunately not possible at the present time to estimate the accuracy of such a system, since no means in the laboratory have as yet been found to measure the coupling between the possible magnetic sensory systems and the magnetic field.

We take as paradigm the migration of the American golden plover (*Pluvialis dominica dominica*). This example was chosen completely randomly with a view toward seeing if one could construct a simple set of instructions which would take the birds from their northern breeding grounds to Brazil. There are, in fact, at least two simple sets of instructions which will do this. As can be seen from the map of total magnetic field intensity given in Fig. 1, the birds will get where they are going if they fly in a southerly direction nearly along the maximum gradient of the geomagnetic field until the earth's magnetic field has been reduced from about 0.6 oersted to 0.25 oersted or alternately until the integrated signal is some specified number. Another set of instructions which is available and will do the job reads something like this: fly in the southeast direction until the integrated signal is a certain number, fly in a southerly direction until the integrated signal is another specified number. This kind of stereotyped behavior is indeed observed in young birds which are caught during their first migration, transported approximately perpendicular to their usual migration route, and then re-leased (Schmidt-Koenig, 1965). They tend to follow the same course as the untransported birds, shifted by the amount that they have been transported. Note that we have not required that plovers have in their possession a world map nor do they need to be able to find their position on command. They need a compass, which the magnetic field can certainly provide, an odometer (Gwinner, 1968), the integrated response, and a guidebook in their genes. We have also provided them with an approximate ground speed indicator in the form of the unintegrated induced electric field.

It is important at this point to consider the temporal variations of the earth's magnetic field. Within a few generations of man, the field can be shifted in both magnitude and direction to a degree sufficient to make any guidebook outmoded (more detail is given in Appendix I). This does not seem to present any serious difficulty to the theory. The birds simply revise their guidebook, much as any first-rate tourist service reviews its recom-mendations yearly, by a process of natural selection. The data are insuffi-cient, but it seems reasonable to suppose that there is enough of a variation in the instructions in a species so that the most popular version of the guidebook will vary correctly according to the variations of the field in the course of time. The selection involved is of the most stringent kind: birds having an inadequate set of instructions simply die in migration.

In the case of carrier pigeons, it is our contention that the crucial experiments which can decide between the reverse azimuth strategy and some other strategy, perhaps true navigation, have yet to be performed. Since we lack knowledge of the threshold and characteristics of magnetic field perception, it is most economical at this point to assume that carrier pigeons obtain the information necessary for homing from the earth's magnetic field during transportation. In order to be able to reject this hypothesis, it would seem necessary to shield the birds from the earth's magnetic field during transportation to an accuracy of perhaps one part in 10^4, corresponding to removing the possibility of their using information contained in local magnetic field anomalies in homing. The problem of the homing of carrier pigeons is of course quite different in kind from that presented to migrating birds. Migrating birds need to get from a given region at one time of year to another given region at some other time of the year. Carrier pigeons need to return from an *arbitrary* position perhaps 700 miles away from the home loft.

As described above, we seem to be dealing with a system which responds to some restricted range of magnetic fields. Magnetic fields which are either too large or too small are rejected by the system. The existence of windows in sensory systems is the rule rather than the exception. The windows we are most familiar with, however, are those concerning frequency or quality rather than intensity. As shown in Appendix III, it is a simple matter to construct a system using only two magnetic receptors which will respond only to a given range of magnetic field intensities and furthermore, with one additional stage of information processing and feedback, will provide whatever long-term accommodation is necessary. We need only suppose that the two receptors have different thresholds and more or less the same saturating levels, although this is not essential, and that the pertinent information is processed, perhaps in the neural connections themselves, as a difference signal. This system will yield a window of any width and any shape depending on the properties of the receptors involved. If a high-intensity, no-difference signal condition is present, a feedback mechanism can adjust the sensitivity of the receptors so as to provide a resetting of the window after some given length of time. If carrier pigeons are to obtain reverse azimuth information upon transportation under varying conditions of magnetic shielding provided by, for example, metal enclosures such as trucks, the accommodation at the low-intensity end of the window would have to be almost instantaneous.

Rapid accommodation of sense organs is known. No accommodation has been found so far which seems to take as long as a few days. Perhaps the resetting of the window at the high-intensity end is governed by hormonal influences.

The question of what the receptors may indeed be is a most puzzling one. There are physical forces acting directly on active nerve cells which correspond in kind to the type of induced fields described above. These are physical forces acting on the action currents produced by Lorenz-type interactions (not K. Lorenz). These forces are commonly called Hall effects or magnetoresistive effects. The finding of Wiltschko and Wiltschko that the magnetic compass of European robins apparently does not distinguish the polarity of the magnetic vector is suggestive of the possibility that the basic physical interaction itself is even in the magnetic field. Magnetoresistive effects have this property, as do magneto- and electrostrictive ones. All, however, seem at first sight to be too small unless some particular structure in the nerve itself such as, for example, regions of large local ion currents or gross changes in ionic mobilities could be assumed to be involved. The possibility of an unrecognized magnetic organ still exists, of course. It is a common observation that all other sensory modalities provide some means for increasing the coupling between the receptors and the external excitatory field, e.g., the structure of the external ear and the optical system of the human eye. We would offhand expect that a magnetic system using a gross magnetic structure rather than direct action would be characterized by a region or regions in the animal or large magnetic susceptibility. A ferromagnetic material would be best, but biological compounds are apparently not commonly ferromagnetic. Suffice it to say that no common biological compound appears promising at first sight.

V. CONCLUSION

We have given a means by which adult golden plovers can perform their fall migrations. The set of instructions involved is remarkably simple. The case of sea birds will undoubtedly involve a more complicated set of instructions, but it does not seem productive here to discuss each case in detail. The point seems to be this: the set of instructions postulated is really a set of elements of behavior. There is no reason to assume that, given the same or similar perceptual mechanisms, different species of birds will behave in the same manner. It is not at all unreasonable to suppose that each species of bird will have its own set of instructions corresponding to different detailed behavior. In fact, one or more aspects of the perceptual mechanisms themselves might well be augmented or deleted as one goes from species to species.

VI. ACKNOWLEDGMENTS

This work was written while the author was a Visiting Scholar in the Department of Zoology at Duke University. I am grateful to the Department and, in particular, to P. H. Klopfer for generously providing the facilities of his laboratories and for his patient and invariably useful advice. Thanks are due also to K. Schmidt-Koenig for helping to introduce a renegade physicist to the mysteries of bird orientation as well as for a most helpful detailed review of the manuscript. I am indebted to D. R. Griffin, C. Walcott, and W. T. Keeton for helpful discussions about work in progress and to the last named for a valuable critique of the manuscript.

VII. APPENDIX I

In this appendix, we present some salient features of the geophysical field: a brief discussion of the major features of the field, its time course in both archeological and geological time, and some of the finer variations manifested as magnetic anomalies.

The earth's magnetic field is a highly complex and subtle force field when considered in detail. Figure 1 is a map of total magnetic intensity obtained in 1955 (Valley, 1965). Superimposed on this map are shown the migration routes of the American golden plover (Dorst, 1962). It is interesting that first-flight birds, that is, young in their first fall migration, take the route normally taken by adults during the spring but of course in the reverse direction. The figure shows clearly that the migration route taken by adults in the fall nearly follows the direction of maximum gradient in the geophysical field. Their destinations are close to regions of relative magnetic extrema in the geomagnetic field.

Within about 20%, the magnetic field of the earth can be represented as a purely dipole field with the dipole located near the center of the earth. Figure 2 shows schematically the distribution of magnetic field over the surface of the earth resulting from this dipole. It is important to note that both the total intensity and the dip angle, that is, the angle that a freely suspended north-seeking magnetic pole makes with the local horizontal, vary with latitude over the surface of the earth. Information as to latitude is therefore available, although we have not suggested that this information is used in orientation. It is entirely possible that the dip angle is an important component in the system the birds actually use. Indeed, experiments by Wiltschko and Wiltschko (1972) suggest that this may be so in the European robin. There are also daily fluctuations in the strength of the earth's magnetic field which amount to perhaps 10 γ (1 γ is defined as 10^{-5} gauss). We have not

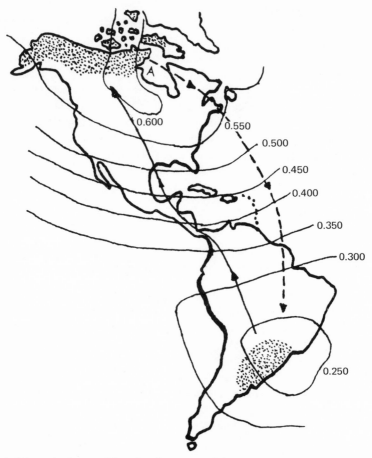

Fig. 1. Schematized map of the migration routes of the American golden plover. The dotted line shows the fall migration of adult birds. The solid line shows spring migration; reversed, the fall migration route of first-flight young. The contours are lines of constant total geomagnetic flux intensity given in gauss. Stippled areas show destinations.

considered these variations as being important, since we have not assumed that birds use the fine features of the earth's magnetic field such as anomalies in orientation.

Figure 3 shows periods of reversals of the earth's magnetic field extending back more than 10 million years. The data were obtained using magnetic patterns from the floor of the sea. The most recent data, that is, up to about 4 million years ago, were obtained from magnetic reversal patterns observed in lava flows on the earth. The dark areas show intervals when the direction of the magnetic field was the same as the present. It can be seen from ex-

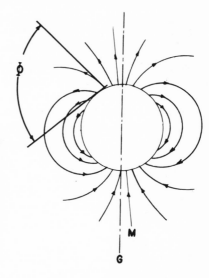

Fig. 2. Schematized distribution of magnetic flux over the surface of the earth resulting from a point dipole at the center. Such a distribution models the field of recent times to within about 20%. Note that information about geographical latitude is available both in the total intensity and in the dip angle Φ. G is the geographical axis. M is the geomagnetic axis. Geographical north is vertically upward.

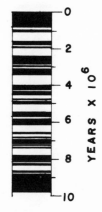

Fig. 3. Last 10 million years' worth of geomagnetic field directions. The black areas indicate a polarity which is the same as that existing today. Magnitude of the field is not known. Data shown are a fair sample of data extending more than 70 million years back, although longer quiet periods appear in the Cretaccous, Paleocene, and early Eocene (Bullard, 1969). Data for the first 4 million years, obtained from geomagnetic remanent magnetization in datable lava flows on land, overlap nicely the rest of the data obtained from measurements of magnetic anomalies on a spreading ocean floor assuming a constant rate of spreading.

amination of the figures that the pattern is quite complex. However, a few salient features probably important for bird migration can be seen. There are fairly long periods of approximately a million years when the direction is constant and then shorter periods of perhaps 5000 years at the minimum when the field reverses. It is during these periods of magnetic field reversal, obviously, that a serious problem in "navigation" using the earth's magnetic field *alone* arises. One may first of all question whether during these times migration was at all necessary. The data obviously need to be correlated in the future with data on the earth's climate at the time. The question simply put is this: supposing birds needed to migrate during these periods of magnetic field reversal, is 5000 years enough to change the set of instructions by natural selection and still have the race survive? There are several bits of information which indicate how this might indeed be possible. In the recent experiment by Wiltschko and Wiltschko, European robins (*Erithacus rubecula*), a migrating species, were shown not to be able to distinguish north from south on purely magnetic data. The authors suggest that clues as to the polar nature of the magnetic field vector arise from the direction of gravity. In addition, as we have discussed above, there is evidence that birds more or less rapidly adjust to changes in the total intensity of the magnetic field within a factor of 2 or 3. It seems possible therefore that a system yet to be discovered is capable of fairly rapidly following geophysical changes. This discussion of course refers to the "navigational system" itself. The birds also have at least 5000 migrations to change their set of instructions. With the present state of knowledge, it seems reasonable to state that the time course of magnetic field reversals presents a problem but does not rule out the use of the magnetic field in migrations during geological times. These difficulties argue for a system whereby the birds have several alternative schemes for finding their way about. This strategy is indeed the most likely one, anyway, on present evidence (Keeton, 1969).

Another clue to how rapidly birds must accommodate to changes in the magnitude of the magnetic field is given by the data presented in Fig. 4. These data show the variation in total intensity of the magnetic field obtained from the examination of lava flows and baked clay pottery during historic times.

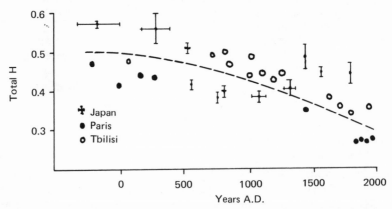

Fig. 4. Total magnetic intensity *vs.* time in recent times at three archeological sites. Data were obtained from basaltic lavas and baked pottery of known ages (Nagata, 1967). The current rate of decrease is about 0.03 % per year. [Redrawn with permission of T. Nagata.]

It is of course not possible to deal with direction, since the pottery does not maintain a fixed orientation with respect to the earth. The data show that a factor of 2 variation in the total intensity of the field in a millennium is not unreasonable. Here again, the birds must somehow reset the total integrated signal from migration to migration. The current rate of 0.03% per year for the secular decrease would not seem to present any serious problem.

Other, more fine-scale variations include magnetic anomalies and fluctuations due to magnetic storms. The former give rise to variations in magnetic field intensity which amount to perhaps 10 γ on a scale of miles. The detail is astonishingly complex. In fact, one is tempted to say that there is far too much detail to be of use to the birds in most cases. Such may not be the situation, however, in the case of oceanic islands (Mason, 1967). Here, the topography of the anomalous field often closely mimics that of the land itself and may thus be of some aid to sea birds. During great magnetic storms, lasting less than a few hours, the field variations at high latitudes may amount to 5% of the main field. The effects are much less at low latitudes (Valley, 1965).

VIII. APPENDIX II

In this appendix, we show that the integrated signal is independent of the birds' ground speed and depends only on details of the migration route.

We suppose that the output, R, of some magnetic receptor is proportional to the local induced electric field, E. Thus

$$R = KE \tag{1}$$

where K is some physiological constant. Note that this is a linear approximation which should hold over some range of ground speeds whatever the detailed response curve of the receptor. We define the integrated signal, I, as follows:

$$I = \int_{\text{time}} R \, dt = K \int_{\text{time}} E \, dt \tag{2}$$

For simplicity, we assume that the receptor is oriented to receive the maximum available induced electric field in level flight, although this is not essential to the result. We may then write the following expression for the induced electric field:

$$E = B(t) \frac{ds(t)}{dt} \sin \theta \, (t) \tag{3}$$

where $B(t)$ is the geomagnetic induction encountered during the course of time, $ds(t)/dt$ is the time-varying ground speed, and $\theta(t)$ is the time-varying angle between the ground velocity and the direction of the magnetic induction and hence is directly related to the dip angle. Substituting (3) into (2), one obtains

$$I = K\int_{t} B(t)\,\sin\theta(t)\,\frac{ds(t)}{dt}\,dt = \int_{t} B(s)\,\sin\theta(s)\,ds \qquad (4)$$

where the final form (4) is an integral over the flight path. Note that the integral depends only on details of the geophysical field over the migration route, s, and not on ground speed. Small, random excursions from the migration route will tend to cancel out, so a nonstop journey is not required.

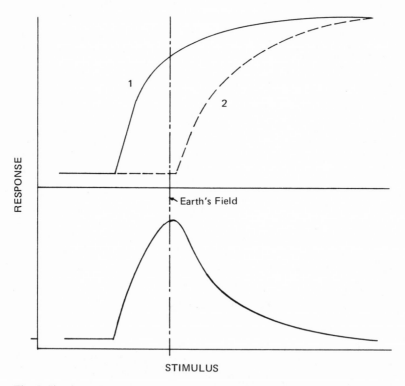

Fig. 5. Simple model of a mechanism of a physiological window. The upper curves show the response of an assumed pair of magnetic receptors, 1 and 2, to a stimulus, the local magnetic field, or to the induced electric field. The difference signal, plotted below, clearly exhibits a response in only a more or less narrow range of magnetic field, i.e., exhibits a window. The window can be shifted by feeding back a no-signal condition from the difference signal to reset one or both receptors.

IX. APPENDIX III

In this appendix, we discuss a simple mechanism by which birds can obtain a physiological "window" of any width centered about any reasonable value of the magnetic field. Figure 5 (upper) shows the response of two assumed receptors to a total applied magnetic field. (It is important to note here that this discussion can easily be extended to include information obtainable from the dip angle of the earth's field. It is merely necessary to assume sets of similar receptors oriented orthogonally.) The assumed response is really quite standard. It consists of a threshold, a more or less linear region, and a region of saturation. The diagram shows two such receptors which are shifted with respect to one another in these parameters. The area between the two curves represents the difference signal from the two receptors and is shown as the curve below. It is clear that adjustment of the relative shift and magnitude of the response parameters can result in a difference signal which has the required characteristics: namely, a window of any width centered about any reasonable magnetic field. We should note here that unsuccessful experiments of the type which attempt to confuse the birds, in particular, homing pigeons, by the application of undetermined externally fixed and time-varying fields during transportation are largely meaningless on this model. The experiments are like trying to jam a communication system without knowing the code or frequency. To then draw conclusions as to the existence or nonexistence of the communication system itself seems highly questionable, to say the least. It is becoming increasingly apparent that the birds use information from a more or less highly selective configuration of magnetic field. We must emphasize again that the model presented here is to be taken as purely speculative. The intention is to show that no obstacles from a point of view of mechanism need stand in the way of the existence of a physiological window. The technique used by the birds has, of course, yet to be discovered.

X. REFERENCES

Adler, H. E. (1963). Psychophysical limits of celestial navigation hypotheses. *Ergeb. Biol.* **26**:235–252.

Barlow, J. S. (1964). Inertial navigation as a basis for animal navigation. *J. Theoret. Biol.* **6**:76–117.

Barnothy, J. M. (1964). Proposed mechanisms for the navigation of migrating birds. In Barnothy, M. F. (ed.), *Biological Effects of Magnetic Fields*, Vol. I, Plenum Press, New York, pp. 287–293.

Bellrose, F. C., and Graber, R. R. (1963). A radar study of the flight directions of nocturnal migrants. *Proc. 13th Internat. Ornithol. Congr.*, pp. 362–389.

Bellrose, F. C. (1967). Radar in orientation research. *Proc. 14th Internat. Ornithol. Congr.*, pp. 281–309.

Bruderer, B. (1971). Radarbeobachtungen über den Frühlingszug im Schweizerischen Mittelland. *Ornithol. Beob.* **68**:89–158.

Bullard, Sir E. (1969). The origin of the oceans. In *The Ocean* (A Scientific American Book), W. H. Freeman, San Francisco, p. 23.

Dorst, J. (1962). *The Migration of Birds,* Houghton Mifflin, Boston, p. 100.

Eastwood, E. (1967). *Radar Ornithology,* Methuen, London, 278 pp.

Emlen, S. T. (1969). Bird migration: Influence of physiological state upon celestial orientation. *Science* **165**:716–718.

Griffin, D. R. (1967). The physiology and geophysics of bird navigation. *Quart. Rev. Biol.* **44**:255–276.

Gwinner, E. (1968). Circannuale Periodik als Grundlage des jahreszeitlichen Funktions wandels bei Zugvögeln. *J. Ornithol.* **109**:70–95. (Suggests that distance may be correlated by the birds with amount of activity expended.)

Hamilton, W. J. (1962). Celestial orientation in juvenal waterfowl. *Condor* **64**:19–33.

Hoffmann, K. (1959). Die Richtungs orientientierung von Staren unter der Mitternachtsonne. *Z. Vergl. Physiol.* **41**:471–480.

Ising, G. (1946). *Ark. Astr. Fys.* **32A(18)**:1.

Keeton, W. T. (1969). Orientation by pigeons: Is the sun necessary? *Science* **165**:922–928.

Keeton, W. T. (1971). Magnets interfere with pigeon homing. *Proc. Natl. Acad. Sci.* **68**:2–106.

Kramer, G. (1953). Wird die Sonnenhöhe bei der Heimfindorientierung verwertet? *J. Ornithol.* **94**:201–219.

Kramer, G. (1955). Ein weiterer Versuch, die Orientierung von Brieftauben durch jahreszeitliche Änderung der Sonnenhöhe zu beeinflussen. Gleichzertig eine Kritik der Theorie des Versuchs. *J. Ornithol.* **96**:173–185.

Kramer, G. (1959). Recent experiments on bird orientation. *Ibis* **101**:399–416.

Lack, D. (1963). Migration across the North Sea studied by radar. *Ibis* **105**:461–492.

Mason, R. G. (1967). In Runcorn, S. K. (ed.), *International Dictionary of Geophysics,* Pergamon Press, New York, p. 895.

Matthews, G. V. T. (1953). Sun navigation in homing pigeons. *J. Exptl. Biol.* **30**:243–267.

Merkel, F. W., Fromme, H. G., and Wiltschko, W. (1965). Nicht-visuelles orientierungsvermögen bei nächtlich zugunruhigen Rotkehlchen: Eine Entgegnung auf eine Arbeit von A. C. Perdeck in Ardea, 51 (1963). *Vogelworte* **23**:71–77.

Nagata, T. (1967). In Runcorn, S. K. (ed.), *International Dictionary of Geophysics,* Pergamon Press, New York, p. 629.

Sauer, E. G. F., and Sauer, E. M. (1960). Star navigation of nocturnal migrating birds. The 1958 planetarium experiments. *Cold Spring Harbor Symp. Quant. Biol.* **25**:463–473.

Schlichte, H. J., Schmidt-Koenig, K. (1972). Homing in Pigeons with impaired vision. *Proc. Nat. Acad. Sci.* **69**:2446–2447.

Schmidt-Koenig, K. (1961*a*). Sun navigation in birds? *Nature* **190**:1025–1026.

Schmidt-Koenig, K. (1961*b*). Die Sonne als Kompass in Heim-Orientierungs-system der Brieftauben. *Z. Tierpsychol.* **18**:221–244.

Schmidt-Koenig, K. (1963*a*). Sun compass orientation of pigeons upon equatorial and transequatorial displacement. *Biol. Bull.* **124**:311–321.

Schmidt-Koenig, K. (1963*b*). Sun compass orientation of pigeons upon displacement north of the Arctic Circle. *Biol. Bull.* **127**:154–158.

Schmidt-Koenig, K. (1965). In Lehrman, D. S. (ed.), *Advances in the Study of Behavior,* Vol. I, Academic Press, New York and London, p. 221.

Valley, S. L. (1965). *Handbook of Geophysics and Space Environments,* Sect. II, Air Force Cambridge, U.S. Air Force.

Walraff, H. G. (1960). Does celestial navigation exist in animals? *Cold Spring Harbor Symp. Quant. Biol.* **25**:451–461.

Wiltschko, W., and Merkel, F. W. (1966). Orientierung zugunruhiger Rotkehlchen in statischen Magnetfeld. *Verh. Deutsch. Zool. Ges.* **1965**:362–367.

Wiltschko, W., and Wiltschko, R. (1972). Magnetic compass of European robins. *Science* **176**:62–64.

Wiltschko, W., Höck, H., and Merkel, F. W. (1971). Outdoor experiments with migrating European robins in artificial magnetic fields. *Z. Tierpsychol.* **29**:409–415.

Yeagley, H. L. (1947). A preliminary study of a physical basis of bird navigation. *J. Appl. Phys.* **18**:1035–1063.

Yeagley, H. L. (1951). A preliminary study of a physical basis of bird navigation. Part II. *J. Appl. Phys.* **22**:746–760.

Chapter 5

DESCRIBING SEQUENCES OF BEHAVIOR[1]

P. J. B. Slater

Ethology and Neurophysiology Group
School of Biology, University of Sussex
Brighton, England

I. ABSTRACT

A review is given of the methods currently available for analyzing sequences of behavior. Simple flow diagrams based on the frequencies or conditional probabilities of individual transitions are considered to be of restricted usefulness except where the sequence is highly ordered and the different patterns occur at similar frequencies. It is more helpful to compare the data with a random model provided that repetitions of the same behavior, and any transitions which cannot occur, are excluded before the expected number of each type of transition is calculated. Such comparisons are most likely to be helpful if the behavior patterns included are closely related and fall into discrete homogeneous categories. The fact that most behavioral data are unlikely to be stationary is considered to be the main factor limiting this approach. It is suggested that first-order transition analysis and correlative techniques are the best current methods for examining such data. The search for higher-order dependencies is useful only in stationary data, where grouping of acts due to changing causal factors can be assumed to be unimportant. Additional difficulties involved in the analysis of sequences of interaction between individuals are briefly discussed. The major complicating factor here is that the behavior of an individual is likely to be dependent both on that of others and on its own previous behavior.

[1]This research was financed by a grant from the Science Research Council.

131

Some ways of improving current techniques to take account of this are put forward. It is emphasized that sequence analysis provides only a description of the behavior under study and that there are dangers in making causal inferences on the basis of such descriptions alone.

II. INTRODUCTION

The analysis of sequences occupies an important position among the methods available to ethologists, particularly those interested in the causation of behavior. By borrowing and adapting mathematical techniques from other branches of science, they have been able to achieve precise and quantitative descriptions of behavior, and on these to base hypotheses about the causal mechanisms involved.

How can sequence analysis help in the study of causation? It is generally considered by ethologists that the occurrence of two behavior patterns close together in time indicates that they have some causal factors in common. This may not always be the case, for one of the behaviors may bring the animal into a situation appropriate to the other, but in many instances the explanation seems plausible. The causal factors shared may be of two types. First, both behaviors may depend on a particular bodily state, such as the presence of a hormone. Second, both may appear in response to the same or related external stimuli and thus be shown only when these are present. Sequence analysis cannot differentiate between these possibilities, but it can indicate in an objective way the groupings in which behaviors occur and so define the relationships which need to be explained.

Its objectivity is perhaps the greatest asset of the approach. Before such methods were developed, observers depended on their own judgment to determine the affinities between behavior patterns. Because activities which are similar in function, such as courtship and grooming movements, often appear close together in time, this subjective classification led to explanation in terms of specific unitary drives (e.g., courtship or grooming drives). The fact that behaviors of different function sometimes occurred in the midst of these groupings could be accounted for only by assuming their causation to be different, and they were labeled as "displacement activities." These early theories thus depended on the acceptance of a rule and the development of theories to account for exceptions to it. The rule, it should be noted, was based on the doubtful assumption that causation and function are equivalent. The inference that functionally related behaviors occurring close together in time were expressions of a common internal drive also neglected the possibility that their grouping was due to the constraints of the external world: courtship movements may occur only when a female is present and thus appear to be grouped. This point was appreciated by Tinbergen (1951), whose model of causation stressed the role of external stimuli in leading to the grouping of behavior patterns, though he still felt it necessary to argue that each functional group shared a particular drive.

With the appreciation that drive concepts are of limited usefulness in the explanation of causation (Hinde, 1970), it has become necessary to arrive at objective ways of examining associations between acts empirically, so that more realistic models of their causation may be constructed (Slater, in press). Sequence analysis was one of the first methods to be employed in this context, and its continuing attractions make an appraisal of its usefulness long overdue. That such methods are preferable to the rather subjective assessments of the interrelations between behavior patterns which were previously common seems beyond dispute, but one must be aware of their limitations and not be overimpressed by their precision. The mechanisms underlying the sequence of acts shown by an animal are likely to be complex, and to analyze them using methods appropriate to simpler situations, such as the sequence of cars passing along a road, will inevitably involve some rather sweeping assumptions. The less valid the assumptions, the less realistic will be the results. But it is perhaps even more important that the questions asked should be framed with the nature of the behavior, rather than the nature of the mathematical tools available, in mind. It is, unfortunately, only too easy for research using sophisticated mathematical methods to give very little insight into behavior.

The purpose of this chapter is to review the methods which have been employed by ethologists to analyze sequences, to assess their validity and usefulness, and to point to the conclusions which may be drawn from their application. Its main concern will be with the role of these methods in helping to outline the relationships between successive acts shown by the same individual, but, as they have also been used in studies of interactions between animals, additional points raised by this application will be more briefly considered.

III. SEQUENCES WITHIN THE INDIVIDUAL

A. Methods of Analysis

The methods used for analyzing sequences vary according to the complexity of the data and how structured the behavior appears to be at first sight. Basically, they may be split into two groups depending on whether or not they involve comparison with a random model.

1. Analysis of Transition Frequencies and Conditional Probabilities

The simplest type of sequencing of events is a deterministic sequence: in this, the events always follow each other in a fixed order, so that the nature of the preceding act defines precisely the nature of that which will follow. Such sequences are rarely studied in normal behavior, partly because two acts always occurring in a particular order tend to be regarded as a single behavior pattern. An instance of a sequence parts of which are deterministic was, however, described by Isaac and Marler (1963). They found the song of a mistle thrush (*Turdus viscivorus*) to be composed of some 21 possible syllables; within a song, each of these could be followed by only a small range of others and often, deterministically, by only one. Fabricius and Jansson (1963) observed that pushing by courting pigeons (*Columba livia*) was always followed by nest demonstration, but it was useful here to regard the two activities as separate because nest demonstration often occurred without pushing preceding it.

Most behavioral sequences are probabilistic rather than deterministic in form, meaning that while the probability of a given act depends[2] on the sequence of those preceding it, it is not possible to predict at a particular point exactly which behavior will follow. Some such sequences are only marginally less precisely ordered than deterministic ones, and these are usually referred to as "chained responses." In these cases, the probability of a particular event is so markedly altered by the nature of that immediately before it that a flow diagram indicating the frequencies with which different transitions occur provides a good impression of the organization of the behavior. This has proved particularly true of goal-directed sequences, such as courtship, in invertebrates and lower vertebrates (e.g., Brown, 1965; Parker, 1972; Baerends et al., 1955; Hinde, 1955/56). Noirot (1969) has applied a similar technique to the maternal behavior of mice: she looked at the order in which different acts occurred during an observation session, disregarding repetitions of the same act.

Where sequences are not so highly ordered, some transitions may be observed between almost every behavior and every other, so that simple flow diagrams become complex and hard to interpret. A way around this is to include only those transitions which have a high probability of occurrence, a method which was used by Fabricius and Jansson (1963). For each sequence of two acts, they worked out the conditional probability (p $B|A$, the probability that B will occur given that A has just occurred), and only where this exceeded 0.1 was an arrow from A to B included in their flow diagram. For each type of act, an impression is thus given of those other behaviors most likely to follow. This method has also been employed to show up differences in sequences between two groups of animals (isolated and socially

[2]The expression "one behavior depends on another" is used throughout this chapter in the statistical sense: the occurrence of the first alters the probability of the second. Where it is intended to imply that the first is in some way causal to the second, this will be referred to as a "sequence effect" rather than a "dependency."

reared guinea pigs, *Cavia porcellus*) by Coulon (1971). It is perhaps best suited to situations like this where comparisons are to be made, for the individual flow diagrams obtained are hard to interpret when behaviors differ strongly from each other in frequency. A large number of transitions, yielding a comparatively complex diagram, may result even from sequences in which the events are independent of one another. This is because the expected conditional probability of a rare act following a common one is low, whereas that of a common act following a rare one is high. In a sequence of 1000 acts, if act A is observed 100 times and B is observed 10 times, and A and B are independent, then one would expect the sequences $A \rightarrow B$ and $B \rightarrow A$ to occur once each (100 × 10/1000), but the expected conditional probabilities are quite different: $p\,A|B = 0.1$ (100/1000); $p\,B|A = 0.01$ (10/1000). In the light of this complication, it is difficult to assess the importance of the different sequences found by Fabricius and Jansson (1963), as they do not give the frequencies of the individual acts.

These simple ways of looking at sequences are thus likely to prove useful only where sequencing is strong and where the different behaviors considered occur at roughly equal frequencies.

2. Comparison with a Random Model

The shortcomings of descriptions based on frequencies and conditional probabilities have led many workers to adopt some of the techniques of Markov chain analysis. A sequence of behavior can be described as a Markov chain if the probabilities of different acts depend only on the immediately preceding act and not on any earlier ones (Cane, 1961). This model is thus appropriate to describe the sequence $A \rightarrow B \rightarrow C$ if the probability that C will follow B is not in any way altered by the nature of A or events prior to A. Such a sequence of events can also, more generally, be referred to as a "first-order" Markov chain, to differentiate this model from ones in which more of the preceding events affect the outcome: in this case, an rth-order Markov chain is one in which the probability of a particular event is significantly affected by the r preceding events (Chatfield, in press). Thus if the probability of C depends on A as well as B, the first-order model is inadequate to account for the structure of the sequence and second-order dependencies can be said to be in operation. If the probability of an event does not depend on any previous events, the events can be said to be independent, and this is sometimes referred to as a "zeroth-order" Markov chain.

Most sequence analysis in behavior is concerned with establishing the existence of, and identifying, first-order dependencies. Here, the matrix of observed transition frequencies is compared with that which would be expected if all acts were independent of one another. For these first-order transitions, the expected values are calculated as for a contingency table (assuming that it is possible for any of the behaviors considered to follow any other), and comparison between the matrices can then be made either in the usual way, using χ^2, or by the use of information theory (Bolles, 1960; Chatfield and Lemon, 1970; Fentress, 1972).[3] If the difference is found to be significant, the hypothesis that the behavior consists of a sequence of independent acts can be rejected. This discovery is not, in itself, surprising; it would be more so if a sequence of independent acts were found, which has not so far been the case. Further analysis is required before any useful description of behavior is obtained, and here two different approaches are possible.

First, the sequence can be analyzed more closely as a whole to decide whether the first-order model is adequate or whether higher-order dependencies affect it. It can be tested, as a start, whether a second-order model is more precise: in the generalized sequence $A \rightarrow B \rightarrow C$, is the probability of obtaining C after B significantly altered by the nature of A? In a first-order Markov chain, this is not the case, and the frequency of $A \rightarrow B \rightarrow C$ is that predicted from the frequencies of $A \rightarrow B$ and $B \rightarrow C$. A difficulty arising here is that N different types of behavior will give N^3 triplet types, and so very extensive data are required before reasonable numbers of each of these are expected. In two cases where triplets have been examined, however, a first-order Markov model has been found to be satisfactory: the frequency of triplets of each type corresponded closely to that predicted from the first-order relationships (Nelson, 1964; Lemon and Chatfield, 1971). As will be seen below, rather special conditions, which are unlikely to arise often in behavioral data, may have to be met before sequences of this type are discovered. The statistical comparison of first- and higher-order models has been thoroughly explored by Chatfield and Lemon (1970; Lemon and Chatfield, 1971; Chatfield, in press).

A second approach to finer analysis is an attempt to detect those first-order transitions which are significantly commoner than their expected values. This is done when the groups in which behaviors occur and the exact relationships between different behavior patterns are the main points of interest. Because such discoveries are behaviorally more interesting than the detection of higher-order dependencies, this approach has greater currency. It carries with it, however, the necessity to examine the discrepancy between the observed and expected values in individual cells of the transition matrix. Because many cells often contain rather low figures and because the observed value in each cell depends on those in others, this is difficult both to carry out and to interpret. The commonest method is a condensation of the whole matrix into a 2×2 table about the cell of interest, followed by a χ^2 test to detect whether that particular transition is more frequent than expected (e.g.,

[3]Strictly speaking, a transition matrix is not a contingency table, as the events included in it are not independent of one another, the second act of each pair being the first act of the next. The use of χ^2 on such data has, however, been validated by Bartlett (1951).

Stokes, 1962; Blurton Jones, 1968). Other techniques have been employed by Andrew (1956), Weidmann and Darley (1971), and Slater and Ollason (1972). In this last case, an attempt was made to overcome the biases involved in massing data from different animals, which is a usual practice where χ^2 is employed in order to boost the figures in individual cells. Perhaps the safest method is a simple inspection of the data, as recommended by Lemon and Chatfield (1971).

Whichever of these tests is used, pairs of behaviors which are commoner than expected can be extracted and a flow diagram constructed along the lines of those used in frequency and conditional probability analyses. This flow diagram has the advantage, however, that the relationships in it are not biased by the frequencies of the individual acts.

A further method may be used if it is of interest to detect asymmetry in particular sequences. If the row and column totals of the transition matrix are the same, then the expected values of the generalized sequences $A \rightarrow B$ and $B \rightarrow A$ will be the same, and tests can be carried out to see if the observed values differ (Blurton Jones, 1968; Slater and Ollason, 1972). While sequence analysis is not the only, or necessarily the best, way of looking for associations between acts, it is certainly the simplest way to detect such asymmetries.

B. Some Comments on the Methods Used

Having given a general outline of the commonest approaches to sequence analysis, it is now appropriate to consider the assumptions underlying them in the wide range of circumstances in which they have been used. Three main points are important here: the validity of using complete matrices, the choice of categories of behavior, and the problem of stationarity.

1. Complete vs. Incomplete Matrices

Comparison with a random model is easy if the matrix considered is a complete one, as shown in Table IA. The expected number of transitions between any pair of behaviors is calculated in the same way as for a contingency table (row total × column total/grand total; Table IB). This then gives the expected number of acts which would succeed each other if all acts were independent of one another. Two considerations are liable to make such a model unreasonable:

a. Repetitions of the Same Act. The complete matrix includes transitions which are repetitions of the same behavior. Figures thus appear on the descending diagonal of the matrix indicating the frequency of transition between *A* and *A, B* and *B,* etc. For each behavior, the observer must develop a criterion with which he can decide when one act ends and the next begins before he can reach a figure for the number of these repetitions. This may be easy for some behaviors: they may almost never be immediately

Table I. Artificial Data Showing the Observed Number of Transitions Between Three Behavior Patterns and the Number Which Would Be Expected If All Acts Were Independent of Each Other

A. Observed Matrix

	A	B	C	
A	10	20	5	35
B	15	4	6	25
C	10	2	8	20
	35	26	19	80

B. Expected Matrix

	A	B	C	
A	15.3	11.4	8.3	35
B	10.9	8.1	6.0	25
C	8.8	6.5	4.7	20
	35	26	19	80

repeated, or, if they are, a consistent time interval may separate succeeding events so that they can be simply distinguished. Other behaviors pose more difficult problems: when, for instance, does one act of locomotion end and the next begin? Here, it would be necessary to select some arbitrary time interval during which the animal did not move as indicating a gap between two acts.

In zebra finches (*Taeniopygia guttata*), other behaviors which highlight this problem are preening and singing (unpublished observations). These birds tend to groom in sessions several minutes long, during which preening of the feathers with the bill is interrupted little by other acts, with the exception of scratching of the head. A preening bird lowers the head to a small area of the body, preens one or a few feathers, and raises the head again. The next series of preening movements is most likely to be directed to the same area of the body as the last. There are thus at least three ways of defining an act of preening:

1. The preening of a single feather, several such acts often taking place between each raising of the head.
2. The series of movements between each raising of the head, which may involve several feathers.
3. The series of movements directed to the same area of the body, which may be interrupted by several instances of head raising, depending on how an area of the body is defined.

In the case of song, a series of almost identical phrases is produced, followed by a gap usually lasting for more than 2 sec before the next such series. An act here might be defined as each individual phrase, or as each series of phrases separated by longer than a certain time.

It is clear from these examples that defining an act other than arbitrarily will often be impossible. For some purposes, this may not be a hindrance, as long as the definition used can be consistently applied, but this is certainly not the case with sequence analysis. Here, two observers might apply different criteria, each in themselves perfectly reasonable, to the same data and so reach quite different conclusions. This is illustrated by Table II, where the same data as in Table I are shown except that the number of $A \rightarrow A$ transitions has been boosted from 10 to 100, as might happen if an experimenter with a different criterion for what constituted an act of A had collected the same data. A comparison of the expected values in Table IIB with those in Table IB shows that this difference, in only one cell of the observed matrix, would lead to a totally different set of expecteds.

The results of sequence analysis in relation to a random model using a complete matrix will usually depend on such arbitrary criteria selected by the experimenter. It is interesting to note, for instance, that each of the behaviors studied by Wiepkema (1961) was recorded as being repeated often, whereas this was seldom the case with the data given by Blurton Jones (1968). This tells one more about the criteria chosen by these workers than about the behaviors which they studied. Most of the matrices published in the literature are rather similar to that of Wiepkema in that the act definitions chosen lead to relatively high figures on the descending diagonal (e.g., Bolles, 1960; Baerends *et al.*, 1970; Lemon and Chatfield, 1971). Carrying out sequence analyses on matrices such as these will tend to show just that these acts occur in bouts rather than more interesting relations between different behaviors.

Table II. The Same Data as in Table I But with the Number of $A \rightarrow A$ Transitions Boosted from 10 to 100

A. Observed Matrix	A	B	C	
A	100	20	5	125
B	15	4	6	25
C	10	2	8	20
	125	26	19	170

B. Expected Matrix	A	B	C	
A	91.9	19.1	14.0	125
B	18.4	3.8	2.8	25
C	14.7	3.1	2.2	20
	125	26	19	170

If arbitrary act definitions are involved or if behavior patterns tend to fall into bouts, the analysis of the complete matrix of transitions will thus be of little help in elucidating the interrelationships between activities. Under these circumstances it is more fruitful to eliminate the entries on the descending diagonal before analysis so that only transitions between different behaviors are considered.

b. Cells Which Cannot Be Occupied. In some sequences of behavior, there may be transitions which cannot occur, particularly because of environmental constraints. Thus the matrix of Slater and Ollason (1972) could not include transitions from feeding to drinking, as locomotion, another category of behavior included in the study, had to occur between the sources of food and water. Likewise, Baerends *et al.* (1971) recognized in their matrix that a herring gull (*Larus argentatus*) could not show "looking down while not on nest" immediately after "sitting on nest" as well as certain other transitions. These constraints may not occur in many sequences, but they should be recognized where they do so.

While many workers eliminate the descending diagonal from their transition matrices and some remove other cells which cannot be occupied, few realize that the calculation of expected values must then be altered so that the row and column totals add up correctly without figures appearing in the empty cells. Goodman (1968) gives a method whereby this may be done, and a simpler, though more approximate, method has been developed by Lemon and Chatfield (1971), whose study of song in cardinals (*Richmondena cardinalis*) was the first to use this technique on sequential data. Table IIIA shows the same data as in Tables I and II but without the descending diagonal. The expecteds in Table IIIB are calculated according to the Goodman method; they add up to the correct row and column totals despite the

Table III. The Same Data as in Tables I and II but with Expected Values Calculated Ignoring Transitions Between Each Behavior and Itself

A. Observed Matrix			
	A	*B*	*C*
A	—	20	5	25
B	15	—	6	21
C	10	2	—	12
	25	22	11	58

B. Expected Matrix			
	A	*B*	*C*
A	—	17.8	7.2	25
B	17.2	—	3.8	21
C	7.8	4.2	—	12
	25	22	11	58

presence of empty cells. These expected values give the number of bouts of each type of behavior which would succeed each other type if all bouts were independent of one another. A "bout" is here defined as a sequence of a given type of behavior not punctuated by the occurrence of any of the other types considered.

As these methods are available, there is no hindrance to the analysis of transition matrices in which repetitions of the same act and transitions which are impossible are not considered. While only Lemon and Chatfield (1971) and Slater and Ollason (1972) have so far used them, it is clear that most data on sequences of behavior would benefit from their application.

2. The Choice of Categories of Behavior

An important preliminary to any study, and especially to the analysis of sequences, is the classification of behaviors into discrete categories. Some acts are almost invariant in form and intensity and can easily be defined and recognized; others are more problematic and may require more or less arbitrary definition. However the categories of behavior are defined, it is important that they should be mutually exclusive and that the observer should "be prepared to treat all members of a given category as equivalent" (McFarland, 1971). Several factors may make these conditions hard to comply with. First, several different behaviors may frequently occur simultaneously, making it difficult to cast the data into a transition matrix. Data of this kind are more appropriately analyzed using correlation techniques (e.g., Delius, 1969), and Golani (unpublished manuscript, cited by Golani and Mendelssohn, 1971) has also applied multidimensional scalogram analysis to this problem. Second, the equivalence of behavior patterns within a category may be doubted where they differ greatly in form (as where all grooming movements are grouped under a single heading) or where there is a wide spread of bout lengths within a category. In this instance, long and short bouts may differ in context so that grouping them in one category clouds their different relationships with other activities.

Studies differ a great deal in the diversity of behavior patterns considered and the ways in which these are classified. It is common, for example, for transition matrices to include mainly behaviors from one functional category (comfort movements: Andrew, 1956; Delius, 1969; Fentress, 1972; courtship behavior: Nelson, 1964; Wiepkema, 1961; Coulon, 1971; song syllable types: Isaac and Marler, 1963; Lemon and Chatfield, 1971). Certain other behaviors may be included, but these are mainly taken as markers indicating the beginning and end of sequences of the group under study (e.g., in studies of comfort movements: resting and locomotion by Andrew, 1956; "other behavior" by Delius, 1969). Where such markers are not included, it is often unclear how a sequence is defined, though the careful study of Nelson (1964) is an exception. He established the time interval between two events for the succeeding to be independent of the nature of that before it and took intervals longer than this to be between, rather than within, sequences. Other matrices avoid the problem of sequence definition by including all behaviors shown in a particular situation regardless of function (e.g., at the nest site: Baerends *et al.*, 1970; at a food source: Blurton Jones, 1968; in isolation from conspecifics: Bolles, 1960; Slater and Ollason, 1972). The range of behaviors included can therefore vary from a wide spectrum of very different acts (e.g., sleep, grooming, and feeding) to a variety of very similar acts (e.g., grooming different areas of the body, different song types). All the behaviors included in a matrix of the latter type may be lumped as a single category in matrices of the former type.

For a number of reasons, the application of Markov chain analysis to the whole matrix is likely to prove more fruitful if the behaviors studied are homogeneous within categories, distinct between categories, and yet all closely related to one another:

a. Cane (1959) comments that "If the states of a Markov chain are grouped, the resulting process is not in general a Markov chain." Homogeneity within each category of behavior will make it less likely that such a grouping has been made.

b. If the behaviors considered are all closely related to each other, it may be possible to split them into categories which are, at least approximately, at the same level of organization. If the categories considered are very dissimilar, this may affect the order of the Markov chain found. Birds, for example, tend to wipe their beaks after drinking, presumably stimulated to do so by water left on the outside. If drinking and beak wiping were included in a matrix with various small bodily movements, some of the latter would often be interpolated between them, making it necessary to invoke higher-order relationships to explain the sequencing.

c. It is useful to work out the order of the dependencies governing a sequence only if the rules are the same for the whole matrix. This is more likely to be the case if the different categories can be taken to be distinct from each other and yet closely related. One deterministic sequence in a matrix of acts which are otherwise independent of one another could lead the observed and expected matrices to differ significantly.

d. Markov chain analysis presupposes that the animal is in a steady state (i.e., that the system is stationary), which is much more likely to be the case during sequences of closely related acts which are considered motivationally similar. The problem of stationarity has more widespread repercussions, however, and these will be discussed in the next section.

3. Stationarity

A stationary sequence is one in which the probabilistic structure does not change with time (Cox and Lewis, 1966; Delius, 1969). Nonstationarity can lead to very complex transition matrices: while the rules underlying sequencing may be simple on the short term, if they are not the same throughout the observed data in a transition matrix the matrix will be a complicated sum of the various simple processes involved and extremely hard to interpret.

Checking to see that the data are stationary is therefore a first requirement of rigorous Markov chain analysis as well as of other methods for examining sequences. It is a condition likely to be met seldom in behavioral data, which is one reason why this type of analysis is not often taken far. The use of concepts such as "drive" and "arousal" to explain behavior (Bindra, 1959; Hinde, 1970) and the discovery of daily cycles (e.g., Palmgren, 1950; Aschoff, 1967) and of short-term cycles (Richter, 1927; Wells, 1950) all speak for the long-standing realization among those studying behavior that the data they collect are not stationary. In certain situations, the steady-state assumption may be valid, but this will not often be the case. The shorter the observation period, for example, the more reasonable the assumption, but the more difficult it becomes to obtain adequate data for analysis.

The most likely source of nonstationarity is the existence of a trend within the data, such as might be caused by daily cycles. The effects of such trends may be minimized by testing only for short periods at a consistent time of day, and methods are also available for detecting trends so that the extent to which they affect the data may be assessed. Perkel *et al.* (1967) suggest breaking the data into segments, analyzing each separately, and then testing to see if they are drawn from the same population. Especially where the types of act considered are heterogeneous, most behavioral data will fail to pass this test: sporadic periods of sleep or long bouts of grooming, which may take place only a few times per day, will tend to occur in some segments but not others.

Moore *et al.* (1966) point out the complexity of rigorous criteria for establishing stationarity in a simple system such as the firing pattern of a single neuron. Very few workers on sequences of behavior attempt such checks, and the frequent interpretation of results in terms of changing motivation suggests that nonstationarity is accepted. Rough tests for stationarity have been performed in two cases. Nelson (1964), working on courtship in fish, was able to carry out Markov chain analysis on one species studied (*Corynopoma*) after excluding intersequence intervals from the data, but he rejected the possibility of sequence analysis on another (*Pseudocorynopoma*), as the probabilities of different acts showed marked temporal fluctuations. Lemon and Chatfield (1971) proceeded with Markov chain analysis after finding no significant differences in the probabilities of different song types between the first and second halves of the records they analyzed.

The fact that these workers were able to satisfy at least some of the criteria for stationarity with their data and the fact that in both cases the behavior was found to follow a first-order Markov chain may be no coincidence. Moore *et al.* (1966) point out that "the distinction between serial dependence (with an overall stationary sample) and actual nonstationarity is an arbitrary one." Thus data which appear to follow a Markov chain of high order could also be interpreted as nonstationary and *vice versa*. In some behavioral data, triplet sequences of the general form $A \rightarrow B \rightarrow A$ have been found to be commoner than expected (Slater and Ollason, 1972; Fentress, 1972), indicating the inadequacy of a first-order Markov model. It is possible here that a second-order model might be satisfactory, but shortage of data precludes testing for this. Another interpretation would be that periods of time exist during which the two acts involved in the sequence are of high probability, while that of all other behaviors is low, thus leading to alternation between them. This interpretation assumes nonstationarity, i.e., fluctuating probabilities.

C. Alternative Methods of Approach

It will be apparent from the above discussion that there are many reasons why the techniques of Markov chain analysis should be applied only with caution to behavioral data. The most obvious reason for this is that the probabilities of different behavior patterns tend to change with time. Given this difficulty, it is worth considering briefly other methods which, while not primarily concerned with sequencing, may be used to detect associations between behaviors.

Factor analysis is particularly interesting in this context, as it has also been employed by ethologists for the treatment of data in transition matrices (Wiepkema, 1961; Baerends and van der Cingel, 1962; van Hooff, 1970; Baerends *et al.*, 1970). Using this method, a small number of hypothetical variables are extracted, the existence of which as causal factors could account for most of the observed correlations between acts, without the acts themselves being directly related. It is assumed that the measured variables, in this case behavior patterns, do not depend causally on each other but only on the postulated factors (Blalock, 1961). This assumption highlights the sharp contrast between this method and Markov chain analysis. The choice of which of these approaches is used appears to depend mainly on whether the research worker believes "sequence effects" or "motivational states" to be the more important in determining the relationships between behaviors. In the former case, it is thought useful to describe the probability of an act at a particular instant in terms of the sequence of acts which preceded it. The animal is assumed to be in a steady state, which is tantamount to ignoring the possibility that motivational changes occur. On the other hand, if interpretation is in terms of motivational states (underlying variables which are postulated as changing more slowly than the switching in overt behavior), the same data can be treated quite differently using factor analysis. An association between acts is here taken to be indicative of common causal factors underlying the behaviors involved, and the factor on which they have a high loading may be argued to represent a motivational state (e.g., "aggressive," "sexual," "nonreproductive": Wiepkema, 1961; "affinitive," "play," "aggressive," etc.: van Hooff, 1970). If factor analysis is based on a transition matrix, two different models may be derived, depending on whether the frequency of acts following each other or preceding each other is taken as a starting point. The method assumes that sequence effects are unimportant, as these will make the two models differ from one another (Slater and Ollason, 1972). As both motivational changes and sequence effects are likely to occur in most behavior, each of these two descriptions is necessarily imperfect, and cases where either is valid in the strictly mathematical sense may be hard to find.

Factor analysis is open to several other objections, some of which have been discussed by Overall (1964), Andrew (1972), and Slater and Ollason (1972). The use of the method supposes it to be useful to describe behavior in terms of a few underlying variables, and yet it is not altogether clear what these may represent. While Wiepkema (1961) referred to his factors as "tendencies," thus identifying them with a drive type of model, Baerends (1970) gives them weaker status as "areas of higher density within the causal network," thus recognizing that their existence may depend on external as well as internal variables. It is doubtful whether the extraction of factors which are themselves of complex causation advances understanding. At the descriptive stage of analysis, it appears preferable to examine the relationships between individual acts, as these can be directly assessed without theoretical implications and do not place constraints on the type of model of organization which may emerge from subsequent experimental work.

Analysis based on transitions between behaviors is not the only way of detecting associations between individual acts: an alternative approach which has much to recommend it is the use of correlation techniques. This involves measuring the frequencies of different behavior patterns within a series of time units and then correlating between each pair of acts to determine whether they are positively or negatively associated. Stationarity remains a problem here, though perhaps a less acute one, while some of the other objections to sequence analysis are not important: two behaviors which occur concurrently are no problem, nor does it bias the results if acts at different levels of organization are included in the analysis. The main difficulty is that the results may depend on the choice of time unit. While two acts may be significantly correlated over 1-hr periods, this is not necessarily true when their correlation is assessed over 10-sec units. To allow for this, it is therefore preferable to analyze the data separately for a number of different time intervals, as has been done by Baerends *et al.* (1970). With this approach, it becomes possible to detect much looser associations than would be achievable with sequence analysis.

Ordinary correlation techniques give no indication of whether acts tend to follow each other in a particular order. If one behavior leads to another, correlations of this type can only indicate that they are in some way associated. A modification of the method can, however, extract such information; this is the cross-correlation procedure which is discussed in detail by Delius (1969). Here the frequencies of different acts are correlated, not just for the same time unit, but with progressive lags introduced between them. Thus if behavior B tends to occur 5 sec after behavior A and the time unit used is 1 sec, a positive correlation will be found when the data for A are moved forward 5 sec in relation to B. Heiligenberg (1973) has carried out the most elegant study to date employing this procedure and, using 1-sec time units, has been able to demonstrate both sequence effects and looser associations between seven behavior patterns in a cichlid fish (*Haplochromis burtoni*).

Correlative techniques can thus give useful information on a number of facets of behavioral organization, including sequence effects. They are particularly suitable where the behaviors under study are diverse and where their associations are liable to be due to common causation rather than sequencing. A full analysis of this sort requires a considerable amount of data (Heiligenberg's results were based on 360 hr of observation), and, unlike the simpler forms of sequence analysis, the amount of calculation involved makes the use of a computer essential. Where an approximate guide to associations is required, or where sequencing is strong, analysis of first-order transitions between acts is clearly more practicable.

IV. SEQUENCES OF INTERACTION BETWEEN INDIVIDUALS

The discussion so far has centered around sequences of behavior within the individual in contexts where the external world is presumed to be relatively unchanging. Some of the examples used were concerned with animals in a social situation (e.g., the studies of fish courtship by Wiepkema, 1961, and Nelson, 1964); in these cases, the assumption was made that the presence of other animals, while perhaps stimulating the behavior under study, had

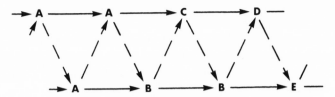

Fig. 1. A simple model of an interactive sequence involving two individuals.

little effect on its sequence. Nelson (1964), for instance, stated that the part played by the female in glandulocaudine courtship is slight, and therefore felt justified in examining sequences within the male as if he were in a constant environment.

Most studies of social sequences have a different primary aim: to demonstrate that the behavior of one animal *is* affected by that of others, and so obtain evidence for the role of different behavior patterns in communication. At first sight, this is an easier task than the analysis of within-individual sequences, for data of this sort can easily be cast into a complete matrix such as Table I, the acts listed down the side being those shown by one individual and those along the top being the subsequent behaviors of another. The objections to including figures on the descending diagonal do not apply: these entries refer to those cases where the performance of a behavior by one animal is followed by the same behavior from the other.

Other difficulties do, however, arise, including those concerned with the choice of categories and with stationarity which have already been discussed. Stationarity may be a particularly serious problem, for as MacKay (1972) points out, many of the most interesting behaviors in communication are those which change in probability as the interaction proceeds. A further difficulty, which is often neglected, is that the behavior of an animal in a social situation is likely to depend partly on the sequence of acts that it has already shown as well as on the behavior of others. Figure 1 shows a simple model of an interaction between two animals designed to take account of both these influences.[4] Studies of within-individual sequences, such as those of Wiepkema (1961) and Nelson (1964), suppose that the effects portrayed by dotted lines are of little importance, whereas some studies of interactions between animals pay only passing attention to the possible influence of the effects illustrated by complete lines (e.g., Hazlett and Bossert, 1965). The most thorough analysis of interactive sequences, that by Altmann (1965)

[4]The model is, of course, too unsophisticated for many purposes. It assumes, first, that each behavior depends only on those shown by the individual and by his companion immediately previously and, second, that the interaction is like a game of chess, the animals making alternate moves.

on social communication in the rhesus monkey (*Macaca mulatta*), does not attempt to differentiate between these two possible influences. His transition matrices are based on the order of events within a group of monkeys, making no distinction between animals; thus two consecutive behaviors may be by the same or by different individuals. Using data of this sort, Altmann has taken Markov chain analysis further than any other worker, his examination extending to quadruplets of acts (third-order relationships). The confounding of these two effects was perhaps inevitable in a field study where several animals might be in view and each could show any one of 120 different behavior patterns, but it means that his results cannot be interpreted easily. In common with a number of other studies, some of sequences within the individual (e.g., Chatfield and Lemon, 1970; Fentress, 1972) and some of interactive sequences (e.g., Hazlett and Bossert, 1965), Altmann's analysis used the techniques of information theory. Such methods can also be used to assess the relative importance of within- and between-individual influences. This can be done by deriving measures of information transfer separately for the transitions shown in Fig. 1 as dotted lines and those shown as complete lines, and comparing the two values so obtained. The study by Dingle (1969) includes such calculations: his conclusion that the two influences were of equal importance in interactions between mantis shrimps (*Gonodactylus bredini*) warns against ignoring the effects of either of them. It should be noted in passing that measures of information transmission, while a useful way of expressing the correlation between events, do not necessarily imply a causal relationship between them (MacKay, 1972).

Consideration of a whole matrix of interindividual transitions can indicate only that the behavior of one animal is in some way linked to that of others. Ethologists usually want to ask more specific questions: e.g., does one animal tend to perform behavior *B* more often than expected after another has shown behavior *A*? If within-individual effects are of no importance, an unlikely finding, this could reasonably be tested by comparing the number of times an event of *B* follows an event of *A* with that expected on a random model. But if, as is quite possible, *A* and *B* tend to fall into bouts, the number of such occurrences may be high purely by chance. If a bout of *B* happens to start just after a bout of *A*, many transitions between the two behaviors will be scored even though the animals may be acting independently of one another. Thus if successive events of *B* are not independent of each other, it is clearly invalid to suggest that each of them was separately stimulated by *A*. It is more valid to ask whether more *bouts* of *B* started during or just after each event of *A* than expected. A bout of *B* must be defined so that the first event in one bout is statistically independent of previous occurrences of that behavior in the same animal. Approximate methods for determining the interval between events necessary for this to be the case have been provided by Duncan *et al.* (1970) and Wiepkema (1968), but have not yet been used in sequence analysis. The closest attempt has been in the study by Wortis (1969) of interactions between ring dove (*Streptopelia risoria*) parents and their chicks. She accepted that only the start of a particular behavior should be scored, but took a gap of only 2 sec as indicating a gap between bouts. This criterion was apparently chosen arbitrarily, and it seems unlikely that 2 sec would be adequate for the second event to be independent of the first. A different way of getting around this problem has been used by Heckenlively (1970), who examined encounters between pairs of crayfish (*Orconectes virilis*). For each encounter, he scored only a single transition, that between the first act of one animal and the response to it of the other, and all transitions could thus be considered truly independent of each other.

The analysis of interactive sequences is clearly a more difficult task than that of sequences within the individual because of these additional complications. Furthermore, in this case other methods do not afford the prospect of achieving an improvement. Sequence analysis seems an obvious first step in the study of communication, for here the suggestion is being made that the behavior of one animal is in some way causal to the behavior of another, while this is not necessarily true of two successive behavior patterns shown by the same individual.

V. DISCUSSION

In this chapter, I have mentioned various ways in which sequences of behavior can be analyzed and some of the assumptions which underlie each of them. It is now pertinent to outline the circumstances in which they are likely to prove helpful and to discuss the conclusions which may be drawn as a result of their application.

The simplest form of sequence analysis, that in terms of frequencies or conditional probabilities, can usefully be applied only in circumstances where sequence effects are strong, the acts following each other in an almost fixed order. If the behavior is less structured than this and, in particular, when the different acts considered occur at markedly different frequencies, a better approach is to compare the observed first-order transition frequencies with those which would be expected on a random model, a method equivalent to the initial stage of Markov chain analysis. This is preferable because it takes into account the frequencies of the individual acts and thus extracts those transitions which are commoner than expected, rather than those which are common in absolute terms. In the past, ethologists have not applied this method as carefully as they might have done, particularly those who have based their analyses on complete matrices of transitions. But, while this methodological problem can be overcome, other difficulties are harder to cope with.

There are several reasons why two behaviors may tend to succeed each other more often than expected. Of these, the two most likely are that they both share causal factors and so occur only when these are present, or that the first act stimulates the second in some way. Data in which behaviors are grouped due to changing causal factors fail to fulfill the statistical requirement of stationarity and will thus be unsuitable for rigorous Markov chain analysis. Despite this, the examination of first-order relationships will provide a rough guide to the groupings into which different activities fall. While first-order transition analysis may therefore be of some usefulness in nonstationary data, the extraction of higher-order dependencies is here pointless, for the statistical dependencies between successive acts may be just a byproduct of their shared causation rather than indicating that rules govern the sequence as such. Correlative techniques using various different time units offer a more effective way of assessing the associations between acts when these are thought to result from changing causal factors.

In situations where internal and external causal factors are thought not to change with time, the data are more likely to pass the test of stationarity, and here it is probable that the sequential dependencies found result from causal dependencies between the successive acts, or between the states of which they are the overt indicators. In these circumstances, the analysis of first-order dependencies is a first step toward discovering the rules governing sequencing. These may then be fully elucidated by a search for higher-order dependencies.

Many behavioral sequences will not fall neatly into one of these categories or the other; perhaps the most profitable approach here is to combine first-order transition analysis to look for associations liable to result from sequence effects, with correlation techniques, to find the looser associations which might result from shared causal factors.

It is important to stress that none of these methods provides more than a description of the behavior under study. That two acts tend to occur in sequence need not imply a similarity of causation, nor that one generates the other in some way: they could be associated by exclusion, or because of particular characteristics of the experimental situation. The choice of descriptive method may bias the researcher toward a particular type of explanation. For example, if $A \rightarrow B$ more than $B \rightarrow A$ in a transition analysis, it suggests that A generates a state appropriate to B, or increases the probability of B. This result can, however, also be explained on a motivational model in which the acts differ in threshold (Bastock and Manning, 1955; Hinde and Stevenson, 1969). Conversely, a symmetrical relationship between two behaviors such that both $A \rightarrow B$ and $B \rightarrow A$ are common could be taken as indicating common causation or that each of the two acts stimulated occurrence of the other. The description of behavior provides hypotheses but not explanations; a frequent mistake made by those using factor analysis is the attempt to provide a causal explanation based on purely descriptive data. On the other hand, the detailed knowledge of the associations between behaviors made possible by the other methods discussed here may prove an invaluable source of hypotheses; the way in which these associations change in response to experimental intervention can then be used to build models of the causation of behavior.

Similar arguments apply to the description of interactive sequences. Even after valid ways have been found of demonstrating that behavior B by one animal tends to occur after behavior A by another more often than expected, this discovery is not evidence that A causes B, though this is one possible explanation. Other possibilities are that a behavior occurring close to A in time stimulates B, or a behavior associated with B, or even that the two behaviors tend to occur synchronously due to some event in the more distant past, such as social facilitation of other activities or the onset of daylight (Andrew, 1972). Experiments are essential if communication is to be demonstrated. Nevertheless, the analysis of sequences of behavior in normal interactions is a useful initial stage before such experiments are carried out to indicate the acts which are likely to be important in communication. The fact that most studies on this topic have failed to distinguish clearly between intra- and interindividual effects does not argue against the usefulness of the general approach.

VI. ACKNOWLEDGMENTS

I am grateful to Professor R. J. Andrew, Dr. P. P. G. Bateson, Dr. C. Chatfield, Mrs. J. C. Ollason, and Dr. L. M. Rosenson, all of whose comments have helped to improve this article.

VII. REFERENCES

Altmann, S. A. (1965). Sociobiology of rhesus monkeys. II. Stochastics of social communication. *J. Theoret. Biol.* **8**:490–522.
Andrew, R. J. (1956). Normal and irrelevant toilet behaviour in *Emberiza* spp. *Brit. J. Anim. Behav.* **4**:85–91.
Andrew, R. J. (1972). The information potentially available in mammalian displays. In Hinde, R. A. (ed.), *Non-Verbal Communication,* Cambridge University Press, Cambridge, England, pp. 179–204.
Aschoff, J. (1967). Circadian rhythms in birds. *Proc. 14th Internat. Ornithol. Congr.,* pp. 81–105.

Baerends, G. P. (1970). A model of the functional organization of incubation behaviour. *Behaviour Suppl.* **17**:263–312.

Baerends, G. P., and van der Cingel, N. A. (1962). On the phylogenetic origin of the snap display in the common heron (*Ardea cinerea* L.). *Symp. Zool. Soc. Lond.* **8**:7–24.

Baerends, G. P., Brouwer, R., and Waterbolk, H. T. (1955). Ethological studies on *Lebistes reticulatus* (Peters): I. An analysis of the male courtship pattern. *Behaviour* **8**:249–334.

Baerends, G. P., Drent, R. H., Glas, P., and Groenewold, H. (1970). An ethological analysis of incubation behaviour in the herring gull. *Behaviour Suppl.* **17**:135–235.

Bartlett, M. S. (1951). The frequency goodness-of-fit test for probability chains. *Proc. Cambridge Philos. Soc.* **47**:89–95.

Bastock, M., and Manning, A. (1955). The courtship of *Drosophila melanogaster*. *Behaviour* **8**:85–111.

Bindra, D. (1959). *Motivation: A Systematic Reinterpretation*, Ronald Press, New York, 361 pp.

Blalock, H. M. (1961). *Causal Inferences in Nonexperimental Research*, University of North Carolina Press, Chapel Hill, 200 pp.

Blurton Jones, N. G. (1968). Observations and experiments on causation of threat displays in the great tit (*Parus major*). *Anim. Behav. Monogr.* **1**:75–158.

Bolles, R. C. (1960). Grooming behavior in the rat. *J. Comp. Physiol. Psychol.* **53**:306–310.

Brown, R. G. B. (1965). Courtship behaviour in the *Drosophila obscura* group. Part II. Comparative studies. *Behaviour* **25**:281–323.

Cane, V. (1959). Behaviour sequences as semi-Markov chains. *J. Roy. Stat. Soc. Ser. B* **21**:36–58.

Cane, V. (1961). Some ways of describing behaviour. In Thorpe, W. H., and Zangwill, O. L. (eds.), *Current Problems in Animal Behaviour*, Cambridge University Press, Cambridge, England, pp. 361–388.

Chatfield, C. Statistical inference regarding Markov chain models. *Appl. Stat.* (in press).

Chatfield, C., and Lemon, R. E. (1970). Analysing sequences of behavioural events. *J. Theoret. Biol.* **29**:427–445.

Coulon, J. (1971). Influence de l'isolement social sur le comportement du cobaye. *Behaviour* **38**:93–120.

Cox, D. R., and Lewis, P. A. W. (1966). *The Statistical Analysis of Series of Events*, Methuen, London, 285 pp.

Delius, J. D. (1969). A stochastic analysis of the maintenance behaviour of skylarks. *Behaviour* **33**:137–178.

Dingle, H. (1969). A statistical and information analysis of aggressive communication in the mantis shrimp *Gonodactylus bredini* Manning. *Anim. Behav.* **17**:561–575.

Duncan, I. J. H., Horne, A. R., Hughes, B. O., and Wood-Gush, D. G. M. (1970). The pattern of food intake in female brown leghorn fowls as recorded in a Skinner box. *Anim. Behav.* **18**:245–255.

Fabricius, E., and Jansson, A-M. (1963). Laboratory observations on the reproductive behaviour of the pigeon (*Columba livia*) during the pre-incubation phase of the breeding cycle. *Anim. Behav.* **11**:534–547.

Fentress, J. C. (1972). Development and patterning of movement sequences in inbred mice. In Kiger, J. A. (ed.), *The Biology of Behavior*, Oregon State University Press, Corvallis.

Golani, I., and Mendelssohn, H. (1971). Sequences of pre-copulatory behaviour of the jackal (*Canis aureus* L.). *Behaviour* **38**:169–192.

Goodman, L. A. (1968). The analysis of cross-classified data: Independence, quasi-independence and interactions in contingency tables with or without missing entries. *J. Am. Stat. Ass.* **63**:1091–1131.

Hazlett, B. A., and Bossert, W. H. (1965). A statistical analysis of the aggressive communications systems of some hermit crabs. *Anim. Behav.* **13**:357–373.

Heckenlively, D. B. (1970). Intensity of aggression in the crayfish, *Orconectes virilis* (Hagen). *Nature* **225**:180–181.

Heiligenberg, W. (1973). Random processes describing the occurrence of behavioral patterns in a cichlid fish. *Anim. Behav.* **21**:169–182.

Hinde, R. A. (1955/1956). A comparative study of the courtship of certain finches (Fringillidae). *Ibis* 97:706–745; 98:1–23.

Hinde, R. A. (1970). *Animal Behaviour: A Synthesis of Ethology and Comparative Psychology*, McGraw-Hill, New York, 876 pp.

Hinde, R. A., and Stevenson, J. G. (1969). Integration of response sequences. In Lehrman, D. S., Hinde, R. A., and Shaw, E. (eds.), *Advances in the Study of Behavior*, Vol. 2, Academic Press, New York, pp. 267–296.

Isaac, D., and Marler, P. (1963). Ordering of sequences of singing behaviour of mistle thrushes in relation to timing. *Anim. Behav.* 11:178–188.

Lemon, R..E., and Chatfield, C. (1971). Organization of song in cardinals. *Anim. Behav.* 19:1–17.

MacKay, D. M. (1972). Formal analysis of communicative processes. In Hinde, R. A. (ed.), *Non-Verbal Communication*, Cambridge University Press, Cambridge, England, pp. 3–25.

McFarland, D. J. (1971). *Feedback Mechanisms in Animal Behaviour*, Academic Press, London, 279 pp.

Moore, G. P., Perkel, D. H., and Segundo, J. P. (1966). Statistical analysis and functional interpretation of neuronal spike data. *Ann. Rev. Physiol.* 28:493–522.

Nelson, K. (1964). The temporal pattern of courtship behaviour in the glandulocaudine fishes (Ostariophysi, Characidae). *Behaviour* 24:90–146.

Noirot, E. (1969). Serial order of maternal responses in mice. *Anim. Behav.* 17:547–550.

Overall, J. E. (1964). Note on the scientific status of factors. *Psychol. Bull.* 61:270–276.

Palmgren, P. (1950). On the diurnal rhythm of activity and rest in birds. *Ibis* 91:561–576.

Parker, G. A. (1972). Reproductive behaviour of *Sepsis cynipsea* (L.) (Diptera: Sepsidae). I. A preliminary analysis of the reproductive strategy and its associated behaviour patterns. *Behaviour* 41:172–206.

Perkel, D. H., Gerstein, G. L., and Moore, G. P. (1967). Neuronal spike trains and stochastic point processes. I. The single spike train. *Biophys. J.* 7:391–418.

Richter, C. P. (1927). Animal behavior and internal drives. *Quart. Rev. Biol.* 2:307–342.

Slater, P. J. B. A reassessment of ethology. In Broughton, W. B. (ed.), *Biology of Brains*, Academic Press, London, in press.

Slater, P. J. B., and Ollason, J. C. (1972). The temporal pattern of behaviour in isolated male zebra finches: Transition analysis. *Behaviour* 42:248–269.

Stokes, A. W. (1962). Agonistic behaviour among blue tits at a winter feeding station. *Behaviour* 19:118–138.

Tinbergen, N. (1951). *The Study of Instinct*, Clarendon Press, Oxford, 228 pp.

van Hooff, J. A. R. A. M. (1970). A component analysis of the structure of the social behaviour of a semi-captive chimpanzee group. *Experientia* 26:549–550.

Weidmann, U., and Darley, J. (1971). The role of the female in the social display of mallards. *Anim. Behav.* 19:287–298.

Wells, G. P. (1950). Spontaneous activity cycles in polychaete worms. *Symp. Soc. Exptl. Biol.* 4:127–142.

Wiepkema, P. R. (1961). An ethological analysis of the reproductive behaviour of the bitterling. *Arch. Néerl. Zool.* 14:103–199.

Wiepkema, P. R. (1968). Behaviour changes in CBA mice as a result of one goldthioglucose injection. *Behaviour* 32:179–210.

Wortis, R. P. (1969). The transition from dependent to independent feeding in the young ring dove. *Anim. Behav. Monogr.* 2:1–54.

Chapter 6

SPECIFIC AND NONSPECIFIC FACTORS IN THE CAUSATION OF BEHAVIOR[1]

John C. Fentress

Departments of Biology and Psychology
University of Oregon
Eugene, Oregon

I. ABSTRACT

In dealing with patterned behavior, an immediate question concerns the degree to which different component activities are separate from one another in their control. Models of integrative specificity stress separate control mechanisms for different behavioral components. Nonspecific models stress mechanisms that are shared among many different activities. Each of these models has certain merits but also certain limitations. Section II of this paper provides a brief outline of some of the major aspects of both the action-specific models and the diffuse activation models of integrated behavioral control. It is suggested that much of the apparent controversy is due to inconsistent application of classificatory criteria which are often not adequately defined. For the purpose of illustration, these issues are discussed in terms of the ethological construct of "action-specific" control and the neuropsychological dimension of "nonspecific" activation. A possible alternative position that stresses overlapping control systems and partial specificity is introduced. Section III concerns aspects of classification and analysis plus some related conceptual issues that appear most relevant to current thoughts about specific vs. nonspecific behavioral control. It is pointed out that the

[1]Preparation of this chapter was supported in part by Public Health Service Grant MH 16955.

most strict criterion for specific control is a system that involves only a single input and a single output and that is entirely independent in its operation from other activities in other control systems, while the most strict definition of nonspecific control is that all inputs affect all outputs, identically. Neither of these criteria is likely to be met, and a more limited operational approach is suggested. It is particularly clear that the problem of input specificity is partially separable from that of output specificity, and that "nonspecific" models at one level of analysis may reflect the operation of "multispecific" mechanisms at a different level of analysis. It is emphasized that input/output analyses provide a logic for the interpretation of intervening processes and do not permit simple S-R models. Section IV provides a detailed literature review that illustrates these and related issues. The data indicate a need to consider separately, and then synthesize, the operation of (1) qualitative variables (which inputs and which outputs), (2) quantitative variables (how much of these inputs and/or outputs), and (3) temporal variables (which part of the input/output sequence). The possible importance of encoding mechanisms (e.g., habit strength) is also briefly considered. Section V reviews the ethological literature on displacement activities as a particularly clear illustration of major issues in specific vs. nonspecific control. It is demonstrated that recent emphasis upon (1) disinhibition as a result of behavioral competition and (2) the operation of specific exogenous and peripheral stimuli provides a useful supplement to models of "nonspecific" endogenous activation. However, it is also suggested that endogenous excitatory processes may be partially shared among competing behavior patterns under certain circumstances and in that sense be "nonspecific." Section VI provides a synopsis of the major issues discussed previously and outlines a "boundary-state" approach that appears to avoid many of the difficulties of strict specific and nonspecific models. This approach stresses both partial and shifting overlap among control systems as a function of intensity and temporal parameters. An analogy is suggested to the center/surround organization of control systems at the neurological level. Integrative pathways are viewed to become more or less focused as a function of the balance between excitatory and inhibitory mechanisms. This balance between excitation and inhibition may shift in a systematic manner as a joint function of qualitative, quantitative, and temporal dimensions employed in a given analysis.

II. INTRODUCTION

When one considers problems of behavioral integration in the intact organism, it is necessary both to dissect ongoing patterns of behavior into component parts and to determine the interrelationships among these component parts. A question that immediately arises is the extent to which these behavioral subunits are separate from one another in their control as opposed to the extent to which they share common mechanisms. This question is traditionally summarized in terms of control specificity. To what extent are there both specific and nonspecific factors in the causation of integrated patterns of behavior? What are the interactions between these two classes of control?

On initial consideration, such questions appear easy to resolve by empirical methods alone. However, as we shall see, they also depend critically on methods of behavioral classification. Questions of behavioral specificity can of course be approached at many different levels, such as the coding of sensory information and organization of complex motivational functions. The emphasis in this chapter is upon integrative systems that mediate various input/output relations in the intact organism. Properties of coding in isolated sensory and motor systems will receive less detailed consideration than will the operation of intervening mechanisms that contribute to integrated behavioral control.

A. Action-Specific Models

Ethologists have traditionally stressed specific factors in the causation of different behavior patterns. Lorenz (1957), for example, has utilized the term "action-specific energy" to emphasize this point in discussion of motivational control of ongoing behavior. Tinbergen (1951) liberalized Lorenz's initial action-specific doctrine in his now famous hierarchical model of behavior, but he too, speaking from a motivational framework, concluded that "there is no question of all the drives being activated at the same time" (p. 2). Such a position is based on the logical framework that different classes of behavior, defined by functional criteria, not only do not occur simultaneously but also can actively interfere with the expression of one another. This position has been amply supported by recent research on behavioral interactions (e.g., van Iersel and Bol, 1958; Sevenster, 1961). I have previously (Fentress, 1968a,b) extended Tinbergen's argument by noting that absolute "nonspecificity" would demand that *all* inputs affect *all* outputs *equally*. As has been pointed out by Hinde (1970), few proponents of nonspecific control of behavior have even examined two input/output relations.

The case, therefore, might seem easily settled. As I have also noted, however (Fentress, 1968a,b, 1972), the strict definition of specificity demands absolute functional independence of different input/output control functions, under *all* conditions. This is tantamount to proving the null hypothesis and is controverted by much of the available data. For example, Tinbergen's statement that different classes of behavior are often antagonistic to one another indicates that at some level of control the underlying mechanisms are *not* independent. This lack of independence implies interaction and thus an influence of one behavioral system on another. As will be seen below, as soon as one has multiple interactions between control mechanisms that are not in themselves examined directly, the overall behavioral consequences may be indistinguishable from "nonspecific" control.

Tinbergen (1952) has argued that the *simultaneous* activation of different classes of behavior is essential to the doctrine of nonspecificity. Many behavioral patterns are mutually exclusive, however (e.g., it is most difficult to stand up and sit down simultaneously). This fact does not in itself preclude the possibility that control systems are shared among these different, incompatible, classes of behavior.

Tinbergen (1952) argues that sleep is known to be dependent on "a drive" and that a generally aroused animal should therefore be expected to sleep more (as well, it is supposed, as to do everything else more). This, he claims, "is contrary to observed facts." He concludes, therefore, that general activating concepts are invalid. As an interesting sideline, it can be noted that there are reports that "excited" organisms do sometimes sleep more (Pavlov, 1927; Levin, 1961). Alternate formulations are possible.

Most ethological categories of behavioral control systems thus are based primarily on the criterion of behavioral elicitation and not on suppression. That is, behavior patterns that are *elicited* by a common factor are viewed as part of the same system, while those behavior patterns which are *blocked* are viewed as part of a different, "separate," system. If not recognized, this can lead to confusion. It is well to recall, for example, that Sherrington (1906) defined reflex control systems in terms of *both* excitatory and inhibitory subfunctions. To illustrate, the normal performance of flexion depends not only on the excitation of flexor motoneurons but also on the simultaneous inhibition of extensor motoneurons. Inhibition as well as excitation can be critical to the definition of a behavioral control system, and the ethologists' emphasis on eliciting factors alone must be evaluated carefully (cf. Norton, 1968).

There are other difficulties for strict action-specific models. Lorenz (1957, pp. 298–299) himself, a long-term proponent of specific control of different behavior patterns, has noted that, "if a breeding pair of the cichlid fish . . . is induced to fight by presenting a dummy, and then, after fighting has attained its fullest intensity, the sham rival is removed, in a few seconds the fish will begin to court with equal intensity." That is, excitatory processes shared between quite diverse categories of behavior are demonstrable. From such observations, Lorenz (1957, p. 299) has noted that the "expression 'action-specific' must not be interpreted too narrowly." For similar reasons, Hinde (1960, p. 211) points out that "an overemphasis on the independence of activities leads to a neglect of, for instance, sensory, metabolic or temperamental factors which affect many activities." Thorpe (1962) also has suggested that the interactions between different classes of behavior might be more complex than indicated in Tinbergen's (1951) initial, and most useful, hierarchical scheme. Thorpe's emphasis on the network, or "nexus" (cf. Hinde, 1970), of interactions underlying integrated behavioral control thus deserves serious consideration in any discussion of specific and/or nonspecific mechanisms.

B. Diffuse Activation Models

It is largely the possibility that excitatory control processes might be shared across a wide range of apparently diverse behavior patterns that has led other workers to stress nonspecific "activation," "arousal," or "drive" (e.g., Berlyne, 1960; Bindra, 1969; Duffy, 1962; Hebb, 1955; Hull, 1943; Lindsley, 1951; Malmo, 1959). One of the major contributions of these workers that I shall return to later is the supposition that intensity variables in integrated behavior may be at least partially separable from directional variables. The idea is not new (e.g., Woodworth, 1918) but has, in recent years, received increasing attention.

Hebb (1955) offers the picturesque analogy of a ship's engine (that controls speed but not direction) and a ship's rudder (that controls direction but not speed) to illustrate this point. This single-engine model has many merits in suggesting research into common mechanisms of integrated behavioral control but also certain limitations. For example, there is much ethological evidence that the prior activation of one behavior pattern can reduce (rather than enhance) the subsequent expression of a second pattern *even when* stimuli appropriate to the second pattern are given (e.g., van Iersel and Bol, 1958; Lorenz, 1957; Sevenster, 1961; Tinbergen, 1952). Such observations themselves must be interpreted carefully, however. For example, as I have argued previously (Fentress, 1968a,b, 1972), the intensity to which different behavioral control systems are activated might be an important determinant in whether they interact with one another in a positive or negative fashion.

There has been a considerable temptation to equate possible "nonspecific" behavioral activation with the operation of particular neurological mechanisms, e.g., the reticular activating system. As is well known, the apparently diffuse activity of the ascending reticular formation (e.g., Moruzzi and Magoun, 1949) immediately made this structure a prime candidate for "nonspecific" drive, etc. While such attempts to relate behavioral phenomena to biological processes are of obvious importance, several potential limitations soon became apparent. First, as indicated above, ethologists and also experimental psychologists have obtained data that challenge the generality of a general drive concept at the behavioral level (for two excellent reviews, see Bolles, 1967; Hinde, 1970). Second, mere correlation between data obtained separately at different levels of analysis does not in itself constitute a sufficient explanation of one level in terms of the other, and subsequent data have indicated that indeed correlations between "reticular arousal" and "behavioral arousal" can be broken. Atropine administration can, for example produce patterns of cortical activity similar to those which occur in sleep although the animal is behaviorally alert (Bradley, 1958; Wikler, 1952), and lesions in the posterior hypothalamus can obliterate behavioral arousal while cortical desynchronization patterns normally characteristic of an alert animal remain (Feldman and Waller, 1962). The admonition by von Holst and von St. Paul (1963) that investigators attempting to relate brain and behavioral functions employ "level-adequate terminology" is important here. Third, as has been pointed out, particularly by Hinde (1959), energizing concepts in behavioral organization can be misleading if not interpreted carefully. He argues, for example, that nonspecific factors may not serve a primary "energizing" role but perhaps serve "merely for the efficient functioning of the control mechanism."[2] Finally, recent analyses of the reticular activating system have shown that its "nonspecific" functions are most readily apparent when the system is considered as a whole but that it can be fragmented into a variety of operational subsystems which can be quite

---

[2]It is interesting that one of the first workers to articulate potential difficulties with the overinterpretation of "energy models" of behavior was C. G. Jung, a disenchanted former pupil of Freud's psychoanalytic school. In his book *Psychological Types* (1923), Jung protested, "Assuredly the learned nominalist of our day is quite convinced that 'energy' is merely a nomen, a 'counter' of our mental calcule; yet in spite of this, our every-day speech refers to 'energy' as though it were something quite tangible; thus constantly sowing among devoted heads the greatest confusion from the standpoint of the theory of cognition" (p. 41). At about the same time, Russell (1927) discussed an analogous problem in the physical sciences: "We must give up what Whitehead admirably calls the 'pushiness' of matter. We naturally think of an atom behaving like a billiard ball; we should do better to think of it like a ghost, which has no 'pushiness' and yet can make you fly" (p. 118). Even today, energizing concepts provide considerable difficulties for the theoretical physicist (e.g., Bohm, 1969). The point that behavioral scientists may not find easy mechanistic explanations of behavioral phenomena by jumping to analogies with the physical sciences has been summed up by Skinner (1962) in "a serious work in jest": " 'It was bound to happen. Look at your equation.' (PROF. S. goes to the blackboard and writes out formula: $E = mc^2$) 'That's why it happened. After all, what is conditioning? It's simply energizing human behavior. Now what is extinction? De-energizing. I've been running the tape backward, trying to undo the effect of the last few feet . . . I had to take the risk. It was a calculated risk. We had to de-energize him, but' (pointing to equation and striking out the E and the m) 'no energy, no mass. You see that smoke? That's c^2. Charley, I sometimes wonder if it's all worthwhile' " (p. 46). In sum, energy models may have their utility if used cautiously to organize a set of data but can easily lead to confusion if overly interpreted in terms of the necessary, and sufficient, operation of mechanism.

separate from one another and limited (i.e., "specific") in their operation (e.g., Anokhin, 1961; Dell, 1963; Jasper, 1963; Rossi and Zanchetti, 1957). This leads to a basic point that is simple but of some importance: a system may be considered quite "nonspecific" at one level of analysis (in which the system is examined in its entirety) but composed of multiple "specific" subfunctions at another level of analysis (cf. Fentress, 1968a,b). Thus easy equations between different levels of analysis are not to be found. Luria's (1966, p. 35) comments are relevant here: "We therefore suggest that *the material basis of higher nervous processes is the brain as a whole* but that *the brain is a highly differentiated system whose parts are responsible for different aspects of the unified whole.*"

The important contributions of "nonspecific" models of behavioral organization are twofold: first, intensity and directional variables in the control of behavior may be at least partially separable; second, causal factors may be at least partially shared among apparently diverse categories of behavior.

C. Partial Specificity ("Wavicle") Approach

To make a short story from a long review, it does appear that certain behavioral control functions are highly discrete ("specific") in their operation at the same time that other control functions appear more generalized ("nonspecific"). One could make a good argument why each of these facts *must* be true. It is imperative, for example, for an organism to keep its signals straight, i.e., separate from one another. If this were not the case, integrated patterns of behavior which depend on complex patterns of sensory input and motor output would not be possible. Stated simply, visual information must be separated from auditory information, and discrete movements can be made which do not involve the entire musculature. However, these separate signals cannot remain separate throughout; otherwise, the organism *as a whole* could not program a patterned *series* of responses on the basis of *combined* information from different sources. To provide another simple illustration, in a noisy room it is useful to combine visual information from a speaker's lips with auditory information while at the same time rejecting other aspects of *both* the visual and auditory worlds. That such interactions occur between superficially separate categories of sensory information has been well documented at the physiological level (e.g., Buser and Imbert, 1961; Fessard, 1961; Gerard *et al.*, 1936; Horn, 1962; Jung *et al.*, 1963; Palestini *et al.*, 1957). We shall see below, however, that simple interpretation of such findings in terms of "diffuse control" can be misleading.

When considering the problem of specific *vs.* nonspecific control of integrated behavior, one is reminded of the still existing controversy between "particulate" and "continuum" models of the physical universe (e.g., Bohm,

1969; Feynman, 1965). In many respects, this may reflect a basic limitation of human conceptual ability to deal with categories of phenomena which are partially homogeneous, partially heterogeneous, and also partially overlapping with one another. By certain measures, basic phenomena such as light, for example, appear to be organized along a diffuse or wavelike dimension, while by other measures these same phenomena appear highly particulate. Some theoretical physicists have coined the term "wavicle" as a semantic covering for this apparent paradox—the best of both possible worlds! The suggestion here, made "in serious jest" (to borrow from Skinner, 1963), is that there are "neurobehavioral wavicles" which the investigator will simply have to deal with rather than provide unitary explanations in terms of *either* absolutely discrete or absolutely diffuse control functions. This, of course, represents a conceptual position and does not imply anything directly about individual underlying mechanisms.

Thorpe (1963, p. 4) stressed a similar point when he emphasized that there appear to be "varying grades of unity in living organisms and varying grades of coordination of actions among these units." Jung *et al.* (1963), studying multisensory convergence, found bisensory responses common but trisensory responses rare, and concluded that localistic or holistic views of brain function "are not mutually exclusive, but they do supplement each other." Working from a broader theoretical framework, Gerard (1960, p. 1941) has similarly argued: "The opposed aspects of neural organization, mass or molar on the one hand and particulate or micro on the other, appear in all phases of structure and function. Pavlov (and Lashley) has been pitted against Sherrington, the gestaltists against the behaviorists, adherents to field mechanisms against those favoring synaptic ones, diffuse against discrete in neural systems and behavioral patterns. Both are, of course, present and useful." In his recent important review of higher brain functions in man, Luria (1966, p. 27) advocates Filmonov's emphasis on "successive and simultaneous gradations in the localization of functions." The electrical stimulation studies by von Holst and von St. Paul (1963) are particularly relevant, since they offer a similar conclusion from an ethological framework. Interactions between stimulated behavior patterns were very subtle, and while the authors found evidence for functional localization these interactions led them to discount any simple "specific center" concept.

There are, of course, two partially separable issues here. The first concerns the degree of *operational specificity* of a given behavioral control system and the second concerns the extent to which a complex control system can be localized in terms of *anatomical centers*. The basic conceptual issues, however, are in many respects analogous and complementary. At this stage, it is useful to examine models and operations used to classify behavioral control systems in greater detail.

III. MODELS AND OPERATIONS: ASPECTS OF CLASSIFICATION AND ANALYSIS OF BEHAVIORAL CONTROL SYSTEMS

A. Model Construction: General Themes

The analysis of behavioral control systems is not separable from methods used both to describe behavior and to classify it (e.g., Hinde, 1970, 1972; Marler and Hamilton, 1966; Simpson, this volume; Slater, this volume). The criteria used to classify behavior, if not explicitly recognized, can be an immediate source of difficulty. The investigator of behavior is faced with a potential paradox in that categories of behavior must be formed (unless one wants to take Heraclitus' ancient challenge that every event is unique!), but these categories are unlikely to be either homogeneous (i.e., indivisible) or mutually exclusive (i.e., functionally independent) on refined analysis. Stated briefly, and without much exaggeration, categories of behavior *must be formed,* but the investigator *must not believe them*!

Categories are necessary abstractions convenient for summarizing data and suggesting subsequent analyses. On subsequent analyses, the original categories may lose their utility. This is particularly true when underlying processes are postulated on the basis of behavioral observations. Hinde (1970), for example, has provided a useful discussion on how "drive" concepts may be helpful at initial stages of analysis but if overly interpreted in terms of homogeneous control systems can actually prevent subsequent examination of mechanism. Miller (1960, p. 846) gives a concrete illustration: "The fact that we have in the English language a word (for example 'hunger' or 'aversive stimulation') to refer to something, does not necessarily mean that the phenomenon is a single unitary variable or controlled by a single center in the brain. The phenomenon may be a heterogeneous cluster of variables."

The first stage in classification of behavioral control systems, or other phenomena, has been well summarized by Ayer (1956, pp. 11–12): "If things resemble one another sufficiently for us to find it useful to apply the same word to them, we are entitled to say, if it pleases us, that they have something in common . . . the resemblance between things to which the same word applies may be of different degrees." Categories, while necessary, inevitably harbor a degree of heterogeneity.

Ashby (1956, p. 39) has stated his position on the necessary oversimplification of models in the natural sciences strongly: "every material object contains no less than an infinity of variables and therefore of possible systems." This is similar to the position earlier expressed by Poincare (1913, p. 181): "If, then, a phenomenon admits of a complete mechanical explanation, it will admit of an infinity of others that will render an account equally well of all particulars revealed by experiment." Using the language of control theory, Ashby points out that all neurobehavioral models are "homomorphic," i.e., major simplifications of the facts they intend to explain, rather than "isomorphic," i.e., *one* to *one* correspondence with all components and functional interrelations between these components. This indeed is a major utility of models, for they permit the investigator to concentrate on particular variables that are thought to contribute importantly to the overall system function. However, as Ashby (1956) notes, comparison between two homomorphic models, each by definition a simplified representation of the universe it is designed to summarize and/or explain, can provide many logical difficulties. One particularly important facet of this is that the criteria used to combine events, structures, etc., into functional groups must be explicitly stated to the fullest extent possible. Different classificatory criteria can give apparently contradictory homomorphic models. Such differences will not be eliminated by empirical research alone (cf. Simpson, this volume).

The importance of explicit, and alternative, criteria in classification of observations has been pointed out nicely by Koestler (1964). He challenges the taxonomist to decide whether a green square should more appropriately be classified with a green triangle or with a red square. Obviously, the solution here is arbitrary, depending whether shape or color is considered to be of prime importance. Such considerations may be considered uselessly academic, but they are not. Cane (1961) has, for example, noted that different measures of ongoing behavior can provide very different (and often apparently contradictory) pictures. Similarly, Hinde (1959, 1970) has noted that most behavioral categories are initially based on the criterion of functional endpoints (e.g., feeding activities and fighting activities) and that, if not recognized, this can lead to confusion when subsequent classifications are made in terms of control mechanisms. For example, the investigator might make the *a priori* assumption that different functional endpoints in behavior (fighting, feeding, etc.) have completely separate causal foundations, although there is no logical necessity that this is true. It takes but a casual glance at much current literature in behavioral biology to recognize the confusion that can often result from such inappropriate assumptions.

One aspect of Hinde's observations deserves elaboration. The distinction between categories of behavior based on functional endpoints and antecedent causation is primarily one of temporal framework. While it is not the intention here to deal with Hume's (1748) challenge that all causal statements are *inferences* based on observed correlations between successive events, it is clear that different sections of ongoing time frames can provide quite different causal inferences. Functional endpoints *vs.* antecedent events are a case in point. I shall demonstrate shortly that, in dealing with problems of specific *vs.* nonspecific control, one must similarly be careful to examine separately input and output functions that operate at different stages and periods of processing. A given mechanism, for example, can be quite diffuse in terms of the operations it controls (i.e., functional endpoints) but highly discrete in terms of the inputs (i.e., causal prerequisites) it receives, or *vice versa*. Such should not be surprising when one thinks of interlocking causal networks that exhibit properties of both convergence and divergence rather than simple causal chains (*cf.* Hanson, 1955).

B. Three Basic Dimensions in Integrated Behavior

We can summarize much of the previous and subsequent discussion in terms of three basic dimensions in the control of integrated behavior: (1) the qualitative dimension (i.e., *what* inputs and *what* outputs), (2) the quantitative dimension (i.e., *how much* input and *how much* output, with the suggestion that intensity and directional variables may be partially separable), and (3) the temporal dimension (i.e., *which part* of the input/output *sequence* is being examined). Each of these three dimensions, and their possible interrelations, will be examined in greater detail below. Each is relevant to the specificity *vs.* nonspecificity problem.

C. A Prototype Model and Its Limitations

It is useful here to mention some of the fundamental issues of behavioral classification in terms of models commonly employed in ethology. Figure 1 is a diagrammatic summary of such a model and its potential limitations. Blocks *A* and *B* represent major categories of behavior defined by functional criteria (e.g., aggressive behavior and feeding behavior).

1. Nonunitary Behaviors

The first point is that neither of these categories of behavior is likely to be unitary when examined in detail at the causal level. This is summarized in

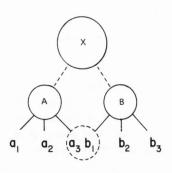

Fig. 1. Schematic representation of two major behavioral systems, *A* and *B*, as they might be classified on the basis of ethological criteria. Three points discussed in the text are summarized: (1) Initial categories are rarely unitary but can be subdivided into components ($a_1, a_2, a_3; b_1, b_2, b_3$). (2) Correlations of subunits in *different* initial categories may, by alternate criteria and/or observations, be greater than correlations of subunits *within* an initial category, thus permitting classification of alternate categories (a_3b_1). (3) Relatively "nonspecific" causal processes may be at least partially shared among initial categories (*X* in diagram).

the diagram by the fractionation of A into a_1, a_2, and a_3, and by the fractionation of B into b_1, b_2, and b_3. For example, grooming behavior in voles can be subdivided into face, belly, and back components each of which is partially separate from the others in terms of its control; i.e., the components can be manipulated more or less independently (Fentress, 1968a, b). Recent research with mice has demonstrated that face grooming itself can be further fractionated into at least eight major components which are partially separable in their control (Fentress, 1972). Similarly, Hinde (1958) has shown that nest building in the domestic canary can be fractionated into component parts such as picking up material, carrying material, and using material to construct the nest, and that each of these activities is under the control of partially separable mechanisms in the sense that their probabilities of occurrence do not fluctuate in unison. Such analyses indicate that the intervening behavioral control system is not unitary and that particular subfunctions may be *more specific* (i.e., separable) than suggested in preliminary models.

There are many other examples of nonunitary behavior, of which but a few will be cited here. Miller (1963) has noted that different measures of hunger do not always correlate closely with one another. Horvath *et al.* (1963) have demonstrated the heterogeneity of "spontaneous motor activity" in rats by separating rearing and locomotion with different doses of amphetamine; 4 mg/kg increased the former and decreased the latter, while 50 mg/ kg had the opposite effect. Similarly, Rowell and Hinde (1963) have emphasized the lack of correlation between various measures of stress in rhesus monkeys. De Lorge and Bolles (1961) have pointed out the inadequacy of using locomotion as a sole measure of exploratory activity, and Mirsky and Rosvold (1963) have broken down attentive behavior into various subdivisions that are not uniformly correlated. Each of the subdivisions made by these authors could probably be further subdivided with still more detailed measures. That the problem exists even at the level of labeling relatively simple motor patterns is demonstrated by Paillard (1960). In each of these cases, the need to postulate at least partially separable ("specific") control subfunctions is indicated.

2. Alternate Behavioral Categories

The second point of the diagram is that different components of behavior *within* the initial behavioral categories may, under certain circumstances, correspond less closely with each other than do the components of different behavioral categories. This means, of course, that alternative categories of behavior which cross over the original ones might be formed (e.g., the category a_3b_1). I have previously shown (Fentress, 1967) that shoulder rubbing (on the ground) in timber wolves can be associated with either "play" situations or inspection of olfactory stimuli. The motor patterns in these two situations appear indistinguishable, and factors which contribute to the elicitation of this behavior in one situation (e.g., olfactory exploration) can facilitate the occurrence of the same behavior in another situation (e.g., social). Thus a shoulder-rubbing category might be formed that includes some but not all aspects of play and olfactory inspection. The basic principle is similar to that discussed by Delgado (1964) on the fractional organization of behavior as measured by brain stimulation. Certain biting movements might be quite similar in feeding, fighting, or even sexual behavior (cf. Hinde, 1970). It is for similar reasons that Thorpe (1963) has argued for a "nexus" of cross-communication between different systems within the lower levels of Tinbergen's (1951) hierarchical model.

3. Shared Causal Factors

The third point of the diagram is that additional factors may exist that can facilitate apparently diverse categories of behavior. This is illustrated by the X in the diagram. The construct here approaches that of "*nonspecific*" control (defined, of course, in terms of the initial categories). For example, I have shown that factors which elicit avoidance and freezing responses can, under appropriate conditions, facilitate predominant patterns such as rodent grooming (Fentress, 1968a,b, 1972). This particular issue is related to ethological constructs of displacement activity, which are discussed in greater detail in Section V.

There are many anecdotal observations of an analogous nature. From his field observations on the North American moose, Geist (1963) notes that after a disturbance there may be an increase in sexual licking in males, nursing in calves, and most commonly feeding by all the animals. In a more quantitative study, Hanby (1972) found that certain environmental disturbances can enhance aspects of mounting behavior in Japanese macaques. At a quite different level of analysis, Havlicék (1963) reports that while adrenalin injections at first decreased feeding and increased defense reactions in the rat, feeding was subsequently enhanced during the "deactivation" phase: "This phenomenon is often so striking that defensive signals call forth 'feeding' reactions which are evidence of the fact that a 'feeding' dominant was set up" (p. 206). In pursuing such cases, of course, and there are many, one should examine carefully both intensity and temporal dimensions of the observed interactions as well as qualitative relations among the behavior patterns themselves. For example, the illustrations cited above concern "residual" effects of apparently irrelevant stimulation. However, at the very least, such observations suggest that apparently diverse behavioral factors can, under appropriate conditions, have widespread excitatory influences.

Berlyne (1960, p. 48) mentions that stimulation of the thalamic reticular system will enhance the intensity of ongoing activity, a statement which would appear to support Hebb's engine concept. Fuster (1958) found a facilitation of perceptual and motor performance with mild brain stem stimulation in monkeys. From a developmental perspective, Anokhin (1961) has reported that "any" external stimulus will increase the rook's act of pecking at the hatching stage and then an identical stimulus will initiate the feeding reaction in a bird already freed from the egg. Anokhin concludes that "the functional system, which from the morphological point of view has been formed earlier than the other system, has a rather low threshold of excitability and therefore acts as a dominant function at that given stage of development. This means that any internal or external stimulus, however small, will indeed first of all stimulate into action the dominant functional system" (p. 643). While this statement is confined to the ontogeny of the organism, it is worth considering the possibility of dynamic (fluctuating) shifts of dominance between functional systems in the everyday life of the organism. This would mean that one behavior might be activated by a particular stimulus at one time and a second behavior at another. Increased emphasis on precise temporal relations would help clarify mechanisms. In a study of fear responses, Brown and Jacobs (1949) do give some evidence that "whatever response is dominant at the moment" (p. 757) is intensified, although not many responses were recorded. More limited data of an analogous nature have been obtained by Fentress (1968a,b). These data appear formally similar to the concept of dominant "set" discussed by Sperry (1955) and Doty (1961). The suggestion in each case is that a given causal factor may, under appropriate conditions, facilitate different behavioral responses. Further illustrations will be given below.

D. Summary of Classification Criteria

Table I summarizes five of the basic criteria that have been used to define the boundaries of a given behavioral control system and thus separation of different control systems. The criteria at the top of the table pose particular difficulties for models of control specificity, while those at the bottom of the table pose particular difficulties for models of nonspecific control.

1. Independence

As indicated above, the strictest criterion for separate (i.e., "specific") behavioral control systems is that each of these systems, defined in terms of input/output functions, is absolutely independent of input/output functions in other control systems (cf. Fentress, 1968a,b). The most pure case is where a single input, and only this input, affects a single output, and only this output. The very fact that many behavior patterns are mutually incompatible,

TABLE I

Specific Control Factors—stringent criteria
(no interactions)

1. Absolute independence of single input–output functions
   (i.e., no cross-excitation or inhibition)
                          (any interactions?)
2. Interactions in one direction only
   (i.e., only cross-excitation *or* cross-inhibition)
                          (any unidirectional interactions?)
3. Identical effects of system inputs on system outputs
   (i.e., no differential in either direction or extent of causal functions)
                          (selected identical effects?)
4. All inputs and all outputs
   (i.e., system that receives *all* inputs and modulates *all* outputs)
                          (inclusive interactions separable in time?)
5. Simultaneous activation of outputs by inputs
   (i.e., sequential or alternate excitation of outputs not included)
                          (all-inclusive simultaneous identical interactions)

Nonspecific Control Factors—stringent criteria

however, (cf. Tinbergen, 1952) suggests that this criterion is rarely met; i.e., many activities have negative interactions with one another. Strong avoidance responses, for example, interfere with grooming behavior in rodents (Fentress, 1968*a,b*, 1972).

2. Same Direction

An alternative, and one which is traditionally applied in the ethological literature, defines a control system in terms of causal functions that operate in the same direction, i.e., elicitation *or* blocking. For example, certain hormones may facilitate components of sexual behavior and thus suppress the expression of components of, for example, feeding behavior. Such observations have been used to suggest that feeding and sexual behavior are controlled by separate "specific" systems. However, since the facilitation of sexual behavior in a given context may be as dependent on the suppression of competing activities (e.g., feeding) as on the direct facilitation of sexual activities, one might alternatively define the control system in terms of *both* the resultant incremental and decremental behavioral functions. While this proposal will probably sound strange to some ethologists, it is logically sound and is identical in principle to techniques used by Sherrington (1906) to define the functional boundaries of a reflex movement control system (e.g., excitation of extensors *and* simultaneous inhibition of flexors). Subsequent and refined analysis will of course often be necessary to determine whether the observed suppression of a given behavioral output is produced, but the logic of initial classification at the behavioral level remains the same.

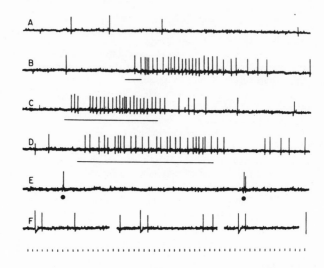

A.

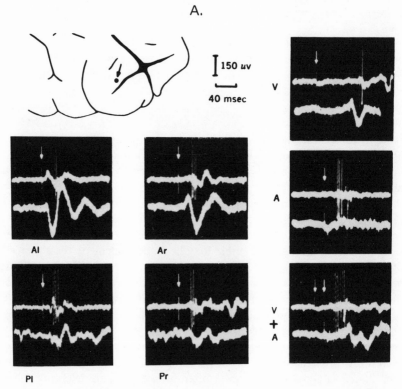

B.

3. Identity of Input/Output Functions

A stricter criterion still is to insist that the input functions included within a given behavioral control system have *identical* effects on each of the outputs in that control system (i.e., differential effects can be used to argue for the possible existence of subspecificities). Figure 2, for example, gives two illustrations of responses of single cells in the reticular formations and perisigmoid cortex, respectively, to different sources of sensory input. These records have been used to argue for nonspecific sensory control (Palestini *et al.*, 1957; Buser and Imbert, 1961). However, it is apparent that the effects of the different sensory inputs are distinguishable, and thus a degree of specificity might be claimed. An alternative, of course, is that the different inputs are given at different effective intensities; i.e., by manipulating each input over a wide range of intensities, identical records might be obtained. Sadly, there appear to be no neurophysiological studies of multimodal cells in which intensity, and source, and temporal variables have been manipulated in a strict parametric manner (Horn, personal communication). A similar critique could be applied to most of the behavioral data at present available.

4. All Inputs and All Outputs

The finding that more than one sensory input can affect the expression of a given behavior or the firing pattern of a particular neuron (Fig. 2) does not permit the conclusion that all inputs will do so. As noted above, for example, Jung *et al.* (1963) have found many cells that respond to *two* sensory inputs (thus indicating a degree of nonspecificity) but many fewer that respond to three or more sensory inputs (thus indicating a degree of specificity). Similarly, a given input may facilitate several but not all outputs. I have reported (Fentress, 1968*a,b*) that an overhead moving disturbance can facilitate either fleeing or freezing behavior in voles as a function of different ongoing activity but is more likely to suppress other behavior patterns.

5. Simultaneous Activation

As indicated previously, Tinbergen (1952) employed the additional demand that nonspecific "drive" mechanisms *simultaneously* increase all

Fig. 2. Two examples of multimodal (polysensory) units traditionally described as "nonspecific." A: Discharge of neuron in reticular formation as spontaneous firing (A), response to ipsilateral forelimb tap (B), response to rubbing back (C), response to touching whiskers (D), response to handclap (E), and response to individual electric shocks to ipsilateral motor cortex (F). Stimulus presentations are represented by lines and dots under unit records. [From Palestini *et al.* (1957).] B: Discharge of neuron in right perisigmoid cortex in response to somesthetic stimulation of each limb, anterior left and right (Al, Ar), posterior left and right (Pl, Pr), light flash (V), click (A), and flash–click sequence (V + A). [From Buser and Imbert (1961).] In each case, note range of responses and stimulus inputs. See discussion in text.

classes of behavior. Since many patterns are mutually incompatible in their expression, this criterion is unlikely to be met in most circumstances. Again, in terms of mechanisms at a refined analysis level such initial observations might reflect actual inhibition, or it is possible that internal excitation of the behavioral programs goes on but one or more of these activated programs are blocked at the output end, such as in certain conflict activities and in coarticulation in speech. (My thanks to Dr. M. Posner for bringing this latter illustration to my attention.) The time periods that different behavior patterns are activated by a given factor may vary, and thus the temporal span examined by the investigator can modify conclusions about organization of control processes (*cf.* Simpson, this volume; Slater, this volume; Fentress, 1968, 1972).

E. Alternative Formulations

The purpose of listing the various criteria above is to reemphasize that conflicts between proponents of specific and nonspecific behavioral control often cannot be resolved easily by empirical methods alone. The conceptual framework employed by different investigators is equally important, although operational criteria for making behavioral categories have rarely been specified with the precision necessary. Four of the most important considerations will be outlined here.

1. Continua vs. Dichotomies

As stated thus far, "specific" *vs.* "nonspecific" factors have been treated primarily in a dichotomous fashion. When used judiciously, such dichotomies can be a very efficient technique for ordering our thoughts (e.g., Attneave, 1959), but this is perhaps only so if such dichotomies are recognized as conceptual conveniences rather than absolute realities. The confusion between these two aspects of heuristic dichotomies can be seen in the often fruitless controversies between "innate" *vs.* "learned" or "endogenous" *vs.* "exogenous" mechanisms in the control of behavior (e.g., Hinde, 1970, 1972; Lehrman, 1970; Marler and Hamilton, 1966). As Hinde (1970, p. 205) notes: "A distinction between specific and nonspecific effects of a stimulus change is therefore often difficult to draw; and the question is more profitably phrased in terms of how specific or how general the effects of each factor may be." One might add to this, "under specified conditions" (cf. below).

2. Separation vs. Isomorphism vs. Overlap

In strict terms, the specific *vs.* nonspecific dichotomy applied to different aspects of ongoing behavior refers to the extent to which the different patterns observed are *entirely separate* from one another in terms of causal mechanisms as opposed to the extent to which they are *indistinguishable* (i.e., isomorphically identical). An alternative is to view different control systems as having varying degrees of overlap. Such a model proposes that different behavioral control systems are *neither* entirely separate from one another *nor* influenced identically by a given variable. Rather, the view is that underlying control systems may be *partially shared* among a defined range of behavior patterns. Such a doctrine receives neurophysiological support. In their recent review of central motor mechanisms, Evarts and Thach (1969, p. 471), for example, state that "overlapping gradients of localization rather than sharp borders between functionally separate zones have been demonstrated for projections of motor cortex pyramidal tract neurons . . . and it would seem likely that gradients rather than sharp boundaries are a general feature of patterns of central organization in the motor system as in other systems: localization is more-or-less rather than all or none" (p. 471). I have published data at the behavioral level (Fentress, 1968a,b, 1972) that suggest analogous principles of organization (cf. von Holst and von St. Paul, 1963; Luria, 1966, cited above).

3. Integrative Dynamics vs. Static Models

As indicated in Fig. 1, most behavioral models are basically static representations of dynamic functions. With respect to the question of specificity and nonspecificity, this may be misleading. There is no *a priori* reason that the degree of specificity within a given control system, and similarly the degree of functional overlap between control systems, need remain fixed at a given point. Rather, these functions may shift with the internal dynamics of the organism. While this possibility has been inadequately explored in the behavioral literature, there is suggestive evidence. I have, for example, shown that apparently irrelevant causal factors may excite a given behavior when these factors are present within a moderate range but inhibit the observed behavior when in excess of this range (Fentress, 1968a,b, 1972). A possibility that I have previously proposed (Fentress, 1972) and that will be pursued in the present chapter is that at moderate degrees of activation output controls may be relatively diffuse (i.e., nonspecific) in their effects, whereas at higher levels of activation these controls may become more tightly "focused" in terms of range of excitatory consequences (defined in terms of different functional classes of behavior affected). There are data at the neurophysiological level that suggest that control fields may shift in their dimensions and/or clarity of boundaries as a function of stimulus parameters (e.g., Adkins *et al.*, 1966; Eccles *et al.*, 1967; Hillman and Wall, 1969; Kozlowskaya and Valdman, 1967; Phillips, 1966). Such data are compatible with the "boundary-state" model of behavioral integration proposed below and provide a possible alternative to static conceptualizations of specific *vs.* nonspecific control. As we shall see, temporal parameters may also be important. Luria (1966) has pointed toward similar possibilities at the neuroanatomical level when he speaks of "dynamic localization of functions." The discussion by Simpson in this volume is also relevant.

4. Operations and Analysis Levels

As indicated previously, it appears important to separate problems of specificity on the input as opposed to the output side of organization, to document the level of analysis at which a given model of behavior is appropriate, and to state explicitly whether excitation, inhibition, or both excitation and inhibition (defined at a given analysis level) are included in definitions of a behavioral control system. Some simple operations relevant to these issues are outlined in the next section.

F. An Operational Approach

Two basic sides in the definition of a behavioral control system, input and output, must be separated. It is necessary to define each of these dimensions in terms of a specific referent; i.e., an output at one stage of a control system may be an input at a subsequent stage of processing. (This of course implies a serial model in addition to any emphasis on parallel operations.) The two basic criteria for including different inputs and different outputs within a given control system are summarized schematically in Fig. 3A: (1)

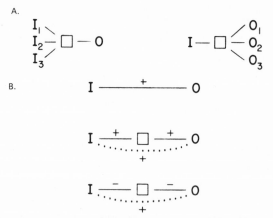

Fig. 3. Schematic for classifying behavioral control systems in terms of input and output functions. A: Initial classification by input convergence and output divergence. B: Direct and two indirect modes of excitation of observed output by a given input. See text for explanation.

If several different inputs (I) affect a given output (O), then it is logically defensible to consider these inputs, at the initial analysis level, as operating through a common system. (2) Similarly, if a given input affects several different outputs, one can conclude, at the initial analysis level, that these outputs are mediated through a common control pathway.

It should be emphasized that this does *not* imply a simple stimulus–response approach. Rather, it is through the separation and subsequent manipulation of specified input/output processes that the most adequate inferences about the operation of critical intervening control systems can be made. Two other points that deserve emphasis are that (1) the approach suggested here is not a simple *linear* representation with single inputs and outputs but rather stresses network processes that involve the interplay between multiple input and output dimensions (cf. Hanson, 1955), and (2) there is no implication that the inputs *drive* the observed outputs in any simple manner—rather, they modify the ongoing state of intervening processes to produce an *alteration* in observed outputs. (My thanks to Dr. J. Adkins for discussion of these issues; cf. Fentress, 1968*a,b*, 1972.)

For simplicity, we can speak of excitatory functions only, although, as we have seen, inhibitory functions can be included in the definition of a given control system as well. If the several inputs are indeed part of the same system, then it should also be possible to demonstrate the combination ("summation") of individual effects. I have, for example, demonstrated that pharmacological treatment can combine with home-pen rearing conditions and with species temperament differences in control of grooming in voles (Fentress, 1968*b*). From the consideration of basic neurological processes of occlusion and facilitation, such summation of individual inputs need not, of course, be strictly linear over all intensity ranges (Fig. 4). It is also suggested here that at the initial stage of analysis the outputs affected by a given input need not occur simultaneously to be included within the same system (cf. previous discussion) but may also occur sequentially or alternately. Hinde (1970), for example, has noted that behavior patterns which occur in close temporal proximity to one another are likely to share causal factors, and I have demonstrated that factors which increase fleeing and/or freezing behavior in rodents may also produce a subsequent increase in grooming (Fentress, 1968*a,b*, 1972).

The essence of these diagrams, while simple, has implications for models of specificity *vs.* nonspecificity. The basic point is this: a given input may have more than one effect; a given output may be affected by more than one input. This permits a tentative combination of these specified input and output functions into a common intervening control system, which, depending on the range of inputs or outputs participating, is defined along the continuum of specific and nonspecific control functions. Subsequent analyses may

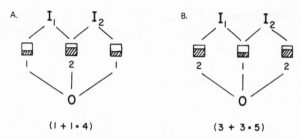

Fig. 4. Summation of two inputs on a given output. A: Facilitation—two inputs give a greater total output than the sum of the two individual effects of these inputs. B: Occlusion—the combined effect of two inputs is less than the sum of their individual effects. For ease of presentation, each control "box" can be thought of as operating in all or none fashion (although this is not necessary). Relative thresholds are indicated by shading in boxes.

of course separate certain parameters of these initially combined input and output relations (cf. Fig. 3B).

It deserves repeating that one cannot on an *a priori* basis combine the multiple input and multiple output features of Fig. 3. It is possible, for example, for I_1, I_2, and I_3 to produce O_1 and I_3 to produce O_2, etc. I_2 might additionally be the only input to produce O_3. The approach suggested therefore allows for partial overlap between subsystems defined in terms of input and output functions. I have previously demonstrated that this can be important on an empirical level (Fentress, 1968a,b). This approach, therefore, represents a departure from that proposed by Miller (1959), in which *both* multiple inputs *and* multiple outputs must be examined in the initial classification of intervening variables. It may be profitable to separate these input and output dimensions in initial analysis.

Figure 3B illustrates that initial classifications of behavior in terms of nonspecific functions may be altered by subsequent and refined analysis. For example, if *I* facilitates an apparently irrelevent behavior (e.g., fear-producing stimuli and grooming; Fentress, 1968a,b), then this might be interpreted as evidence for some type of "general drive." Two alternatives immediately suggest themselves, however. First, the input may increase stimuli that are specifically associated with the apparently irrelevant behavior, such as autonomic factors that increase peripheral irritation, and thus increase grooming, in a "fear" situation (e.g., Andrew, 1956a,b). Second, the input may inhibit an activity (e.g., exploration) that normally inhibits the facilitated output. The facilitation of an apparently irrelevant output such as grooming might thus be interpreted within a "disinhibition" rather than "general activation" framework (e.g., van Iersel and Bol, 1958; Sevenster, 1961; Rowell, 1961). The relevance of these comments to the ethological construct of "displacement activities" will be examined in greater detail below. However, at this time it should be pointed out that neither the presence of "specific" relevant stimuli nor the suppression of normally conflicting behaviors in itself negates the possible influence of other relatively "nonspecific" excitatory factors (e.g., Fentress, 1968*a,b*; Hinde, 1970; Lorenz, 1971; Rasa, 1971; Wilz, 1970*a,b*).

At this stage, it is useful to review some experiments directly related to the issue of specific and nonspecific control of integrated behavior patterns as outlined here.

IV. INTEGRATIVE SPECIFICITY LITERATURE: AN OVERVIEW

The problems we have examined thus far can be summarized in terms of functional boundaries. What is the range of input, output, and intervening functions that should be included within the boundaries of any given control system? To what extent do the boundaries of different control systems overlap, or are they mutually independent? Do different behavioral control systems interact with one another in only a negative fashion ("inhibition"), or are there also positive ("excitatory") interrelations between control systems?

For the neurophysiologist, such questions can be approached in a direct manner through analysis of basic sensory and motor functions. However, for the behavioral scientist the questions are particularly critical for functions that integrate different input/output combinations in the intact organism. These intervening functions must usually be inferred by systematic manipulation of different input/output variables (cf. Fig. 3).

A. Attentional–Motivational Variables

The studies of Broadbent and others in human experimental psychology on attentive mechanisms (e.g., Attneave, 1959; Broadbent, 1958, 1971; Garner, 1962; Keele, 1968; Posner, 1966; Posner and Boies, 1971) provide an elegant demonstration that such an approach can yield precise inferences about underlying control functions. In terms of the question of control specificity, these studies are of particular interest, for they indicate that quite different sensory and motor functions may compete for the same limited-capacity processing channel to a considerable degree. These models are similar to those in the neurophysiological literature on occlusion and thus indicate that integrative functions are not restricted (i.e., "specific") to any single given task. Concentration on one set of stimuli, for example, may interfere with the simultaneous processing of other stimuli, and the performance of complex motor responses may also interfere with processing of sensory input from various modalities (e.g., Broadbent, 1958, 1971). The extent to which single-channel models are adequate at different stages and for different types of processing is a subject of active current research. These models may be of considerable importance to problems of behavioral integration in ethology, but to date they have not been applied in a systematic fashion (cf. Fentress, 1965; McFarland, 1966). (It should also be emphasized that attention models involve the concept of "choice." That is, they concern factors which lead to one set of activities—e.g., processing certain visual information—as opposed to another set of activities—e.g., processing different visual information, auditory information, etc. In this sense, they bear a close relationship to ethological models of motivation, in which a given ongoing behavior is examined in the context of alternative actions that the animal might make at any given point in time.)

Of particular interest to ethologists are fluctuations in the relative strengths of different input/output relationships during ongoing behavior. These changes of responsiveness to stimulation, which are reversible (although usually lasting more than a few seconds) and which cannot be attributed to peripheral factors such as sensory and motor fatigue, are classed within the broad framework of motivational control (e.g., Hinde, 1970). The question of specificity, as indicated earlier, is the extent to which various motivational control systems are independent of one another in their action. Since it is well known that many different classes of behavior are mutually antagonistic (e.g., Tinbergen, 1952), the emphasis here will be on positive interactions and functions that may be shared among differently defined motivational control systems: that is, do factors which increase the intensity of one class of behavior also increase the intensity of another class of behavior? To what extent is it legitimate to separate intensity and directional variables in behavioral control?

B. Intensity Dimensions

Intensity itself deserves brief comment, for it is unlikely to be a unitary factor. At least six, partially separable, dimensions can be noted.

1. Duration

Probably the most common criterion that the strength of a given motivational control system has increased is that the behavior patterns assumed to be under control of this motivational system increase in both their probability and duration. During reproductive seasons, for example, animals spend an increased proportion of their time in courtship, copulatory, and parental activities. These activities do not occur simultaneously, however, and are partially independent of one another in their control (e.g., Lehrman, 1965). The component activities may also be negatively correlated in certain instances; for example, large doses of testosterone might be expected to reduce the time spent in preliminary courtship, with copulatory activities thus coming with a shorter latency. Thus the total duration of sexual activities can be *decreased* when causal factors are sufficiently strong. One model is that the different components have different thresholds, with higher-threshold behavior suppressing the expression of lower-threshold behavior. Thus duration measures that include various appetitive stages in behavior may, under certain conditions, be *negatively* correlated with inferred "drive strength" (cf. Duncan and Wood-Gush, 1972; Fentress, 1965).

2. Completeness

One important approach to this problem is to differentiate components of a motivational sequence and to ask *how complete* the sequence is (e.g., Tinbergen, 1952). This thus presents an alternative, or at least a supplement, to duration measures. Conclusions may still not be entirely unambiguous, however. Beach *et al.* (1955), for example, found that electroconvulsive shock decreased sexual "arousal" in male rats as judged by an increase in intermating intervals but also decreased the number of intromissions necessary for ejaculation. Different measures might therefore suggest that the apparently irrelevant shock increased or decreased sexual motivation. Part of the solution, of course, is that sexual behavior is not a unitary dimension, and there is no reason why a given stimulus situation should affect all components equally. This obviously argues for greater specificity, but at the same time the fact that apparently irrelevant stimulation such as shock can facilitate certain components of sexual behavior, particularly those found in complete rather than aborted bouts, suggests that relatively "nonspecific" factors may also be operative (cf. Barfield and Sachs, 1968; Larsson, 1963—discussed below).

The studies by Tugendhat (1960*a,b*) are particularly interesting in this context. She examined the effects of electric shock on subsequent feeding in three-spined sticklebacks and found that after shock the number of completed feeding responses per unit time increased (as in conditions of food deprivation) but the total time spent feeding decreased (unlike food-deprivation conditions). Thus the number of prey animals eaten (which could also be used as a measure of "motivational strength") might either increase or decrease depending on the time samples examined.

3. Time Samples

The critical importance of time samples examined is also illustrated clearly in the study by Sevenster (1961) on the relationship between aggressive and sexual components of behavior in the stickleback. When one takes long time periods, such as reproductive season *vs.* nonreproductive season, there is positive correlation between these two classes of behavior (Fig. 5). This permits the conclusion, therefore, that causal factors may be shared (i.e., be "nonspecific") between the two major classes of behavior. However, short-term interactions between these activities is in the negative direction. This is not merely due to the incompatibility of showing both patterns simultaneously, for prior strong activation of either tendency reduces subsequent expression of the other. Surely, the patterns are not independent (and thus might be included in a common control system, above), but do they share common excitatory factors?

4. Activation Level and Direction of Interaction

It is possible that this depends on the extent to which either system is activated. For example, might low levels of previously activated sexual behavior *increase* (rather than decrease) subsequent expression of aggressive behavior, and *vice versa*? I have provided evidence that the apparent level of activation of any given system might be more important in the direction of interaction between control systems than previously supposed in the ethologi-

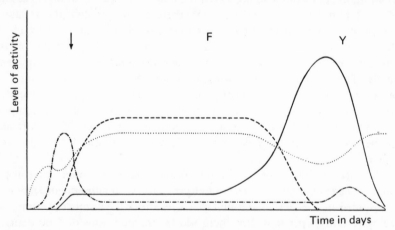

Fig. 5. Schematic representation of breeding cycle activities in the male stickleback (*Gasterosteus aculeatus*). Long-term fluctuations in (– · –) building activity, (– – –) sexual activity, (· · ·) aggressive activity, and (—) amount of fanning. ↓, First occurrence of creeping through (marking the end of the nest-building phase); *F*, fertilization (marking the transition from the sexual into the parental phase); *Y*, hatching of the young. [From Sevenster (1961).]

cal literature (Fentress, 1968a,b, 1972). Supporting this view, Strongman (1965) found that brief shocks (3 sec) increased food intake in rats, whereas longer shocks (30 and 300 sec) decreased food intake. (An additional point of interest in Strongman's data is that when food was treated with quinine none of the experimental shock levels increased; this argues for an additional incentive component in determining direction of motivational interactions.) Related examples will be given below. The main point is that intensity and temporal variables may largely determine whether positive or negative correlations between different patterns are observed and thus influence conclusions of control specificity based on mutual excitation of different classes of behavior.

5. "Vigor"

We have seen that while "intensity" of behavior itself may be difficult to define with precision, two measures—duration and completeness—are frequently used (and under certain circumstances they may be negatively correlated!). It is important to note two other criteria that are commonly employed. The first concerns the apparent "vigor" of a behavior as defined by amplitude and/or frequency of the observed patterns (which again need not correlate perfectly). "Vigor," for example, was the broadly used criterion of ethologists when they spoke of the "frantic" nature of displacement activities (e.g., Tinbergen, 1952). Such criteria can be of considerable value, although they have been neglected in most subsequent research on displacement activities. Fentress (1968a,b, 1972), for example, has noted that rodent grooming may appear particularly "vigorous" (e.g., harried) in situations analogous to those that have been used by ethologists to study displacement behavior (see below). This might imply increased activation of grooming control systems by apparently irrelevant stimuli. In this case, "vigor" correlates quite nicely with increased duration of grooming activities, as well as with their completeness. In a recent study, Duncan and Wood-Gush (1972) found that the duration of individual preening components in domestic fowl can be shortened by experimental thwarting conditions (food placed beneath perspex). As they note, this abbreviated component duration may contribute to the vigorous or frantic appearance of the movements.

6. Ease of Interruption

An additional measure of behavioral intensity concerns the ease with which an ongoing pattern of behavior is interrupted. For example, I have shown that "vigorous" grooming in rodents is more difficult to interrupt than "relaxed" grooming (Fentress, 1972, unpublished). There are numerous other examples in which "drive strength" (i.e., intensity) is inferred from the amount of irrelevant stimulation needed to interrupt an ongoing pattern of behavior (e.g., reviews by Bolles, 1967; Hinde, 1970).

7. Overview

Fortunately, these various measures of "intensity" often correlate sufficiently closely to permit some evaluation of amount of activation. We can thus continue our exploration of specific and nonspecific causal factors. However, the importance of individual measures and the possibility that the views expressed here will need modification should be clearly recognized.

The various dimensions of behavioral "intensity" dealt with rather cursorily here refer to observed output functions. Input, or stimulus, intensity is often more easily, and less ambiguously, determined (e.g., measurement along a physical scale). There is still potential room for confusion, however, if the particular measures of input "intensity" are not stated explicitly. For example, train duration, interpulse interval, pulse width, and current are not necessarily interchangeable in neurophysiological investigations (Fentress, in preparation). Also, the "effectiveness" of a given stimulus (as measured by output parameters) need not be directly correlated with stimulus "strength" (i.e., under appropriate conditions we may show a greater startle to a slight click than to a loud bang). Furthermore, the effectiveness of a given stimulus can be modulated as a function of the organism's behavioral state. This, of course, is the essence of much motivational research in ethology (e.g., Hinde, 1970). It is thus sometimes useful to speak of "effective intensity" of a given stimulus (e.g., Sevenster, 1961; Fentress, 1972), which combines relevant input variables with the current state of the organism as judged by the experimenter. This can be particularly true when interactions between different classes of behavior are to be determined, since the direction of interaction between these classes of behavior may be partially a function of the "intensity" at which each behavior would have been expressed separately (Fentress, 1968a,b, 1972). Such equations involving "effective intensity" must of course be determined prior to interaction studies by separate analysis of the individual constituent elements and various combinations thereof. It is likely that much of the current confusion about behavioral interactions will be resolved only through such a stepwise procedure.

C. General Drive Constructs: Some Specific Dimensions

An introduction to general drive constructs was presented in Section I of this chapter. This will now be elaborated. Since "drives" are usually defined in terms of relatively short-term excitatory functions only, the emphasis in this section will be evidence for positive associations between apparently different classes of behavior. The question can be phrased operationally: do factors which normally elicit one class of behavior also facilitate other classes of behavior? For the moment, we shall assume that concepts of facilitation, intensity, etc., can be defined satisfactorily.

The appropriate experimental situation is shown in Fig. 6. A given input (I_1) is first found to facilitate a given class of outputs (O_1). A second input (I_2) is then found to facilitate a second class of outputs (O_2). The question is then phrased as follows: under appropriate conditions, can presentation of I_1 facilitate O_2; can I_2 facilitate O_1? What is the effect of an apparently "irrelevant" input on the production of a given output or, conversely, the effect of a given input on an apparently irrelevant output? An alternative phraseology is as follows: what is the divergence of a given input as measured by facilitation of diverse outputs; the convergence of different inputs on a given output? (Cf. Fig. 3.)

In an attempt to examine the construct of general drive as developed by Woodworth (1918) and particularly Hull (1943), Kendler (1945) examined the influence of water deprivation (0, 3, 6, 12, 22 hr) on operant responses under food deprivation during extinction trials. Under the four lower irrelevant "thirst" conditions, there was a monotonic rise in resistance to extinction on the food-deprivation task. Water deprivation of 22 hr reversed this trend, however. Kendler's experiment thus appeared to support Hull's concept of generalized activation from an irrelevant "drive" but only under certain ranges of this irrelevant drive. McFarland (1965) has confirmed that strong thirst motivation can suppress feeding responses (which in turn facilitate thirst), so the question becomes whether less severe thirst can have the opposite effect, as suggested by the Kendler study.

To check possible facilitation of moderate thirst on a food-reinforced task, Bolles and Morlock (1960) examined the alley-running behavior of rats which were subjected to different combinations of food and water deprivation. They too found that low levels of thirst in a hunger task increased performance, whereas high levels of thirst (e.g., 48-hr deprivation) antagonized performance on a hunger-motivated task. They also noted, however,

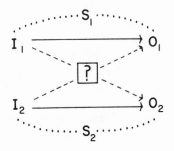

Fig. 6. Operational method for determining functional interrelationships between two previously defined input/output control systems. See text for explanation.

that the relationships between hunger and thirst were not symmetrical; even low levels of hunger disrupted performance of thirsty animals running to water (perhaps the opposite result from that which would be expected from McFarland's data).

The two main points here can be summarized briefly. (1) Low levels of irrelevant "drive" may facilitate performance of behavior patterns primarily associated with another "drive," whereas stronger activation leads to suppression of these same patterns. (2) Interactions between two motivational systems need not be symmetrical, and, to quote from Bolles (1967, p. 275), "It seems that the conditions under which, and the extent to which, different drive conditions may be substituted for each other must be determined the hard way."(!)

The study by Miles (1958) on interactions between a food-rewarded operant task in rats and irrelevant activation produced by reduced temperature and/or cocaine injections adds an additional dimension that may be of some importance. He found that the "irrelevant" states facilitated bar pressing for food when an animal was in a small box but not when it was in a large box. In the former case, the animal had a limited set of behavior patterns which it could perform, but in the large box a greater variety of potentially competing activities could occur. I have recently obtained similar results on the effects of "irrelevant" stimuli on grooming in mice. A variety of disturbances will increase grooming in mice confined in a small enclosure but are much less likely to increase grooming when the mice are in a larger enclosure with more behavioral alternatives available to them. Indeed, under certain circumstances preliminary evidence suggests that stimuli which increase grooming in a small restricted environment may decrease grooming in a larger and more complex environment. Again, interactions between different behavioral control systems may be either positive or negative depending on the exact circumstances investigated. The fact that they can be positive, however, suggests that activation factors which are not specific to a single class of behavior patterns are relatively common.

A point that should be made clear is that most demonstrations of positive interactions between different classes of behavior are based on "after-effects" of the apparently irrelevant stimulation. For example, Barfield and Sachs (1968) and Larsson (1963) found increases in sexual behavior *after* the administration of shock and handling, respectively. Similarly, Tugendhat's study of feeding in three-spined sticklebacks (1960*a,b*) concerned consequences subsequent to the termination of shock. One might ask what would happen *during* the presentation of a lower-intensity disturbance, but few data are available.

I have reported one such case (Fentress, 1972). As C57BL6 mice recover from anesthetic doses of Nembutal, they typically go through a sequence of hind leg scratching, gnawing, and face grooming. Mild pinches on the tail (an "irrelevant stimulus") facilitate *each* of these activities at the appropriate stage of recovery (i.e., increase probability, duration, and apparent "vigor" of these activities during the period of recovery in which they are most likely to occur anyway). Stronger tail pinches *block* the spontaneous expression of these activities. However, *after* a strong pinch on the tail the mice show an increase in either hind leg scratching, gnawing, or face grooming as a function of the recovery stage that they are in. That is, direct effects of a stimulus may be either incremental or decremental as a function of stimulus intensity, and negative effects *during* the presentation of a strong stimulus may be followed by more diffuse activation. These data are systematic and permit precise predictions on a given trial. Similar results have been obtained on the facilitation and blocking of perseverant movement stereotypes in zoo-caged mammals by systematic manipulation and observation of stimulus parameters (Fentress, 1965).

There are other data suggesting that irrelevant stimuli may facilitate a given response during the presentation of these irrelevant stimuli. Webb and Goodman (1958) and Siegel and Sparks (1961) have reported that satiated rats that were previously trained on an operant feeding task increased their bar-pressing rate when the apparatus was flooded with water, and Sterritt (1962) found that feeding in pigeons could be increased during the presentation of a 5-sec electric shock. I have previously reported preliminary evidence for a momentary enhancement of different ongoing behavioral patterns prior to their suppression by an external disturbance (Fentress, 1968*a,b*, 1972).

The Webb and Goodman study also introduces another potentially important dimension in the consideration of "nonspecific" drive functions. When they presented two bars, only one of which had been reinforced with food reward, flooding of the cage was found to produce particularly marked facilitation of responding to the previously rewarded bar. Thus habit strength may influence whether a given response is facilitated by irrelevant stimulus conditions. Faidherbe *et al.* (1962) gave methylphenidate (an amphetamine-like compound) to cats that had previously been trained to press a bar for food. Well-established bar-pressing responses were increased by this procedure, whereas less well-established behaviors were reduced. In this case, the animals did not respond to the food; rather, their behavior appeared to be more of a "motor discharge." Other well-established and relatively stereotyped activities were also increased by methylphenidate injections. These authors conclude: "The drug generates a state of generalized excitation, in which a bit of behavior—whether it is a highly automatized conditioned response or anything else from the animal's natural repertoire, such as licking its paw or moving its head forward rhythmically—tends to recur and repeat itself for a long time" (Faidherbe *et al.*, 1962, p. 523). Michelson and Schelkunov (1963) gave moderately large doses of amphetamine (0–8.0 mg/kg) to rats and report (p. 26) that "The food CR's were completely abolished for a long time by these doses. At the same time certain subcortical automatisms were revealed: the rat stereotypically repeats the same movement, characteristic of smelling, or scratching, or licking and so on." The point here is that apparently irrelevant stimulation may facilitate well-established activities at the same time it blocks those that are less well established. Thus the specificity of a given control system, defined in terms of range of facilitatory inputs, may vary as a function of factors such as complexity and degree of establishment. These factors will be seen to be potentially important in the subsequent discussion of displacement activities. For example, Bindra and Mendelson (1963) and Fentress (1968*b*) found that moderate doses of methylphenidate and amphetamine, respectively, can produce an increment in relatively stereotyped and predominant activities such as grooming, while other more complex and novel behavior patterns are blocked. These facts add an additional consideration to questions of intensity and temporal relationships in the possible "nonspecific" control of behavior, i.e., complexity and degree of encoding. In support of this suggestion, I have found (Fentress, 1965, in preparation) that well-established perseverant movement stereotypes in rodents and Canidae can be facilitated by a considerable variety of stimulus inputs over a wide range of intensities, whereas less well-established movement stereotypes are blocked by these same stimuli. Voles which had been isolated in individual covered cages for a long period of time, for example, developed persistent circling movements around the waterspout. In initial stages of development, these movements were easily suppressed by external disturbances, whereas later they were facilitated by separate and combined disturbances over a wide range of intensities. Berkson and Mason (1964) also report an increase in persistent movement stereotypes in monkeys as a function of increased "arousal."

D. Summation of Heterogeneous Factors

Overall, there have been relatively few attempts to demonstrate synergistic relationships between two or more "input" factors that normally underlie different motivational control systems, although this is important to any construct of "general drive." Two relevant examples of apparent summation are (1) the study by Hall (1956) in which simultaneous environmental stimulation (flashing lights and noise) plus food deprivation increased activity as measured on a revolving drum more than did either stimulus source alone, and (2) the experiment by Jerome *et al.* (1957), who found that light plus food deprivation had a similar additive effect on activity in a maze situation. A third well-known example is provided by Brown (1961). He found that startle response to a sudden auditory stimulus was a resultant of combined food deprivation and "fear" (produced by previous electric shock in the test situation). Such observations are obviously compatible with models of behavioral integration that stress a certain degree of nonspecificity among mechanisms normally contributing to different classes of behavior, at least under certain conditions. I have previously shown (Fentress, 1965, 1968b) that summation of heterogeneous factors can be demonstrated in rodent grooming (e.g., Fig. 7B) and particularly clearly in perseverant stereotypes, although, of course, strict linear combinations may not always be found.

Such studies indicate that it is possible to demonstrate synergistic effects of diverse inputs on a given behavioral output. At this point, it is well to recall that recent electrophysiological studies have demonstrated the convergent effects of different sensory modalities on the output firing of single neurons in both cortical and subcortical structures (e.g., Buser and Imbert, 1961; Fessard, 1961; Horn, 1962; Jung et al., 1963; Palestini et al., 1957). Processes mediating different sensory modalities are not as separate as had been formerly believed. Furthermore, it is now well documented that activation of movement control systems can influence processing of sensory information (e.g., Adkins et al., 1966; Jasper, 1963), as can, of course, stimulation of central structures underlying motivation per se (e.g., MacDonnell and Flynn, 1966). Obviously, models which stress the complete separation of behavioral subsystems can be brought into question by such facts.

The opposite emphasis is on complete nonspecificity, e.g., "nonspecific arousal." As is hopefully now apparent, this is not a necessary (nor sufficient!) alternative to specific models of behavioral integration. Rather, gradations between these two extreme forms of control appear most likely. We have previously seen, for example, that while different sensory inputs may affect the activity of a given neuron in the brain, the inputs do not necessarily have an identical ("nonspecific") effect (e.g., Fig. 2). There are numerous related data concerned with behavioral activation (e.g., Hinde, 1970).

E. Central–Peripheral Factors

A further consideration important to the interpretation of specific and nonspecific factors in the interaction between separately classified motivational systems is the balance between central and peripheral factors. While "general drive" implies central control, it is now well established that many motivational control functions are mediated to a large extent by relatively specific responsiveness, and alterations in responsiveness, to peripheral stimulation. Beach and Whalen (1959*a,b*), for example, have shown the importance of innervation from the glans penis in sexual behavior in male rats, and Hinde and Steel (1966) have documented changes in the brood patch in canaries, and thus sensitivity to tactile stimulation, as a function of the animals' hormonal state. Similarly, MacDonnell and Flynn (1966) have shown that attack behavior of cats on rodents produced by hypothalamic stimulation is partially attributable to alterations in responses to tactile stimulation applied to the face. This recognition of peripheral contributions to motivated behavior is important (but, as noted by Stellar, 1960, should not be overly interpreted at the exclusion of central states). The reason this is mentioned at the present time is that if an animal can be shown to be responding to "specific" peripheral cues produced by apparently irrelevant stimulation this may modify conclusions regarding "nonspecific" sources of endogenous motivation.

The influence of "emotionality" on water and food intake is particularly instructive here. The situation is similar to those in which the ethologist might employ the concept of "displacement activity," i.e., activation of an apparently irrelevant response by a given stimulus.

Siegel and Stuckey (1949) found that faradic stimulation increased subsequent drinking in rats, and they interpreted this in terms of general cellular dehydration and local dryness of the buccal cavity. However, close inspection of the data shows that 27 of the 40 animals in the experiment did not drink at all, a fact which casts some suspicion on the explanation offered. Amsel and Maltzman (1950) obtained similar results when shock was administered in a situation different from that in which the animals drank, but they preferred explanation in terms of central motivational factors. The crucial test was performed by Siegel and Brantley (1951). Here, feeding rather than drinking was employed. The peripheral dehydration hypothesis would predict a decrease in dry food intake following emotional stimulation, while a hypothesis involving more central factors might predict an increase in feeding. An increase was found. The authors conclude: "We interpret our findings as contradictory to the tissue dehydration hypothesis and as confirming prediction from Hull's hypothesis of general drive" (Siegel and Brantley, 1951, p. 306). While this is probably an overinterpretation, the main point that increased feeding is unlikely to be due solely to peripheral factors appears valid.

Ullman (1951) found that rats shocked in an apparatus where they had been previously trained to eat fed less for the first 2 days of shock but then fed more than in the preshock trials. This again emphasizes the importance of temporal (and perhaps "intensity") factors. The enhancement occurred even when the animals were satiated, and the author interprets his finding in terms of "a generalized tension-reducing response." On a more anecdotal level, Bauer (1956) notes that enhanced feeding may occur in aggressive mice, and Geist (1963) finds that moose in the wild will often feed intensely after the passing of a disturbance. Raber (1948) reported that either feeding or drinking would occur in turkeys depending on the availability of food or water. (This report is interesting to compare to the recent study by Valenstein *et al.*, 1968, on the excitation of both feeding and drinking responses by hypothalamic stimulation under appropriate conditions.) The increased completion ratio of feeding responses in sticklebacks after shock noted by Tugendhat (1960*a,b*) has particular difficulty being squeezed into a dry mouth only model. Epstein and Teitelbaum (1962) have bypassed oropharyngeal and olfactory receptors by employing direct intragastric self-injections in rats, and they found regulation of food intake to occur normally for long periods of time. It would be interesting to combine this type of technique with feeding or drinking in studies of "irrelevant" motivation. Such should be workable surgically (Epstein, personal communication).

The point of this brief review is to emphasize that while "specific" peripheral stimuli may indeed play an important role in the mediation of different classes of motivated behaviors, as well as interactions between these classes, it is premature to designate all such functions as due to peripheral factors only. Central control factors which are not "specific" to any single behavioral class may also play a role. (Of course, it must be remembered that demonstration of central, e.g., excitatory, effects shared among more than one class of behavior does not indicate that these effects are absolutely "nonspecific"; rather, a cautious interpretation is that they may have more than one "specific" effect. It is, however, this criterion of multiple specificity that leads to the inference of nonspecific control.)

The recent electrical stimulation studies by Valenstein *et al.* (1968) are of some interest at this juncture. These authors found that stimulation of the hypothalamus in rats could lead to an increase in either gnawing, feeding, or drinking behavior as a function of the external stimuli present. Thus the particular behavior elicited appeared to be determined *both* by the specific environmental stimuli available and by the central stimulation. Since the animals could be made to respond strongly to more than one type of environmental stimulus (e.g., food, water), the authors correctly concluded that effects of stimulation do not appear strictly "specific." However, the range of activities that can be facilitated by this hypothalamic stimulation remains to be tested; it is very unlikely that the results of such stimulation are absolutely "nonspecific."

Studies employing electrical stimulation of the brain must of course be interpreted with some caution, since the activity pattern produced in neuronal firing is highly abnormal. In this respect, the study by Barfield and Sachs (1968) is relevant in that these authors applied electric shock to the skin (and muscles) of rats rather than to the brain. Following such shocks, they noted an increase in copulatory behavior in male rats and thus concluded that the shock produced nonspecific "arousal." This is a potentially surprising result (but see Beach *et al.*, 1955; Larsson, 1963), and the authors are indeed correct in pointing out that the peripheral shock does not produce a single specific effect on behavior. However, potential limitations of the alternative "nonspecificity" model should be noted. For example: (1) the copulatory behavior occurred *after* the shock, not during it (thus aftereffects of a stimulus may be partially separable from direct effects of the stimulus; e.g., Fentress, 1968*a,b*, 1972); (2) the applied shock was not varied over a wide range of intensities (could, for example, lower-intensity shocks facilitate copulatory activity *during* the shock presentation?); (3) there is, of course, no evidence that the shock would increase *all* classes of behavior even when appropriate stimuli were present. These comments are not made primarily as a critique but rather to show that there may be limitations to which the proposed construct of "nonspecific arousal" is applicable. Certainly, the authors have demonstrated that shock may have effects other than those expected, and thus strict "specific" models of motivational control are thrown into some question.

F. Approach–Avoidance and Activation Level

One of the interesting conclusions shared by theories of "nonspecific" activation is that a wide variety of stimuli at moderate intensity will produce approach reactions, while the same stimuli at greater intensity will lead to avoidance (cf. review by Schneirla, 1965). Sokolov (1963) has provided a valuable review of his own and other Russian work that supplements the arousal models by the general finding that orientation "reflexes" are replaced by "defensive" reflexes (measured by, for example, cephalic and peripheral vasodilation and vasoconstriction) with the increasing intensity of a wide variety of stimuli. (Sokolov is careful to point out potential limitations of his model, however. For example, under certain conditions the introduction of a novel stimulus can result in the replacement of the defensive with the orientation reaction. This certainly suggests that nonspecific "arousal" is not the whole answer.) Internal and experiential factors are also important. Menzel (1962), for example, found that rhesus monkeys would initially avoid large objects of various types more than small ones but that familiarity with these large objects would decrease avoidance and increase contact.

The potential relevance of such approach–avoidance models for ethological investigations that emphasize specific behavioral control functions has been illustrated by Hirsch *et al.* (1955) and Schleidt (1961) in a critique of the classical study of young birds giving specific avoidance responses to overhead objects that resemble a hawk as opposed to a goose (cf. Tinbergen, 1951). As these authors point out, inadequate controls for possible habituation in the original study may have confounded responses to quantitative dimensions of the stimulus with responses to qualitative dimensions. While several studies (e.g., Melzack *et al.*, 1959) indicate that qualitatively different stimuli do preferentially elicit different degrees of avoidance behavior when habituation factors are controlled, the major point of previous critiques (and related studies) makes an important point: the *intensity* of an input can affect the *direction* of an output (e.g., approach or avoidance behavior). Therefore, if stimuli are given at different "effective intensities" (see above), specific directional factors may be confounded with more general quantitative factors.

G. Integrative Efficiency and Activation Level

There have been many studies which suggest that efficiency of behavioral processing is positively correlated with an intermediate level of behavioral activation (e.g., Berlyne, 1960; Bindra, 1959; Malmo, 1959). Perhaps the most famous early formulation of this principle was that of Yerkes and Dodson (1908). One important implication of this position is that the relationship between intensity of stimulation or internal activation and performance is a nonmonotonic one; i.e., either too much or too little activation reduces performance efficiency. There have been numerous studies that support this contention, and it coincides well with everyday experience.

It possibly has some relevance also to the direction of interactions between control systems. For example, suggestive evidence has been presented that moderate activation of apparently irrelevant motivational inputs can facilitate a given output, while excessive activation of these same inputs can block or reduce expression of this same output. This could be of obvious importance when considering problems of control specificity. If, for example, the specificity of a given control system is defined in terms of the range of inputs that will excite, rather than inhibit, a given output, then this definition of specificity is potentially dependent on the intensity to which the inputs under investigation are activated.

The potential relevance of such considerations to questions of specific *vs.* nonspecific control of displacement activities will be examined in the next section. At this point, it is worth suggesting briefly that since displacement activities often occur in states of relatively extreme arousal, during which processing efficiency appears to be low (cf. Welford, 1962), this may be partially due to the fact that such activities are more easily programmed than are other activities. This in turn may be due partially to the fact that displacement activities are typically well established in the organism's behavioral repertoire and relatively stereotyped in their expression (cf. Fentress, 1965, 1968a,b).

The basic point thus far is that concepts of "nonspecificity" as determined by multiple interactions between input and output variables, plus partial separation of intensity and directional dimensions of behavior, may provide useful complements to models that stress strict specificity in behavior, in which directional and intensity variables are considered synonymous, and in which the only interactions between different classes of behavioral control are considered negative. Of course, the converse is also true: models of specific control provide an important complement to those stressing nonspecificity. Indeed, it is now well established that different measures of "nonspecific" activation can be separated from one another and thus that this dimension is unlikely to be unitary (e.g., Lacey, 1967). Even given this, however, the broad concept may have something to offer. This possibility will now be examined with respect to a classical model of behavioral specificity in the ethological literature: displacement activities.

V. DISPLACEMENT ACTIVITIES: A CASE STUDY

A. Overview

Displacement activities in the ethological literature were initially defined as activities which appear to be "out of context" and facilitated by causal factors normally associated with other behaviors (cf. reviews by Fentress, 1965; Hinde, 1970; McFarland, 1966; Tinbergen, 1952; Zeigler, 1964). For example, during a bout of aggressive activities different bird species often peck at the ground or preen their feathers (Tinbergen, 1952); rodents which have been disturbed by an overhead moving object may groom excessively (Fentress, 1968a,b, 1972).

Displacement activities have often been described as "irrelevant," since they appear in contexts which are at first surprising (e.g., why should a fighting bird suddenly stop and preen?). This criterion of irrelevance is of course based on presumed behavioral goals and, as has been pointed out by Bindra (1959), may reflect implicit criteria set by the observer. For example, instead of asking why a fighting bird should stop and preen one might ask why a preening bird should stop and fight! However, most ethologists are able to resolve such issues to their satisfaction by observing the broad context (environmental, behavioral) in which displacement activities occur.

The important question (for the present purposes) concerns *causation*. It is perhaps some confusion between functional endpoints and causal mechanisms that led early ethologists to suggest that displacement activities are also causally irrelevant. However, as pointed out by Kruijt (1964) and Fentress (1965), the construct of causal irrelevance is a logical anomaly. If a factor (or set of factors) has an effect on behavior, it is relevant. If a factor has no (measurable) effect, it is not causal. What, then, are the most relevant causal factors in displacement activities?

The term "displacement" itself is a loaded one, for while it is derived initially from descriptive data, it also implies a mechanism of causal control. Consistent with the ethological emphasis on "specific" motivational factors, it was suggested (e.g., Kortlandt, 1940; Tinbergen, 1940) that specific drive variables were "displaced" from one channel of expression (e.g., fighting) into another (e.g., preening). Displacement activities were considered to be caused by factors other than those associated with them in "normal" contexts.

As has been pointed out previously (e.g., Fentress, 1965; Zeigler, 1964), there was an obvious alternative explanation. Rather than "specific" factors being shunted from one channel to another, displacement activities might indicate the presence of relatively "nonspecific" motivational control functions. This view is compatible with the survey of specific and nonspecific causal factors outlined above.

Subsequent ethological research, however, took two different tacks. First, it was suggested that patterns such as displacement preening may reflect response of the animals to relevant "specific" peripheral stimuli. Andrew (1956a,b), for example, noted that various conflict situations (in which displacement activities are particularly common) might actually increase autonomic stimulation, which in turn would increase the probability of displacement preening activities (cf. Morris, 1956). Similarly, Sevenster (1961) demonstrated that peripheral factors relevant to "normal" fanning behavior in sticklebacks (e.g., CO_2 in the water, which is usually associated with the presence of eggs, toward which fanning is directed) also facilitated fanning in "displacement" contexts. Rowell (1961) demonstrated that peripheral irritation of the plumage in chaffinches could increase both "normal" and "displacement" preening. More recent studies (e.g., Feekes, 1971; Kruijt, 1964; Rasa, 1971) support the basic conclusion that "specific" peripheral stimuli can play an important, although certainly not exclusive, role in various displacement activities.

The second approach taken by ethologists to the displacement behavior problem was to emphasize negative interactions between different classes of behavior (see previous discussions). The resulting model was summarized under the term "disinhibition hypothesis" (e.g., Andrew, 1956a,b; van Iersel and Bol, 1958; Rowell, 1961; Sevenster, 1961). Through an elegant series of studies, it has been shown that expression of one given behavioral tendency can be blocked by strong activation of another behavioral tendency. Thus in a situation of conflict between two predominant tendencies (e.g., approach and avoidance behavior), each of which normally has a strong inhibitory effect on activities such as grooming and *also on one another,* the inhibitory actions of the two predominant tendencies may cancel one another out, thereby "disinhibiting" the displacement behavior. The displacement behavior can therefore again be interpreted in terms of its own "specific" excitatory mechanisms. A similar model has been applied with considerable precision at the level of attention mechanisms by McFarland (1966).

These two strategies thus provide a powerful alternative to the initial "surplus" model of displacement activities (e.g., Tinbergen, 1952) and preserve the notion of "specific" behavioral control. The strategy is directly analogous to that outlined in Fig. 3B and might usefully be employed in psychological investigations of "general drive." That is, the data suggest an alternative to a general drive model. But is the issue settled?

As indicated above, negative interactions between patterns do not permit the conclusion that these behavior patterns share no ("nonspecific") causal factors. For example, it is obvious that the behavior patterns studied by ethologists in the context of displacement activities are not independent from one another in their control; i.e., they inhibit one another. Might these different behavioral tendencies also excite one another under appropriate conditions? It is important to note that disinhibition theorists do not deny this possibility. However, they suggest that this excitation of one behavioral system by another can *only* take place either (1) through an increase of appropriate "specific" peripheral stimuli (e.g., via autonomic mechanisms, above) or (2) through an intermediate inhibition of a normally antagonistic activity. Such a model is difficult to test, since it can quickly approach the tautological statement that mutually exclusive activities occur between others. Also, the model is obviously correct in its broad outlines; peripheral stimuli are important to patterned behavior, and strong activation of one behavior suppresses the expression of another.

As I drum my fingers on the table while trying to compose the next section, however, I cannot help but wonder whether it is really true that all such "irrelevant" patterns are either activated all along (and thus merely disinhibited in this strange context) or are solely the response to peripheral cues—my fingers do not seem to itch.

B. Shared Excitation vs. Disinhibition

Two potential chinks in the armor of the disinhibition model do permit some further analytical exploration of it. First, the demonstration that strong simultaneous activation of different patterns produces mutual inhibition does not preclude the possibilities that (1) weaker activation of these behavior patterns can produce some degree of mutual facilitation and (2) the *aftereffects* of even relatively strongly activated tendencies can produce relatively diffuse excitatory functions that are shared among more than one class of behavior patterns. Some justification for such proposals is obtained from the literature sampled above.

The suggestions appear supported in a recent series of experiments I have conducted (Fentress, 1968*a,b*, 1972). For example, initial experiments (Fentress, 1968*a*) with two species of vole (*Microtus agrestis* and *Clethrionomys britannicus*) produced the potentially paradoxical finding that presentation of an overhead disturbance *increased* grooming over control periods in the former and *decreased* grooming over control periods in the latter species. By other criteria, it was shown that *Microtus* is the more docile and *Clethrionomys* the more timorous of the two species. An immediate suggestion, therefore, is that an "optimal level" of disturbance increases grooming, while higher levels of disturbance decrease grooming. "Fear" behavior might either facilitate or inhibit the control of grooming partially as a function of the intensity of "fear."

Another relevant point is that increases in grooming were observed *after* the stimulus presentation had ceased. Since it is well known that motivational effects of a stimulus can outlast the presentation of the stimulus and then gradually decline (e.g., Hinde, 1970),[3] by the "optimal arousal" model grooming should first increase and then decrease during the course of recovery from the disturbance. That the data for *Microtus* fit this explanation is seen in Fig. 7A. The recovery period for *Clethrionomys* by these criteria is extended.

In such studies, interpretation can be aided by examination of behavior other than the one of primary concern. Figure 7A also shows the progressive increase in walking following the disturbance. Here, an important point can be made: grooming occurs between freezing and subsequent locomotion. Now the disinhibition theory would also predict this result. It could be argued that freezing inhibits grooming and that locomotion inhibits grooming, and thus that when these two activities are in "conflict" or in balance, grooming is "disinhibited." There is much merit in this model, as we have seen, but the

---

[3]A simple illustration of such "motivational inertia" is provided in Fentress (1968*a,b*). I found, first, that animals which were walking at the time of the stimulus presentation were likely to flee from the stimulus, whereas those sitting still were more likely to show only freezing behavior. However, animals which *had been* walking within 10 sec of the stimulus presentation, but which were immobile at the instant that the stimulus was presented, were almost equally likely to flee as were animals that displayed concurrent locomotion, and were very significantly different from animals immobile for a longer period of time. These effects can be termed "motivational," since they do not depend on momentary feedback from ongoing locomotor movements, etc. Also, of course, these data indicate that a given stimulus can have more than one "specific" effect: i.e., qualitatively different outputs can be facilitated by the stimulus as a function of ongoing endogenous activity.

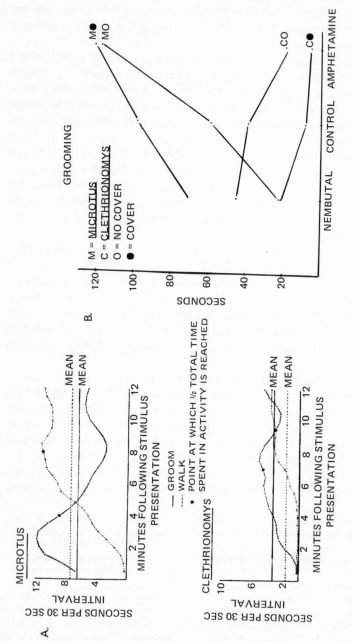

Fig. 7. A: Smoothed mean grooming and walking plotted against time following cessation of presentation of overhead visual disturbance. The ordinate is time spent in each activity per 30-sec interval after the overhead stimulus is stopped. B: Mean grooming scores for two species as a function of both home-pen rearing conditions (covered vs. noncovered enclosures) and injections of 15 mg/kg Nembutal, water control, and 0.8 mg/kg amphetamine. Animals were exposed to an overhead disturbance as in A. Species studied were *Microtus agrestis* and *Clethrionomys britannicus* (voles). [From Fentress (1968a) and Fentress (1968b), respectively.]

question is whether it is sufficient as a sole explanation for the observed phenomena. Closer inspection of the data suggests that it is not. First, the correlation between grooming and subsequent locomotion in individual animals was less consistent than the correlation between the disturbance and subsequent grooming; i.e., subsequent locomotion was not a necessary prerequisite for the occurrence of grooming. If the disinhibition model is expanded to include all behavior that follows grooming (i.e., "quiet sitting"), it has some danger of stating the obvious, namely, that a behavior pattern occurs between others. Second, if one examines intervals of a given length that occur during transitions between stimulus-produced freezing and subsequent locomotion, the total duration of grooming during these intervals could be increased by manipulation of various independent variables (e.g., speed of overhead stimulus, home-pen housing conditions, amphetamine injections). This fact would not be predicted from the disinhibition hypothesis without additional assumptions. Third, the experimental manipulations could also increase the apparent vigor and completeness of grooming sequences over that observed in control animals. Finally, on initial presentation of the overhead disturbance *during* ongoing grooming, individual movements appeared momentarily accelerated before inhibition.

In the same study (Fentress, 1968*b*), I was also able to demonstrate that qualitative different independent variables such as species, home-pen rearing conditions, and administration of activating or depressant drugs would summate in their effects on grooming (Fig. 7B). This type of summation among diverse inputs is another important criterion for "nonspecificity" (cf. Fig. 3 above). The basic model proposed, therefore, was that, under appropriate conditions, relatively "nonspecific" activation or arousal functions may be shared among apparently diverse classes of behavior. Consideration of these functions can provide an important *supplement* to the disinhibition hypothesis. The proposed activation functions appear to operate in a nonmonotonic fashion; i.e., at moderate levels they increase grooming, but at higher levels they decrease grooming. The basic position has received subsequent support in research reported by Rasa (1971) and Wilz (1970*a,b*). In a recent personal communication, Lorenz has argued that such studies have led him to conclude that "arousal" may be a very important "independent" dimension in the organization of displacement activities (cf. Lorenz, 1971). Duncan and Wood-Gush (1972) have confirmed that experimental thwarting (food placed beneath perspex) can accelerate the completion (reduce the duration) of individual preening components in domestic fowl, thus giving the behavior its hurried (vigorous) appearance.

The conclusions must be formulated carefully, for none of the experiments cited indicate that the basic tenets of the disinhibition model are incorrect. Rather, they suggest that the model might be usefully supple-

mented by consideration of excitatory causal factors which, under appropriate conditions, can be shared among more than one class of behavior. The
extent to which excitatory as opposed to inhibitory interrelations among
different classes of behavior occur in this context appears to be a function of
(1) the intensity to which these behaviors are activated, and (2) the temporal
relations between the various input/output variables under consideration. A
diagrammatic revision of the traditional disinhibition model suggested by
these data is provided in Fig. 8.

The question still remains, why should an animal perform an activity
such as grooming? Obviously, not *all* behavior patterns are facilitated in
their expression by the overhead disturbance, since many of them cannot be
expressed simultaneously: increase in the duration of one activity necessitates
the decrease in the duration of all those with which it is incompatible.

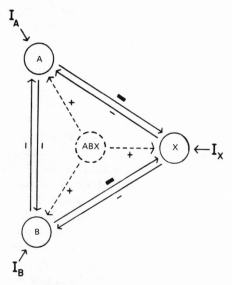

Fig. 8. Modified diagram typically used in discussions
of the "disinhibition" hypothesis for displacement
activities. Two major classes of behavior, such as
approach and avoidance, are represented by A and B.
The apparently irrelevant behavior is represented by
X. Each behavior is assumed to have excitatory factors
specific to it (I_A, I_B, I_X). Each behavior has inhibitory
effects on the other two. The inhibition of A and B
on X is strongest; the inhibitory effect of X on A and
B is weakest. The diagram suggests that there may be
other positive causal factors (ABX) that are at least
partially shared among the three classes of behavior.
See text for discussion.

C. Central vs. Peripheral Factors

Here, the second facet of recent research on displacement activities—peripheral stimulation—becomes relevant. The easiest answer to the question of why an animal grooms is that it itches. The experiments by Rowell (1961), Sevenster (1961), and others have documented the important influence of peripheral stimuli on displacement activities. Furthermore, McFarland (1965, 1966) has provided elegant demonstrations that conflict, etc., can lead to a shift of attention to such peripheral stimuli. However, it is obvious that demonstration that peripheral factors are relevant to a given behavior does not permit the conclusion that these are the *only* relevant factors. This is logically similar to the fact that demonstration of the importance of genetic factors in behavior does not exclude the importance of experiential factors (cf. Hinde, 1970; Lehrman, 1970).

A brief reexamination of data provided by Rowell (1961) is relevant here. From his experiments with chaffinches in which peripheral stimuli were increased (1) by dirtying the bill and (2) by spraying the bird with water, Rowell reported that the proportion of bill wiping increased from 41% of total grooming to 56% of total grooming in the former case and decreased from 38% to 36% in the latter case. However, reanalysis of the data demonstrates an *absolute* (rather than percentage) increase of approximately 15% in grooming oriented to the bill in the former case and of approximately 30% in grooming oriented to the bill in the latter case. Thus there must be some central linkage among different grooming components as well (cf. Andrew, 1956b).

Preliminary evidence (Fentress, 1965) for the importance of central factors in rodent grooming was obtained in which (1) injections of adrenalin and and hexamethonium (an autonomic blocking agent) had relatively little effect on probability, duration, or organization of grooming behavior under controlled conditions; (2) grooming was conditioned by repeated exposures to a test tin which had previously been associated with ammonia fumes; (3) *face* grooming was potentiated by moderate applications of peripheral disturbance to the animals' *backs*; i.e., the animals began grooming sequences with face-grooming strokes in the usual manner before moving to their backs (cf. discussion of Rowell, 1961, above). Furthermore, perseverant motor stereotypes (e.g., weaving around the waterspout in previously isolated animals) could be shown to increase and decrease in a roughly analogous fashion to the increases and decreases of grooming in control animals. Such data certainly suggest the importance of central control mechanisms in well-established motor activities. (Strongly developed stereotyped weaving behaviors in the voles persisted after removal of the waterspout, even though the animals would occasionally lose their balance by this manipulation. Such observations strongly support the concept of central programming.)

As has long been known (e.g., Lorenz, 1957; Tinbergen, 1952), displacement activities are usually (1) relatively stereotyped in their expression and (2) well established in the organism's behavioral repertoire. Each of these factors might be relevant to *why* particular activities should occur in conflict and related situations.

More recent experiments on grooming in mice indicate that indeed a remarkable degree of central programming might underlie this behavior. The major findings are reported in Fentress (1972) and can only be summarized very briefly here. First, face grooming survives lesions of sensory branches of the trigeminal nerve (thus denervating the face). Most interestingly, the face grooming is *least* affected by these lesions when the animals, by other criteria, are strongly aroused (e.g., placed in novel cage). Thus in situations similar to those discussed in the literature on displacement activities, animals may be particularly *insensitive* to changes in the peripheral environment, an opposite suggestion to that provided by McFarland (1966). Second, basic components of grooming survive lesions of the forelimb dorsal roots (thus removing proprioceptive inputs from the forelimbs to the central nervous system) to a remarkable degree, whereas other, more complex and novel motor activities are abolished. Third, amputation of the forelimbs in infant mice does *not* prevent the development of basic grooming movements involving the shoulders, nor even the "central" coordination between limb movements and licking movements of the tongue (which now occur in midair) (cf. Fentress, 1933).

D. Summary and Conclusions

To summarize, two fundamental supplements to the traditional models of displacement activities are suggested by these findings. First, under conditions of moderate activation two normally antagonistic behavioral systems may share excitatory processes. The model suggests that the influence of "irrelevant" factors on the production of a given behavior is nonmonotonic. They are facilitatory when activation is low, *after* higher levels of activation, or during the very *initial* stages of irrelevant stimulation. Second, central (rather than just peripheral) encoding mechanisms may be of considerable importance in determining *which* activities occur, and when. The facts that displacement activities are particularly well established in the animal's behavior repertoire and that they are also relatively stereotyped in their expression are likely to be important considerations (e.g., Bindra, 1959; Zeigler, 1964).

This last point cannot be elaborated on in detail in the space available, but certain broad suggestions deserve brief summary. First, during displacement situations central integrative machinery is likely to be operating at less than maximum efficiency (cf. the concept of "noise" in experimental psychology). Previous suggestions that optimal efficiency in behavior may occur during intermediate "arousal" conditions have already been mentioned (e.g., the Yerkes–Dodson law). Many displacement activities occur not only in conflict situations and under conditions that appear to be moderately stressful but also in the context of low "arousal" states such as those which precede and follow sleep (e.g., Fentress, 1965; Delius, 1970). One function of displacement activities, which are usually relatively stereotyped in their form, may be to increase the level of organization of integrative processes. Although this construct is difficult to test empirically, there is reason to suggest that the highly organized outputs of displacement activities serve the critical *input* function of increased organization (my thanks to Dr. J. Adkins for discussion of these issues). Second, well-established and relatively simple behavior patterns are most likely to be facilitated by suboptimal conditions, whereas more complex and novel activities are more likely to be suppressed. This suggestion fits well with the common observation in ethology that displacement activities are usually well established and relatively stereotyped, and with much recent research in human experimental psychology (e.g., review by Broadbent, 1958, 1971). I have written an earlier review of these considerations (Fentress, 1965). Available space permits brief mention of only two additional points relevant to mechanism at this point. First, it may at first seem paradoxical that activities such as grooming which are usually inhibited easily by other patterns and which exert only weak inhibitory influence on these activities in return (e.g., van Iersel and Bol, 1958; Sevenster, 1961) should occur in displacement contexts that result in the blockage of these other more complex and apparently "tougher" behavior patterns. However, recent studies in human experimental psychology suggest that (1) it is *because* behavior patterns are relatively noncomplex and strongly encoded that they occupy little attention (and thus processing capacity) and interfere minimally with other activities, and (2) these "simple" behavioral patterns in turn require less complex information processing and thus can occur in situations of lowered efficiency which prevent the expression of other more complex and/or less well-established behavior (e.g., Broadbent, 1958, 1971; Posner, 1966; Welford, 1958).

Second, it is likely that well-established species-characteristic movement patterns such as grooming are controlled largely by subcortical (e.g., brainstem) structures (e.g., Wang and Akert, 1962; Woods, 1964) and thus are relatively immune to disruption of cortical functions. It may of course be true that in situations of conflict, stress, etc., tonic inhibition from the cortex on subcortical structures is blocked, thus "releasing" these subcortically controlled species-characteristic behavior patterns. Such possibilities are currently being pursued in our laboratory. However, this is using the construct of "disinhibition" at a very different level from that employed in the ethological studies reviewed thus far and therefore is not directly relevant to the arguments presented.

There are obviously also "specific" (i.e., unshared) factors that determine *which* form of displacement behavior will occur in *which* behavioral context (e.g., Feekes, 1971; Kruijt, 1964; McFarland, 1966; Tinbergen, 1952; Wilz, 1970*a*,*b*). An important point is thus that one set of explanatory factors does not in itself negate the importance of other factors. Perhaps a major difficulty with models of displacement activities is that investigators have been overly tempted to replace one unitary explanation with another rather than to use any systematic approach based on converging operations.

With these points in mind, a tentative and necessarily "rough" approach to some of the basic problems of "specific" and "nonspecific" factors in integrated behavioral control will be sketched below.

VI. SYNOPSIS AND EXTRAPOLATION: A "BOUNDARY-STATE" APPROACH

A major task of ethology has been to delineate control systems that integrate patterns of ongoing behavior in the intact organism. What is the range of inputs (and their interactions) that influences a given output; the range of outputs (and their interactions) affected by a given input? The question of control specificity is therefore raised as a problem of fundamental importance. As this review has shown, however, there are conflicting views with respect to this issue.

A. Synopsis: Major Specificity/Nonspecificity Dimensions

1. Basic Input/Output Dimensions

Behavior in the intact organism can be subdivided into a variety of component activities. As we have seen, these behaviors can be classified by various criteria, and these criteria in turn can affect models of behavioral organization. Each behavioral component is likely to be affected, directly or indirectly, by a variety of factors, either singly or in different combinations. Each of these factors, termed "inputs" in the present discussion, is also likely to affect more than a single class of behavioral outputs. The amount of convergence of different inputs on a behavioral output and the amount of divergence of a given input as measured by its effects on different output classes are at the heart of the problem of specific *vs.* nonspecific control.

The inputs are most appropriately viewed as modifying the ongoing state of the animal, rather than in simple S–R terms. For example, a given input can produce quite different outputs as a function of the ongoing state of the animal (Fentress, 1968a,b). Since at present it is not possible to observe *directly* all intervening processing stages, these must be inferred from systematic manipulation of selected inputs and observed alterations in behavioral output. The degree of separateness *vs.* commonality of different input/output combinations can be determined empirically and offers an operational approach to the question of integrative specificity at the level of analysis employed. The input/output language is one of operational convenience. Intervening factors can always be assumed (cf. Hume, 1748), and it is precisely these intervening factors that the input/output approach is designed to clarify. How specific *vs.* nonspecific are the intervening processes in integrated behavioral control? That is the issue of concern here.

It has been suggested above that it may be useful to separate input and output dimensions in the initial approach to this problem of control specificity. For example, a given mechanism may be affected by many inputs as defined in a particular investigation but in turn influence only a small range of output, or *vice versa*. We have also seen that it is essential to specify the criteria used to define input/output variables to be included in a control system (e.g., both inhibitory and excitatory functions, excitatory functions only, identical effects of different inputs on a given output—or of a given input on different outputs, simultaneous effects only, etc.). A third critical feature is that the level of analysis for which a model is formulated must be defined and maintained. For example, apparent "nonspecificity" at one level of analysis may be defined more appropriately in terms of "multiple-specific" operations at a refined level of analysis. Indeed, nonspecificity is an *inference* based on observation of multiple specific effects. Specificity, however, is also an inference which may be based on examination of a too limited range of input/output functions. Further, as elegantly demonstrated by the disinhibition hypothesis, excitatory functions at one level of analysis may be mediated by inhibitory mechanisms at a more refined level of analysis. Fourth, it has been suggested that as a dichotomy "specific" *vs.* "nonspecific" may be less useful than an approach which allows for more gradations in the range of particular control functions. As noted, for example, rarely does a *single* input affect only a *single* output with *complete independence* from other input/output functions. Nor is it likely that a case will be found where *all* inputs of a given control system under investigation will affect *all* outputs equally.

2. Three Major Interaction Classes

One of the fundamental contributions of ethology to problems of integrated behavioral control has been examination of interactions between different classes of behavior. These classes of behavior are usually defined in terms of functional endpoints, and thus problems about their *causal* interrelationships demand empirical investigation (cf. Hinde, 1970). There are three main possibilities relevant here: (a) the systems are independent of one another, (b) the systems are mutually antagonistic (symmetry being assumed for the sake of simplicity), and (c) the systems share excitatory functions. Under different conditions, of course, one might obtain evidence for each of these possibilities.

3. Overlap and Integrative Dynamics

This introduces the problems of (a) *overlap among control processes* and (b) *static vs. dynamic* models. First, it is suggested that "different" control systems in behavior typically show varying degrees of overlap rather than being either absolutely independent of one another or isomorphically superimposable. This fits well with current neurological research (e.g., Evarts and Thach, 1969; Gerard, 1960; von Holst and von St. Paul, 1963; Luria, 1966) but is a rather radical departure from many behavioral models (cf. Norton, 1968). That is, the approach stresses the concept of *partially shared* (and thus partially unshared) control processes. To summarize this basic idea, a Venn diagram approach to ethological models (partially intersecting circles) might be more useful than conventional diagrammatic procedures (cf. disinhibition models, e.g., Fig. 8).

The next consideration is at the heart of the matter. Static representations of dynamic behavioral processes may be misleading (cf. Simpson, this volume). I suggest that we consider that specificity within a control system can shift as a function of the internal dynamics of the organism, and thus the degree of operational overlap between control systems can also shift. I also postulate that there may be relatively systematic rules that contribute to many such shifts among "behavioral boundaries." The term "boundary-state" summarizes this position.

4. Intensity and Temporal Dimensions; Encoding

The discussion thus far has concentrated on the qualitative or directional dimension in behavior (i.e., *what* inputs and *what* outputs). The preceding review suggests two additional dimensions of primary importance: a quantitative dimension (i.e., "intensity" of input/output functions) and a temporal dimension (i.e., during what time frame do inhibitory and/or excitatory relations between input and output occur?). These latter two dimensions are not entirely separate from that of direction in behavior, and, most importantly for the present argument, they may modulate specificity as defined in terms of excitatory and/or inhibitory input/output effects. The approach here is fundamentally different from most previous formulations in that, for example, it does not assume that intensity and directional models are synonymous (ethological "specific" control models) or that they are independent (certain neuropsychological "nonspecific" control models) but rather suggests that the *level of activity of a given control process may partially determine its specificity.* Temporal variables are viewed in a similar light. These latter may be partially synonymous with the intensity dimension (cf. warmup, fatigue, and after-discharge in Sherringtonian reflex studies). In addition to the three main variable classes summarized thus far (qualitative, quantitative, temporal), there is a fourth dimension suggested by the previous review of psychological and ethological data: the extent of encoding, organizational complexity, and relative predominance of different control functions.

B. Extrapolation: "Boundary-State" Model

1. The Basic Model

I suggest the analogy to receptive fields in the neurophysiological literature as one approach to this problem. Many such receptive fields have been documented, for example, that have excitatory centers and inhibitory surrounds. We shall concentrate on this type of organization for simplicity of expression. The suggestion is that "integrative control fields" may also be viewed in terms of a core of excitatory functions and a surround of inhibitory functions. This of course must be recognized as an abstraction and does not in any way imply local "centers" of anatomical control. The basic *abstraction* is presented schematically in Fig. 9A. The degree of specificity of the system at any one cross-section is represented by the diameter of the inhibitory and excitatory rings at that level. There is a formal similarity to the purely operational approach in Fig. 3, but now *both* excitatory and inhibitory functions are dealt with explicitly.

Different control systems are here viewed as sharing different degrees of these excitatory and inhibitory subfunctions. A multidimensional abstraction that employs basic Venn diagram methods for interlocking spaces can be employed to conceptualize various possible interrelations among different control systems at any one point in time. These are amenable to direct test.

The schema as thus far presented is a static representation. It deals with the qualitative dimension of different inputs and different outputs. The next question is whether consideration of intensity and/or temporal variables can be used to superimpose a dynamic framework with respect to the question of control specificity.

From the literature reviewed above, I suggest the following as an initial. approximation for a relatively wide range of integrative phenomena. For the sake of simplicity, I shall initially emphasize the output functions of the

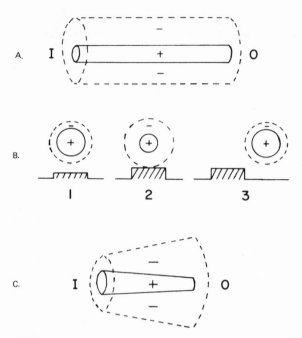

Fig. 9. A: Schematic representation of integrative control field with central core of excitatory processes and outer surround of inhibitory processes. B: Possible fluctuations in focus of excitatory and inhibitory processes (measured by influence on defined outputs) as a function of excitation level and temporal parameters. C: Schematic suggestion that by certain criteria excitatory outputs become more tightly focused with increasing activation but that input sensitivity may expand. This and a possible alternative based on occlusive mechanisms are discussed in the text.

system. First, with moderately low levels of activation the excitatory functions of a given control system are relatively diffuse (i.e., many behavioral outputs potentially excited) and the band of inhibition is relatively narrow and weak (Fig. 9B-1). Second, with higher levels of activation the excitatory output functions become more vigorous *and* more tightly focused (i.e., few outputs excited but these strongly so) and the band of inhibition expands in range and strength (Fig. 9B-2). Third, *after* high-intensity activation the system gradually returns to a lower activation level (e.g., behavioral "inertia") at which time relatively broad fields of output excitability and more narrow bands of inhibition are again seen (cf. Sherrington's, 1906, "after-discharge") (Fig. 9B-3). Finally, at the *very initial* stage of strong activation there may be a relatively broad beam of excitation which then quickly tightens (cf. Sherrington's "warmup"). Literature reviewed below indicates that while outputs may become constricted with increasing activation, animals may become *more* sensitive to an increasing range of inputs (Fig 9C); thus input/output functions may follow somewhat separable rules.

2. Experimental Operations and Predictions

What predictions does this make about interactions between control systems? The basic operational strategy for determining such interactions is summarized in Fig. 6. First, one control system (S_1) is defined in terms of range of inputs (I_1) affecting range of outputs (O_1) by the criteria outlined in Fig. 3. A second control system (S_2) is similarly defined in terms of appropriate inputs (I_2) and outputs (O_2). The question can then be asked: what is the influence of inputs in system 1 (I_1) on outputs in system 2 (O_2); what is the influence of inputs in system 2 (I_2) on outputs in system 1 (O_1)?

The predictions which follow include the following: (a) At low levels of activation of one control system by inputs appropriate to it, some facilitation of the outputs of the other system will be seen moderately often. (b) At high levels of activation of one control system by inputs appropriate to it, this facilitation will have a much lower probability, while inhibition will have a higher probability. (c) *After* high levels of activation of one control system by inputs appropriate to it, facilitation, rather than inhibition, of the other control system will again increase in probability. (d) At the *very initial* introduction of a high-intensity stimulus relevent to one control system, there may be a momentary facilitation of outputs in the other control system, followed by inhibition. The range over which these predictions are valid is subject to direct experimental test. They are summarized in schematic form in Fig. 10.

3. Neural Analogies

At another level, it might be asked whether this schema has any potential neurological reality. For example, if the analogy is made between receptive fields and "integrative control fields" is there any evidence that the size of receptive fields can alter with either the internal state of the organism or with dimensions of sensory input? There are indeed such examples, a few of which were indicated above. Pribram (1971) summarizes several additional examples of apparent shifts in receptive field size and has offered useful speculations on their possible meaning. Perhaps a more appropriate question at this stage is whether there is any evidence that "integrative fields" at the neuronal level can also shift. Here, for obvious reasons, clear experimentation

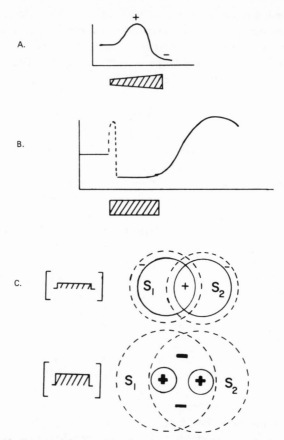

Fig. 10. Proposed integrative dynamics involving influence of one control system on expression of another as a function of excitation level and temporal factors. A: Enhancement and suppression of control-level expression of behavior as a function of intensity of "irrelevant" input. Input increases in strength from left to right. B: Temporal consequences of presentation of "irrelevant" input at high intensity. There may be a momentary enhancement of ongoing behavior and then decrement during stimulus presentation, followed by a second increment *after* the stimulus presentation. C: Schematic of two overlapping output control fields under low and high levels of activation. See text for explanation.

is much more difficult, but one suggestive illustration is provided in Fig. 11. In their monograph on cerebellar organization, Eccles *et al.* (1967) speak directly of a "focusing action" produced on granule fiber excitation by repeated mossy fiber stimulation. The effect is mediated by inhibitory connections from Golgi cells on the granule cell–parallel fiber system. In their

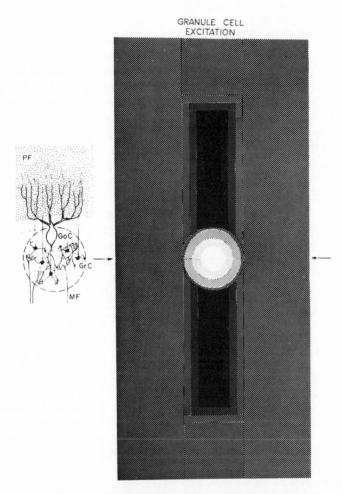

Fig. 11. Diagram of postulated action of impulses in mossy fiber bundle (MF) that innervates a focus of granule cells (GrC) in the cerebellum. Golgi cell (GoC) distributed to that focus and transverse cross-section of parallel fibers (PF) are also shown. Band of excitation is shown in white, while inhibition is represented by gray shading. [From Eccles *et al.* (1967).]

words, "this focusing action will occur only when there is a repetitive mossy fiber input. . . . As a consequence of the focusing action of the Golgi cell inhibition, it is justifiable to postulate that repetitive impulse discharge along a small bundle of mossy fibers comes restricted to an effective excitatory action on a small focus of granule cells" (pp. 221 and 223). Of course, it is dangerous to equate such very different levels of analysis at the present time. Further, it is difficult to extrapolate the consequences of altered inhibition–excitation ratios at one processing stage to the operation of "the system as a whole" or even to relative changes in inhibition and/or excitation at some other functionally interconnected processing stage. However, a basic "boundary-state" approach appears compatible with operation of integrative principles at the neuronal level. Also, the approach can suggest direct experimentation at the neuronal level. For example, are interactions between two sensory inputs on a multimodal cell more likely to be inhibitory when placed close together and during strong as opposed to weak stimulation? With a Sherringtonian reflex preparation, might inhibitory relations between reflexes (e.g., inhibition of scratch reflex by flexion reflex be replaced by mutual facilitation at low levels of activation? This illustrates the type of prediction that can be made from the present model. Predictions at higher levels are necessarily difficult, since, as so elegantly demonstrated by the disinhibition model, inhibition at one stage can lead to excitation at the next (and *vice versa*). Also, it is necessary to make clear that actual mechanisms of inhibition, as suggested here, are often difficult to distinguish from spread of activation to competing programs. Such details depend on finer levels of analysis for their resolution.

C. Related Behavioral Literature

1. Ethological and Related Models

More realistically, one can scan for similarities with broad outlines of behavioral control as outlined in the ethological literature. Here, for example, one might note that "appetitive" phases of behavior, which are defined in terms of a relatively wide range of expressed outputs, typically occur early in a behavioral sequence ("appetitive phase") and during conditions of moderate levels of motivation (e.g., Tinbergen, 1951). As the animal gets closer to its goal, behavioral outputs become progressively restricted to activities relevant to a narrow range of specific functional endpoints ("consummatory phase"). The intensity of behavior within these narrow functional endpoints appears to increase by most measures (e.g., Hinde, 1970; Miller, 1959; Tinbergen, 1951). Activities that are primarily relevant to other functional endpoints of behavior are then suppressed. The study by Bolles (1963) in which rats were found to groom more during active periods of the day, but less when "specifically" motivated by hunger, appears formally similar: i.e., grooming was facilitated by unfocused activity but suppressed as the activity became focused toward a particular goal.

2. Behavioral Anticipation and Interim Activities

An interesting further test of the principles suggested here would involve analysis of behavioral "anticipation" in which consummatory responses and/or stimuli are available only at certain time intervals. One would predict, for example, that early in the interval (i.e., shortly after preceding consummatory response and/or stimulation) behavioral activity would be relatively low, later in the interval a variety of different activities might be facilitated ("diffuse" activation), and still later, as the consummatory response and/or stimuli approach more closely, the animal's behavior would become more narrowly focused (i.e., activities irrelevant to specific goal-directed responses would become suppressed).

There is some evidence to suggest that this basic pattern does indeed occur. Extensive studies on schedule-induced polydypsia by Falk (1969), for example, have shown that marked increases in drinking behavior occur in rats between food reinforcements given at experimentally determined intervals. Falk reviewed considerable evidence that the excessive drinking thus observed cannot be attributed to peripheral factors alone. By interpreting the interval between food reinforcements, deprivation conditions, and type of food reinforcement as indices of "thwarting" of food-directed responses, Falk concluded that "too little or too much thwarting is inimical to the elicitation of adjunctive behavior" (i.e., drinking, etc.) (p. 584) (see Fig. 12). Falk notes possible similarities to the ethological literature on "displacement activities" and accepts the basic interpretation of "disinhibition."

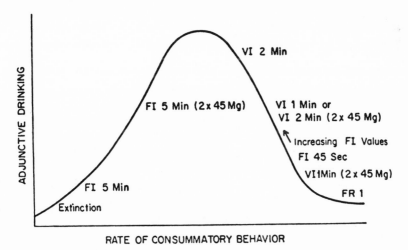

Fig. 12. The relationship between consummatory feeding and adjunctive drinking as a function of reinforcement schedules in an operant test situation in rats. [From Falk (1969).]

In a more recent and expanded review, Staddon and Simmelhag (1971) examined a variety of "interim" activities that occur in inter-reinforcement intervals in pigeons and concluded that schedule-produced thwarting situations produce an "*elevation* (rather than merely disinhibition) of motivational states other than the blocked one" (p. 39). They propose an argument, similar to the one suggested here, that "displacement activities" might usefully be interpreted in the same light. Each of the authors indicates that such data illustrate complex interactions (*vs.* independence) of different motivational systems. This matter lies at the heart of specific *vs.* nonspecific models of motivation (cf. above). Similarly, Wilton *et al.* (1969, p. 378) have concluded: "The general picture that emerges . . . is that frustration has energizing or depression effects according to its intensity . . . which increases with reward magnitude, and as it occurs near to the time of reinforcement." Here, the experiments reported by Staddon and Simmelhag (1971) are of further interest, for they demonstrate that many "interim" behaviors bear a nonmonotonic relationship to the interval between successive reinforcements; i.e., they increase and then decrease during interstimulus intervals. Subsequent studies within the context of "displacement activities" in which these temporal and intensity variables were manipulated in a systematic way would be valuable.

Space precludes a more detailed review of related studies at this time. The basic point that activation of one control system can have nonmonotonic influences on the expression of another control system (defined in terms of functional endpoint) is clear, however, That is, the "boundaries" of a given behavioral control system appear to shift with external factors and/or the animal's internal state.

3. Sensory Processing and Behavioral Performance

Thus far, we have concentrated on the output side of our behavioral equation. The input side may follow somewhat different rules. For example, while very "hungry" animals may restrict their behavioral outputs to food-seeking activities, the range of stimuli which will elicit feeding often increases, as is well known (cf. Hinde, 1970). Similarly, Baerends *et al.* (1955) found that with increasing levels of "sexual motivation" male guppies would respond to progressively smaller females. Using electrophysiological stimulation techniques, MacDonnell and Flynn (1966) found that the size of the receptive field on the face of a cat that would elicit tactile-induced biting responses increased with stimulation of certain hypothalamic structures relevant to "attack." Does the range of outputs decrease with higher levels of activation at the same time the range of inputs that will elicit these outputs increases? This possibility, diagrammed in Fig. 9C, has not been fully explored and stresses the potential value in separating input from output functions in discussions of control specificity. It would be of considerable interest, using MacDonnell and Flynn's basic procedures, to see whether stimulation of "attack" regions in the hypothalamus *reduces* the animals' sensitivity to stimuli that normally subserve other nonattack, functional categories of behavior.

Research by McFarland (1966) does suggest that during intervals in which animals show a variety of different patterns of behavior (e.g., "displacement activities") they may be particularly responsive to different classes of sensory information. However, during strong activation of a given set of behavioral outputs there is also evidence that organisms are *less* sensitive to sensory information relevant to other behavioral outputs (e.g., Broadbent, 1958, 1971; Fentress, 1972; Posner, 1966; Welford, 1968).

Recent investigations on human attention and task performance reviewed by Posner and Boies (1971) are of interest here. These authors note that current studies of "attention" can be subdivided into basic dimensions of alertness, selection, and processing capacity and that these dimensions may differentially influence different stages of a given task. Of particular interest here were experiments in which reaction times to an "irrelevant" auditory probe were measured during different phases of an experiment which involved a visual warning ("alerting") stimulus, a first letter, and a second letter which the subjects were to respond to as "same" or "different" from the first. Reaction times to the auditory probe were significantly *reduced* from intertrial interval times after the warning stimulus prior to the first letter and even shortly after presentation of the first letter. Auditory probe reaction times then increased during the remaining interval between first and second letters and remained high for some time after presentation of the second letter, when the subjects were actively engaged in the specific task. Thus the subjects' attention appeared to become progressively more "focused" during the task period, a fact which bears formal similarities to the animal studies and schemata reviewed above. The fractionation of attention into component stages suggested by Posner and Boies suggests an approach that might have value in ethological studies usually pursued with the framework of motivational control. For a broadly similar, although not identical, attempt to relate motivational ("arousal") and attention constructs, see Gray and Smith (1969).

The construct of a limited-capacity information-processing system in behavioral integration as outlined so elegantly in recent studies on human attention and performance (previous references) suggests a supplement to the schemata presented thus far here. It may have struck the reader that the stress in this chapter on increasing focus of behavior during high levels of activation is superficially the reverse of investigations originating with Sherrington (1906) that demonstrate progressive "irradiation" of behavior

with increasing stimulus strength, etc. However, note that the emphasis here
has been on facilitation *vs.* suppression of *different classes* of behavior (i.e.,
activities subserving different functional endpoints), while the clearest demonstrations of irradiation have concerned response components *within* a
given class of activities (e.g., scratching components). If activities *within*
a behavioral class are expanded, then one might well expect activities in other
classes of behavior to be reduced through occlusive and related mechanisms.
At this point, the ethologists' claim that examination of behavior in terms of
functional endpoints may assist analysis of causal mechanisms (e.g., Lorenz,
1957; Tinbergen, 1952) appears particularly worthwhile. At present, it remains difficult to separate a model that stresses expanding inhibition from
one that stresses expansion of excitation and subsequent occlusion due to
competition of excited pathways for subsequent processing through a limited-
capacity mechanism (cf. Posner and Boies, 1971). Each of these models,
however, stresses dynamic fluctuations in processing fields, unlike more
traditional static conceptualizations (e.g., Norton, 1968).

An example of a neuronal model that stresses principles of irradiation
relevant here is that proposed by Gerard (1960). In his "neuron reserve"
hypothesis, Gerard notes that "the physiologically available neurons rather
than the anatomically existent ones must determine behavioral complexity
at any one time." Gerard's argument is that when too many neurons are
activated there is a decrease in signal/noise ratio and a diminishment in neural
channels available for the specific task due to occlusive mechanisms. A very
similar view has been proposed on the basis of human experimentation by
Welford (1962). It must be emphasized again that at the present time it is
difficult to separate precisely such irradiation models from the progressive
"focus" approach outlined here. Broadbent (1971, p. 51) has recently noted
that "It is not clear whether there is simply a deterioration due to high arousal
itself, or whether at high levels arousal ceases to be important and gives way
to the effects of expectance," a position more directly analogous to the one
I have outlined. These issues can only be settled by further research. Each
approach, however, has the fundamental advantage of examining higher
integrative functions within a dynamic rather than static framework.

Considerable evidence at both behavioral and neurophysiological levels
suggests that the demands incurred in the programming of behavioral outputs
can also interfere with the processing of sensory inputs; i.e., input and output
processing are not independent (e.g., Broadbent, 1958, 1971; Jasper, 1963;
Welford, 1968). Thus increasing the range of output demands might be
expected to decrease the range and/or complexity of input processing possible, and *vice versa*. The relative unresponsiveness of animals to changes in
sensory input during highly activated behavior patterns and the increased

stereotypy of movement sequences during high attention demands (Fentress, 1972) could be indicative of such principles.

4. Information vs. Energy Models

One advantage of the overall approach proposed here is that by dissecting behavioral control systems into various input/output parameters, quantitative information measures can be applied to basic problems of motivational organization. The techniques appear particularly valuable when applied to the dissection of ongoing behavioral sequences as studied by ethologists. For example, I have (Fentress, 1972) recently provided quantitative support for hierarchical control in rodent grooming, thus supporting at a descriptive level basic principles of organization outlined by Tinbergen (1951). Recent unpublished data from my laboratory also suggest that these different descriptive levels are controlled by mechanisms that are at least partially separable (i.e., one can modify activities at one level of control with little if any modification of organization at a different level). Similar recent uses of information and Markov measures for ethological analyses of integrated behavior sequences can be found in Cane (1961), Chatfield and Lemon (1970), Delius (1969), Golani and Mendelssohn (1971), Lemon and Chatfield (1971), Nelson (1964), Slater and Ollason (1972), and Vowles (1970). See also Slater's article in this volume. A basic hope, of course, is that such information measures may help avoid vague "energy" concepts that have long made motivational analysis so difficult (cf. Hinde, 1960, 1970).

Weiner (1948) was an early pioneer of this approach to neuronal function and concluded that "the bookkeeping which is most essential to describe their function is not one of energy" (p. 42). "The mechanical brain does not secrete thought 'as the liver does bile,' as the earlier materialists claimed, nor does it put it out in the form of energy, as the muscle puts out its activity. Information is information, not matter or energy. No materialism which does not admit this can survive at the present day" (p. 132). This does not of course mean that intensity measures of behavior are invalid or that concepts such as "arousal" do not have a certain utility at this stage in our knowledge. It is rather a caution, as has been reemphasized by Hinde (e.g., 1970), that as analysis proceeds detailed statements of *patterning* (whether of behavior or neuronal activity) will prove to have greater analytical power than the extrapolation of blanket "drive" variables viewed within a simple energetics framework. It is, after all, the patterning of behavior we wish to explain!

5. Epilogue on Displacement Activities

This brings us to the final epilogue on displacement activities. Obviously, behavior patterns within this category do not represent a unitary dimension of behavior, but it is to the credit of early workers such as Lorenz (1957) and Tinbergen (1952) that they stressed the fact that most displacement activities are well established in the organism's behavioral repertoire and relatively stereotyped in their expression. Since most of the contexts in which displacement activities occur involve some form of "conflict" or "thwarting," one can presume that the processing capacity of the organism is strained, and thus activities which demand little information processing are most likely to occur. While this broad theme appears to have some general applicability, it is still incumbent on the research worker to determine why *this* rather than *that* displacement behavior occurs in *a particular* context. This is a question of *specific* causation (cf. Feekes, 1971; van Iersel and Bol, 1958; Kruijt, 1964; Sevenster, 1961; Wilz, 1970*a*).

It is encouraging to see that ethological studies on the integration of behavioral sequences are now at a stage where tentative construction of dynamic functions involving both relatively specific and nonspecific factors, plus their interactions, can be formulated in a manner which is compatible with an increasing range of literature in both experimental psychology and the neurosciences. While the views presented here will obviously need some modification in light of future experiments, they hopefully will help suggest the experiments that will lead to this modification. Thus the present set of arguments should be viewed as an approach to and not as a solution of problems of control specificity in organized behavior. These "solutions" will for some time remain imperfect approximations—but that is where the excitement, both general and specific, lies!

VII. RÉSUMÉ

1. Both specific and nonspecific models of behavioral integration have certain merits but also certain limitations.
2. Review of the literature indicates that disagreement between specific and nonspecific models is largely due to inconsistent application of classification criteria.
3. The most strict definition of specificity is that a system involves only a single input and a single output and is entirely independent in its operation from other input/output combinations that operate in parallel. The most strict definition of nonspecificity is that all inputs affect all outputs, identically. Neither of these criteria is likely to be met.
4. Partial specificity (and partial nonspecificity) is a more difficult conceptualization but also is likely to be more accurate. This approach implies partial overlap (and partial separateness) among

control processes that underlie different input/output combinations.

5. Dynamic, as opposed to static, models may also be required to account for possible shifts in specificity within systems and overlap between systems as a function of interactions between different variables.

6. Ethological models traditionally classify a given integrative system in terms of factors of elicitation only. However, suppression of one behavior with the elicitation of another is equally important and implies lack of independence of control systems defined by elicitation only. An alternate view is that both elicited and suppressed activities are components of the same system.

7. Multiple interactions between control processes may be indistinguishable from nonspecificity at the initial analysis level. At refined analysis levels, "multispecific" models may be more adequate than "nonspecific" models.

8. With respect to a given stage of processing, it is often useful to distinguish between input specificity and output specificity. Many inputs may affect a limited number of outputs, or many outputs may be affected by a limited number of inputs. An operational approach can avoid confusion between these dimensions.

9. The direction of interaction between factors that contribute to different behavior patterns may be either positive or negative as a function of degree of activation, temporal relationship between activation and the behavior pattern measured, and the structural complexity and perhaps degree of encoding of the behavior pattern in question.

10. Excitation at one level of analysis can be produced by inhibition at another (e.g., via "disinhibition"). Recognition of this can affect the formulation of specific vs. nonspecific models of integration.

11. Central activation is often supplemented by the operation of peripheral factors. This can affect conclusions about specificity of central control mechanisms.

12. Ethological research on "displacement activities" is an illustration of how multiple inhibition (and disinhibition) among factors underlying the control of different behavior patterns, plus excitation of particular behavior patterns by specific external and/or peripheral factors, can provide an alternative explanation to generalized central activation. However, the data currently available indicate that the "disinhibition hypothesis" as usually stated may be valuably supplemented by further consideration of relatively nonspecific and centrally operating control processes.

13. A proposed "boundary-state" model suggests that the relative

specificity of a behavioral control process, and similarly the degree of overlap among control processes, may shift in a systematic manner as a function of intensity, temporal, and encoding dimensions. As a guide for future thought and experimentation in behavioral integration, an analogy is suggested to the center/surround operation of certain sensory and motor control systems at the neurological level.

14. With current techniques, it remains difficult to separate proposed spread of inhibition from diffuse excitation with subsequent blocking due to competition for limited-capacity mechanisms (occlusion) near the output side of processing. This presents a major task for future analyses. Application of information measures may further reduce emphasis on ambiguous "energizing" constructs in motivational analysis and thus assist resolution of balance between specific and nonspecific factors in patterned behavior.

VIII. ACKNOWLEDGMENTS

I appreciate the many helpful discussions by members of the University of Oregon BioSocial Research Center on the material covered here. I also thank Mr. H. Howard for photographic assistance, Ms. H. Parr and Ms. V. Stickrod for typing, and Ms. F. P. Stilwell for preparing the final figures and proofreading.

IX. REFERENCES

Adkins, R. J., Morse, R. W., and Towe, A. L. (1966). Control of somatosensory input by cerebral cortex. *Science* **153**:1020–1022.
Amsel, A., and Maltzman, I. (1950). The effect upon generalized drive strength of emotionality as inferred from the level of consummatory response. *J. Exptl. Psychol.* **40**:563–569.
Andrew, R. J. (1956a). Some remarks on behaviour in conflict situations, with special reference to *Emberiza* sp. *Brit. J. Anim. Behav.* **4**:41–45.
Andrew, R. J. (1956b). Normal and irrelevant toilet behaviour in *Emberiza* sp. *Brit. J. Anim. Behav.* **4**:85–91.
Anokhin, P. K. (1961). Contributions to general discussion: I. Inborn and reflex behavior. In Delafresnaye, J. F. (ed.), *Brain Mechanisms and Learning*, Oxford University Press, New York, pp. 642–644.
Ashby, W. R. (1956). *An Introduction to Cybernetics*, Chapman and Hall, London.
Attneave, F. (1959). *Applications of Information Theory to Psychology*, Holt, New York.
Ayer, A. J. (1956). *The Problems of Knowledge*, Penguin, Edinburgh.
Baerends, G. P., Brouwer, R., and Waterbolk, H. T. (1955). Ethological studies of *Lebistes reticulatus* (Peters). I. An analysis of the male courtship pattern. *Behaviour* **8**:249–334.
Barfield, R. J., and Sachs, B. D. (1968). Sexual behavior: Stimulation by painful electrical shock to skin in male rats. *Science* **161**:392–395.

Bauer, F. J. (1956). Genetic and experiential factors affecting social reactions in male mice. *J. Comp. Physiol. Psychol.* **49**:359–364.

Beach, F. A., and Whalen, R. E. (1959a). Effects of ejaculation on sexual behavior in the male rat. *J. Comp. Physiol. Psychol.* **52**:249–254.

Beach, F. A., and Whalen, R. E. (1959b). Effects of intromission without ejaculation upon sexual behavior in male rats. *J. Comp. Physiol. Psychol.* **52**:476–481.

Beach, F. A., Goldstein, A. C., and Jacoby, G. A. (1955). Effects of electroconvulsive shock on sexual behavior in male rats. *J. Comp. Physiol. Psychol.* **48**:173–179.

Berkson, G., and Mason, W. A. (1964). Stereotyped behaviors of chimpanzees: Relation to general arousal and alternative activities. *Percept. Motor Skills* **19**:635–652.

Berlyne, D. E. (1960). *Conflict, Arousal, and Curiosity,* McGraw-Hill, New York.

Bindra, D. (1959). *Motivation: A Systematic Reinterpretation,* Ronald Press, New York.

Bindra, D. (1969). An interpretation of the "displacement" phenomenon. *Brit. J. Psychol.* **50**:263–268.

Bindra, D., and Mendelson, J. (1963). Training, drive level, and drug effects: A temporal analysis of their combined influence on behavior. *J. Comp. Physiol. Psychol.* **56**:183–189.

Bohm, D. (1969). Some remarks on the notion of order. Further remarks on order. In Waddington, C. H. (ed.), *Towards a Theoretical Biology,* Vol. 2, Aldine, Chicago, pp. 18–60.

Bolles, R. C. (1963). The effect of food deprivation upon the rat's behavior in its home cage. *J. Comp. Physiol. Psychol.* **56**:456–460.

Bolles, R. C. (1967). *Theory of Motivation,* Harper and Row, New York.

Bolles, R. C., and Morlock, H. (1960). Some asymmetrical drive summation phenomena. *Psychol. Rep.* **6**:373–378.

Bradley, P. B. (1958). The central action of certain drugs in relation to the reticular formation of the brain. In Jasper, H. A., Proctor, L. D., Knighton, R. S., Noshay, W. C., and Costello, R. T. (eds.), *Reticular Formation of the Brain,* Little, Brown, Boston, pp. 123–149.

Broadbent, D. E. (1958). *Perception and Communication,* Pergamon Press, London.

Broadbent, D. E. (1971). *Decision and Stress,* Academic Press, New York.

Brown, J. S. (1961). *The Motivation of Behavior,* McGraw-Hill, New York.

Brown, J. S., and Jacobs, A. (1949). The role of fear in the motivation and acquisition of responses. *J. Exptl. Psychol.* **39**:747–759.

Buser, P., and Imbert, M. (1961). Sensory projections to the motor cortex in cats: A microelectrode study. In Rosenblith, W. A. (ed.), *Sensory Communication,* Wiley, New York, pp. 607–626.

Cane, V. (1961). Some ways of describing behavior. In Thorpe, W. H., and Zangwill, O. L. (eds.), *Current Problems in Animal Behaviour,* Cambridge University Press, Cambridge, England, pp. 361–388.

Chatfield, C., and Lemon, R. E. (1970). Analyzing sequences of behavioral events. *J. Theoret. Biol.* **29**:427–445.

Delgado, J. M. R. (1964). Free behaviour and brain stimulation. *Internat. Rev. Neurobiol.* **6**:349–447.

Delius, J. D. (1969). A stochastic analysis of the maintenance behaviour of skylarks. *Behaviour* **33**:137–178.

Delius, J. D. (1970). Irrelevant behaviour, information processing and arousal homeostasis. *Psychol. Forsch.* **33**:165–188.

Dell, P. (1963). Reticular homeostasis and critical reactivity. In Moruzzi, G., Fescard, A., and Jasper, H. H. (eds.), *Progress in Brain Research,* Vol. 1, Elsevier, Amsterdam, pp. 82–114.

de Lorge, J., and Bolles, R. C. (1961). Effects of food deprivation on exploratory behavior in a novel situation. *Psychol. Rep.* **9**:599–606.

Doty, R. W. (1961). Conditioned reflexes formed and evoked by brain stimulation. In Sheer, D. E. (ed.), *Electrical Stimulation of the Brain: An Interdisciplinary Survey of Neurobehavioral Integrative Systems,* University of Texas, Austin, pp. 397–412.

Duffy E. (1962). *Activation and Behavior,* Wiley, New York.
Duncan, I. J. H., and Wood-Gush, D. G. M. (1972). An analysis of displacement preening in domestic fowl. *Anim. Behav.* **20**:68–71.
Eccles, J. C., Ito, M., and Szentagothai, J. (1967). *The Cerebellum as a Neuronal Machine,* Springer-Verlag, New York.
Epstein, A., and Teitelbaum, P. (1962). Regulation of food intake in the absence of taste, smell, and other oropharyngeal sensations. *J. Comp. Physiol. Psychol.* **55**:753–759.
Evarts, E. V., and Thach, W. T., Jr. (1969). Motor mechanisms of the CNS: Cerebrocerebellar interrelations. *Ann. Rev. Physiol.* **31**:451–498.
Faidherbe, J., Richolle, J., and Schlag, J. (1962). Non-consumption of the reinforcer under drug action. *J. Exptl. Anal. Behav.* **5**:521–524.
Falk, J. L. (1969). Conditions producing psychogenic polydipsia in animals. *Ann. N. Y. Acad. Sci.* **157**:569–593.
Feekes, F. (1971). "Irrelevant" ground pecking in agonistic situations in Burmese red jungle fowl (*Gallus gallus spadiceus*). *Rijkuniversiteit te Granigen.*
Feldman, S. M., and Waller, H. J. (1962). Dissociation of electrocortical activation and behavioural arousal. *Nature* **19**:1320–1322.
Fentress, J. C. (1965). Aspects of arousal and control in the behaviour of voles. Ph. D. dissertation, Cambridge University.
Fentress, J. C. (1967). Observations on the behavioral development of a hand-reared male timber wolf. *Am. Zoologist* **7**:339–351.
Fentress, J. C. (1968a). Interrupted ongoing behaviour in two species of vole (*Microtus agrestis* and *Clethrionomys britannicus*). I. Response as a function of preceding activity and the context of an apparently "irrelevant" motor pattern. *Anim. Behav.* **16**:135–153.
Fentress, J. C. (1968b). Interrupted ongoing behaviour in two species of vole (*Microtus agrestis* and *Clethrionomys britannicus*). II. Extended analysis of motivational variables underlying fleeing and grooming behaviour. *Anim. Behav.* **16**:154–167.
Fentress, J. C. (1972). Development and patterning of movement sequences in inbred mice. In Kiger, J. (ed.), *The Biology of Behavior,* Oregon State University Press, Corvallis, pp. 83–132.
Fentress, J. C. (1973). Development of grooming in mice with amputated forelimbs. *Science* **179**:704–705.
Fessard, A. (1961). The role of neuronal networks in sensory communications within the brain. In Rosenblith, W. A. (ed.), *Sensory Communication,* Wiley, New York, pp. 585–606.
Feynman, R. (1965). *The Character of Physical Law,* MIT Press, Cambridge, Mass.
Fuster, J. M. (1958). Effects of stimulation of brain stem on tachistoscopic perception. *Science* **127**:150.
Garner, W. (1962). *Uncertainty and Structure as Psychological Concepts,* Wiley, New York.
Geist, V. (1963). On the behavior of the North American moose (*Alces alces andersoni* Peterson, 1950) in British Columbia. *Behaviour* **20**:377–416.
Gerard, R. W. (1960). Neurophysiology: An integration. In Field, J., Magoun, H. W., and Hall, V. E. (eds.), *Handbook of Physiology,* Sect. I, Vol. III, American Physiological Society, Washington, D.C., pp. 1919–1965.
Gerard, R. W., Marshall, W. H., and Sault, L. J. (1936). Electrical activity of the cat's brain. *A.M.A. Arch. Neurol. Psychiat.* **36**:675–738.
Golani, I., and Mendelssohn, H. (1971). Sequences of precopulatory behavior of the jackal (*Canis aureus* L.). *Behaviour* **38**:169–192.
Gray, J. A., and Smith, P. T. (1969). An arousal–decision model for partial reinforcement and discrimination learning. In Gilbert, R. M., and Sutherland, N. S. (eds.), *Animal Discrimination Learning,* Academic Press, London and New York, pp. 243–272.
Hall, J. F. (1956). The relationship between external stimulation, food deprivation, and activity. *J. Comp. Physiol. Psychol.* **49**:339–341.
Hanby, J. P. (1972). The sociosexual nature of mounting and related behaviors in a confined troop of Japanese macaques. Ph. D. dissertation, University of Oregon.

Hanson, N. R. (1955). Causal chains. *Mind* **64**:289–311.

Havlicék, V. (1963). Some aspects of the central effect of adrenalin. In Votava, Z., Horváth, M., and Vinar, O. (eds.), *Psychopharmacological Methods*, Pergamon Press, London, pp. 197–208.

Hebb, D. O. (1955). Drives and the CNS (Conceptual Nervous System). *Psychol. Rev.* **62**:243–254.

Hillman, P., and Wall, P. D. (1969). Inhibitory and excitatory factors influencing the receptive fields of lamina 5 spinal cord cells. *Exptl. Brain Res.* **9**:284–306.

Hinde, R. A. (1958). The nestbuilding behaviour of domesticated canaries. *Proc. Zool. Soc. Lond.* **131**:1–48.

Hinde, R. A. (1959). Unitary drives. *Anim. Behav.* **7**:130–141.

Hinde, R. A. (1960). Energy models of motivation. *Symp. Soc. Exptl. Biol.* **14**:199–213.

Hinde, R. A. (1970). *Animal Behaviour: A Synthesis of Ethology and Comparative Psychology*, 2nd ed., McGraw-Hill, New York.

Hinde, R. A. (1972). *Social Behavior and Its Development in Subhuman Primates*, Condon Lectures, University of Oregon Press, Eugene.

Hinde, R. A., and Steel, E. (1966). Integration of the reproductive behaviour of female canaries. *Symp. Soc. Exptl. Biol.* **20**:401–426.

Hirsch, J., Lindley, R. H., and Tolman, E. C. (1955). An experimental test of an alleged innate sign stimulus. *J. Comp. Physiol. Psychol.* **48**:278–280.

Horn, G. (1962). The response of single units in the striate cortex of unrestrained cats to photic and somaesthetic stimuli. *J. Physiol.* **165**:80–81.

Horváth, M., Frantík, E., and Formánek, J. (1963). Experimental studies of higher nervous functions in pharmacology and toxicology (methodological implications). In Votava, Z., Horváth, M., and Vinar, O. (eds.), *Psychopharmacological Methods*, Pergamon Press, London, pp. 131–150.

Hull, C. L. (1943). *Principles of Behavior*, Appleton-Century-Crofts, New York.

Hume, P. (1748). An enquiry concerning human understanding. In Hendel, C. W., Jr. (ed.), *Hume Selections*, Scribner's, New York, pp. 107–193.

Jasper, H. H. (1963). Studies of non-specific effects upon electrical responses in sensory systems. In Moruzzi, G., Fessard, A., and Jasper, H. H. (eds.), *Progress in Brain Research*, Vol. 1, Elsevier, Amsterdam, pp. 272–293.

Jerome, E. A., Moody, J. A., Connor, T. J., and Fernandez, M. B. (1957). Learning in multiple-door situations under various drive states. *J. Comp. Physiol. Psychol.* **50**:588–591.

Jung, C. G. (1923). *Psychological Types*, Pantheon, New York (1959 printing).

Jung, R., Kornhuber, H. H., and Forseca, J. S. (1963). Multisensory convergence on cortical neurons: Neuronal effects of visual, acoustic, and vestibular stimuli in the superior convolutions of the cat's cortex. In Moruzzi, G., Fessard, A., and Jasper, H. H. (eds.), *Progress in Brain Research*, Vol. 1, Elsevier, Amsterdam, pp. 207–240.

Keele, S. W. (1968). Movement control in skilled motor performance. *Psychol. Bull.* **70**:387–403.

Kendler, H. H. (1945). Drive interaction: I. Learning as a function of the simultaneous presence of the hunger and thirst drives. *J. Exptl. Psychol.* **35**:96–109.

Koestler, A. (1964). *The Act of Creation*, Hutchinson, London.

Kortlandt, A. (1940). Wechselwirkung zwischen Instinckten. *Arch. Néerl. Zool.* **4**:401–442.

Kozlowskaya, M. M., and Valdman, A. V. (1967). In Valdman, A. V. (ed.), *Pharmacology and Physiology of the Reticular Formation*, Vol. 20, Elsevier, Amsterdam, pp. 93–127.

Kruijt, J. P. (1964). Ontogeny of social behaviour in the Burmese red jungle fowl (*Gallus gallus spadiceus*). *Behaviour*, Suppl. 12.

Lacey, J. I. (1967). Somatic response patterning and stress. Some revisions of activation theory. In Appley, M. H., and Trumbull, R. (eds.), *Psychological Stress: Issues in Research*, Appleton-Century-Crofts, New York, pp. 14–42.

Larsson, K. (1963). Non-specific stimulation and sexual behaviour in the male rat. *Behaviour* **20**:110–114.

Lehrman, D. S. (1965). Interaction between internal and external environments in the regulation of the reproductive cycle of the ring dove. In Beach, F. A. (ed.), *Sex and Behavior*, Wiley, New York.

Lehrman, D. S. (1970). Semantic and conceptual issues in the nature–nurture problem. In Aronson, L. R., Tobach, E., Lehrman, D. S., and Rosenblatt, J. S. (eds.), *Development and Evolution of Behavior*, W. H. Freeman, San Francisco, pp. 17–52.

Lemon, R. E., and Chatfield, C. (1971). Organization of song in cardinals. *Anim. Behav.* **19**:1–17.

Levin, M. (1961). Sleep, cataplexy, and fatigue as manifestations of Pavlovian inhibition. *Am. J. Psychotherap.* **15**:122–137.

Lindsley, D. B. (1951). Emotion. In Stevens, S. S. (ed.), *Handbook of Experimental Psychology*, Wiley, New York, pp. 473–516.

Lorenz, K. (1957). The past 12 years in the comparative study of behavior. In Shiller, C. H. (ed.), *Instinctive Behaviour: The Development of a Modern Concept*, Methuen, London, pp. 288–310.

Lorenz, K. (1971). Preface to Rasa, O. A. E., *Appetence for Aggression in Juvenile Damsel Fish*, Paul Parey, Berlin.

Luria, A. R. (1966). *Higher Cortical Functions in Man*, Basic Books, New York.

MacDonnell, M. F., and Flynn, J. P. (1966). Control of sensory fields by stimulation of the hypothalamus. *Science* **152**:1406–1408.

Malmo, R. (1959). Activation: A neuropsychological dimension. *Psychol. Rev.* **66**:367–386.

Marler, P., and Hamilton, W. J. (1966). *Mechanisms of Animal Behavior*, Wiley, New York.

McFarland, D. J. (1965). Hunger, thirst, and displacement pecking in the Barbary dove. *Anim. Behav.* **13**:293–300.

McFarland, D. J. (1966). On the causal and functional significance of displacement activities. *Z. Tierpsychol.* **23**:217–235.

Melzack, R., Penick, E., and Beckett, A. (1959). The problem of "innate fear" of the hawk shape: An experimental study with mallard ducks. *J. Comp. Physiol. Psychol.* **52**: 694–698.

Menzel, E. W. (1962). The effects of stimulus size and proximity upon avoidance of complex objects in rhesus monkeys. *J. Comp. Physiol. Psychol.* **55**:1044–1046.

Michelson, M. I., and Schelkunov, E. L. (1963). The action of chlorpromazine, chloracizine, and amphetamine on the food and avoidance conditional reflexes. In Votava, Z., Horváth, M., and Vinar, O. (eds.), *Psychopharmacological Methods*, Pergamon Press, London, pp. 28–30.

Miles, R. C. (1958). The effect of an irrelevant motive on learning. *J. Comp. Physiol. Psychol.* **51**:258–261.

Miller, N. E. (1959). Liberalization of basic S–R concepts: Extensions to conflict behavior, motivation, and social learning. In Koch, S. (ed.), *Psychology: A Study of a Science*, Vol. 2, McGraw-Hill, New York.

Miller, N. E. (1960). Motivational effects of brain stimulation and drugs. *Fed. Proc.* **19**: 846–854.

Miller, N. E. (1963). Some reflections on the law of effect produce a new alternative to drive reduction. In Jones, M. R. (ed.), *Nebraska Symposium on Motivation*, University of Nebraska Press, Lincoln, pp. 65–112.

Mirsky, A. F., and Rosvold, H. E. (1963). Behavioral and physiological studies in impaired attention. In Votava, Z., Horváth, M., and Vinar, O. (eds.), *Psychopharmacological Methods*, Pergamon Press, London, pp. 302–315.

Morris, D. (1956). The feather postures of birds and the problem of the origins of social signals. *Behaviour* **9**:75–114.

Moruzzi, G., and Magoun, H. S. (1949). Brain-stem reticular formation and activation of the EEG. *Electroencephalog. Clin. Neurophysiol.* **1**:455–473.

Nelson, K. (1964). The temporal patterning of courtship behaviour in the glandulocaudine fishes (Ostariophysi, Characidae). *Behaviour* **24**:90–146.

Norton, S. (1968). On the discontinuous nature of behavior. *J. Theoret. Biol.* **21**:229–243.

Paillard, J. (1960). The patterning of skilled movements. In Field, J., Magoun, H. W., and Hall, V. E. (eds.), *Handbook of Physiology*, Sect. I: *Neurophysiology*, Vol. III, American Physiological Society, Washington, D.C., pp. 1679–1708.

Palestini, M., Rossi, G. F., and Zanchetti, A. (1957). An electrophysiological analysis of pontine reticular regions showing different anatomical organization. *Arch. Ital. Biol.* **95**:97–109.

Pavlov, I. (1927). *Conditioned Reflexes. An Investigation of the Physiological Activity of the Cerebral Cortex*, Oxford University Press, New York.

Phillips, C. G. (1966). Changing concepts of the precentral motor area. In Eccles, J. C. (ed). *Brain and Conscious Experience*. Springer Verlag, New York, pp. 389–421.

Poincare, H. (1913). *The Foundations of Science*, Science Press, New York.

Posner, M. I. (1966). Components of skilled performance. *Science* **152**:1712–1718.

Posner, M. I., and Boies, S. (1971). Components of attention. *Psychol. Rev.* **78**:391–408.

Pribram, K. H. (1971). *Languages of the Brain*, Prentice-Hall, Englewood Cliffs, N.J.

Raber, H. (1948). Analyse des Balzverhaltens eines domestizierten Truthans (*Meleagris*). *Behaviour* **1**:237–266.

Rasa, O. A. E. (1971). *Appetence for Aggression in Juvenile Damsel Fish*, Paul Parey, Berlin.

Rossi, G. F., and Zanchetti, A. (1957). The brain stem reticular formation: Anatomy and physiology. *Arch. Ital. Biol.* **95**:199–438.

Rowell, C. H. F. (1961). Displacement grooming in the chaffinch. *Anim. Behav.* **9**:38–63.

Rowell, T. E., and Hinde, R. A. (1963). Responses of rhesus monkeys to mildly stressful situations. *Anim. Behav.* **11**:235–243.

Russell, B. (1927). *An Outline of Philosophy*, Allen and Unwin, London.

Schleidt, W. (1961). Über die Auslosung der Flucht vor Raubvogeln bei Truthuhneren. *Naturwissenschaften* **48**:141–142.

Schneirla, T. C. (1965). Aspects of stimulation and organization in approach/withdrawal processes underlying vertebrate behavioral development. In *Advances in the Study of Behavior*, Vol. I, Academic Press, New York.

Sevenster, P. (1961). A causal analysis of displacement activity (fanning in *Gasterosteus aculeatus* L.). *Behaviour*, Suppl. 9.

Sherrington, C. S. (1906). *The Integrative Action of the Nervous System*, Yale University Press, New Haven.

Siegel, P. S., and Brantley, J. J. (1951). The relationship of emotionality to the consummatory response of eating. *J. Exptl. Psychol.* **42**:304–306.

Siegel, P. S., and Sparks, D. L. (1961). Irrelevant aversive stimulation as an activator of an appetional response: A replication. *Psychol. Rep.* **9**:700.

Siegel, P. S., and Stuckey, H. S. (1949). The effect of emotionality on the water intake in the rat. *J. Comp. Physiol. Psychol.* **42**:12–16.

Skinner, B. F. (1963). A Christmas caramel, or a plum from the Hasty Pudding. *Worm Runners Digest* **5**:42–46.

Slater, P. J. B., and Ollason, J. C. (1972). The temporal pattern of behaviour in isolated male zebra finches. *Behaviour* **42**:248–269.

Sokolov, Y. N. (1963). *Perception and the Conditioned Reflex*, Pergamon Press, London.

Sperry, R. W. (1955). The neural basis of the conditioned response. *Brit. J. Anim. Behav.* **3**:41–44.

Staddon, J. E. R., and Simmelhag, V. L. (1971). The "superstition" experiment: A reexamination of its implications for the principles of adaptive behavior. *Psychol. Rev.* **78**:3–34.

Stellar, E. (1960). Drive and motivation. In Field, J., Magoun, H. W., and Hall, V. E. (eds.), *Handbook of Physiology*, Sect. I: *Neurophysiology*, Vol. III, American Physiological Society, Washington, D.C., pp. 1501–1527.

Sterritt, G. M. (1962). Inhibition and facilitation of eating by electric shock. *J. Comp. Physiol. Psychol.* **55**:226–229.

Strongman, K. T. (1965). The effect of anxiety on food intake in the rat. *Quart. J. Exptl. Psychol.* **17**:255–260.

Thorpe, W. H. (1963). *Learning and Instinct in Animals,* 2nd ed., Methuen, London.

Tinbergen, N. (1940). Die Übersprungsbewegung. *Z. Tierpsychol.* 4:1–40.

Tinbergen, N. (1951). *The Study of Instinct,* Clarendon Press, Oxford.

Tinbergen, N. (1952). "Derived" activities; their causation, biological significance, origin, and emancipation during evolution. *Quart. Rev. Biol.* 27:1–32.

Tugendhat, B. (1960a). The normal feeding behavior of the three-spined stickleback (*Gasterosteus aculeatus*). *Behaviour* 15:284–315.

Tugendhat, B. (1960b). The disturbed feeding behavior of the three-spined stickleback: I. Electric shock is administered in the food area. *Behaviour* 16:159–187.

Ullman, A. D. (1951). The experimental production and analysis of a "compulsive eating syndrome" in rats. *J. Comp. Physiol. Psychol.* 44:575–581.

Valenstein, E. S., Cox, V. C., and Kakolewski, J. W. (1968). Modification of motivated behavior elicited by electrical stimulation of the hypothalamus. *Science* 159:1119–1121.

van Iersel, J. J. A., and Bol, A. C. A. (1958). Preening of two tern species. A study of displacement. *Behaviour* 13:1–88.

von Holst, E., and von S. Paul, U. (1963). On the functional organization of drives. *Anim. Behav.* 11:1–20.

Vowles, D. M., (1970). Neuroethology, evolution, and grammar. In Aronson, L. R., Tobach, E., Lehrman, D. S., and Rosenblatt, J. S. (eds.), *Development and Evolution of Behavior,* W. H. Freeman, San Francisco, pp. 194–215.

Wang, G. H., and Akert, K. (1962). Behavior and the reflexes of chronic striated cats. *Arch. Ital. Biol.* 100:48–85.

Webb, W. B., and Goodman, I. J. (1958). Activating role of an irrelevant drive in absence of the relevant drive. *Psychol. Rep.* 4:235–238.

Weiner, N. (1948). *Cybernetics,* Wiley, New York.

Welford, A. T. (1962). Arousal, channel-capacity, and decision. *Nature* 194:365–366.

Welford, A. T. (1968). *Fundamentals of Skill,* Methuen, London.

Wikler, A. (1952). Pharmacologic dissociation of behavior and EEG "sleep pattern" in dogs: Morphine, *N*-allylnormorphine, and atropine. *Proc. Soc. Exptl. Biol. Med.* 79:261–265.

Wilton, R. N., Strongman, K. T., and Nerenberg, A. (1969). Some effects of frustration in a free responding operant situation. *Quart. J. Exptl. Psychol.* 21:367–380.

Wilz, K. J. (1970a). Causal and functional analysis of dorsal pricking and nest activity in the courtship of the three-spined stickleback *Gasterosteus aculeatus. Anim. Behav.* 18:115–124.

Wilz, K. J. (1970b). The disinhibition interpretation of the "displacement" activities during courtship in the three-spined stickleback *Gasterosteus aculeatus. Anim. Behav.* 18:682–687.

Woods, J. W. (1964). Behavior of chronic decerebrate rats. *J. Neurophysiol.* 27:635–644.

Woodworth, R. S. (1918). *Dynamic Psychology,* Columbia University Press, New York.

Yerkes, R. M., and Dodson, J. D. (1908). The relation of strength of stimulus to rapidity of habit formation. *J. Comp. Neurol. Psychol.* 18:459–482.

Zeigler, H. P. (1964). Displacement activity and motivational theory: A case study in the history of ethology. *Psychol. Bull.* 61:362–376.

Chapter 7

SOCIAL DISPLAYS AND THE RECOGNITION OF INDIVIDUALS

M. J. A. Simpson

Sub-Department of Animal Behaviour
University of Cambridge
Madingley, Cambridgeshire, U. K.

I. ABSTRACT

Ethologists often single out displays in their studies of social behavior. This chapter opens by considering the ways in which conspicuous social actions may be interpreted, and it suggests a framework of interpretation which attends particularly to the sequential and temporal relationships among social actions. Although this interpretation is contrasted with more widely used approaches to displays, it is not an exclusive alternative. It does, however, lead us to pay special attention to social interactions as processes occupying time which can lead to progressive and sometimes irreversible change in relationship. Much of this chapter is concerned with showing how some current approaches to animal behavior can make it difficult for us to describe progressive social exchanges and how these approaches limit the kinds of description that emerge. This limitation is to be expected of current approaches referring specially to physiological levels of description. This chapter is a plea for more complete descriptions of social processes at the behavioral level. One measure of the success of such descriptions will be the new questions they pose for physiological and developmental studies.

II. INTRODUCTION

Students of social behavior often distinguish displays among the actions of the animals they study. Approaches concerned with displays are admirably summarized in the several review articles in Hinde (1972a), and by Hinde (1970), Marler and Hamilton (1966), and Bastock (1967). Another review would be superfluous at this time.

In this chapter, I shall explore the possibilities of an approach to social interactions which does not single out "displays" as special kinds of social actions. Instead, my approach is concerned with the relationships between individuals in their social groups, as expressed in their social interactions with each other. To what extent can such relationships be interpreted in terms of the past histories of interactions which involved the individuals concerned? Such histories may, of course, include events which ethologists have called "displays." It will become clear that this chapter emphasizes observations and experiments that could be carried out rather than work that has been done.

To place a social action, including a display, in a history of actions in the life of an individual is to interpret that social action. Gregory Bateson's (1969) comparison between sequences of animal behavior and reports of human dreams provides a starting point from which to develop a concept of "interpretation." Bateson's article included the questions: "How is a dream put together inside itself?" and "Could animal behavior be put together in the same way?" Both dreams and sequences of animal behavior pose common problems of interpretation for observers. Dreams, like episodes in an animal's life, can be explained in terms of situations and conditions which make their occurrence more likely, such as indigestion or the prevalence of a particular concentration of androgens in the blood. Interpretations in terms of stimulating and driving conditions are perhaps most common in causal studies of animal behavior, and I return to consider such interpretations later in this chapter. Here, I introduce two further kinds of interpretation.

Conspicuous aspects of both dreams and animal behavior sequences seem to lend themselves to codes of translation: for one species of animal, a display peculiar to that species may be correlated using evidence about the situation in which it occurs with other activities that are regarded as equivalent across many species. Examples are acts of social approach, aggression, and sequences which usually lead to mating. Using appropriate methods of correlation, the species' peculiar display may be placed in wider classes of social activities, such as "social," "aggressive," or "courtship" activities. In the dreams of certain classes of individuals in certain cultures, particular activities dreamed of, such as climbing steps, may be shown to be connected with the movements of copulation. Thus when an individual from such a culture describes a dream which includes a stair-climbing episode, we may suspect that such an individual is concerned about sex. To such a "dream-book" approach, Freud (1900) made some contribution, but his real achieve-

ment was to develop methods for the application of the third kind of interpretation, which is the one to be explored most fully in this chapter.

This third framework of interpretation can be introduced with reference
to an imaginary study of mounting behavior in a group of rhesus monkeys
which includes several adult females. Most female rhesus monkeys mount
other females on occasion. A study made within the *first* kind of framework
of interpretation might be concerned with the hormonal state of mounting
females; the *second* framework of interpretation might be concerned to show
whether females mounted during aggressive episodes and whether it was usually the higher-status female who did the mounting (e.g., Bernstein, 1970).
In nice contrast to the concern with sex of some human beings who dream
about mounting stairs, mounting in female rhesus monkeys might be shown
to be an aggressive action. The *third* framework of approach recognizes that
both reports of single dreams by an individual human being and conspicuous
episodes in an animal's social life, such as its displays, must be considered
within a broader context of the *individual's* life, not merely his life as a member of a culture or species. Of the mounting episode, we may want to know
which individual mounted, who was mounted by her, what both had been
doing before, what they did afterward, and what social activities characterized
her in her social group. Of the dreamer, we may want to know whose dream
it was, to whom he was reporting it, and what other waking activities and
dreams he has talked about and refrained from talking about.

To attempt to place a behavioral episode in a broader context of an
individual's social history is to suppose that the attempt may reveal the
episode to be part of some kind of pattern or meaningful structure. In his
book *The Interpretation of Dreams,* Freud (1900) regarded dreams as "psy-
chical structures which have meaning and which can be inserted at an as-
signable point in the mental activities of waking life." "Structure" implies
some kind of relationship among component parts, and "meaning" implies
some relationship between that structure and the events in the dreamer's
waking life. To discover that relationship, in cooperation with the dreamer,
is to "interpret" the dream. (See Seaborne Jones, 1968, for a discussion about
how such "interpretations" may be tested.)

The emphasis of the third kind of approach to episodes in social behavior is on the meaning or structure that may be discoverable in sequences of
events, and combinations of simultaneous events, of which a behavioral
episode or display, such as a female monkey's act of mounting, may only be
one. In our imaginary group of rhesus monkeys, there may be three adult
females, A, B, and C. When A mounts B, B, then grooms A. However, when
A mounts C, it is usually A who grooms C. Episodes involving A's mounting
B have a different structure from episodes where A mounts C, when grooming
behavior is taken into account. At other times, and especially when food

sources are spatially restricted so as to crowd the monkeys, A regularly supplants B, but A and C supplant each other equally often, though rarely. If a study of many such groups confirms that in any pair of female rhesus monkeys where the mounter is then groomed by the mountee the mounter is dominant in terms of supplanting ability, while in pairs where it is the mounter who then does the grooming the supplanting relationship between the two is uncertain, we can suggest that the "structure" mount-then-groom-first has a different "meaning" in terms of its correlation with the status relationship of the pair than the "structure" mount-then-be-groomed-first. This kind of approach begins by looking closely at the behavior of named individuals in known social groups, and it searches for patterns connecting their various social actions, without imposing categories like "sex" or "aggression." Although it starts by scrutinizing episodes involving individuals closely, it aims at generalizations applicable to the species being studied, such as the foregoing generalization about the relationship between individuals where the mounter is then first to groom.

This example presents the individuals involved and the information about their mounting, grooming, and supplanting behavior as if those "facts" were easily distinguishable in the stream of events which constitutes their interaction. In real life, many other social actions in addition to grooming, mounting, and supplanting may have preceded and followed the pair of events, mounting and grooming, which was said to constitute the "structure" of the social episode. Other members of the group may also be interacting with, and milling around, the pair of animals we select. It follows that to suggest that particular combinations of particular kinds of social action by named individuals constitute episodes with "structure" relevant to those individuals at other times and in other contexts is a pointless exercise, unless methods for "recognizing" such "structures" in the stream of behavior can be suggested. In our observations of behavioral episodes involving the animals living in social groups which we know well, we may believe ourselves able to place the events of a particular episode in the wider context of the animals' current and preceding situations. Yet those observers who do *not* know the species and this particular social group may find the particular episode as confusing and inconsequential as a dream, especially if they prevent themselves from forcing anthropocentric, "functional," or quasi-causal and physiological interpretations on the events they see. To be newly placed amid a group of chimpanzees can be to experience a marvellous waking dream, until we regain our grasp of our habitual methods for containing such confusion.

Studies of social behavior commonly start to contain the confusion of real life by distinguishing approaches concerned with causation from approaches concerned with function (e.g., Tinbergen's 1969 introduction). I suggest that such a distinction can hinder studies of social interaction. For in any social exchange, an action by one animal may, by definition of social exchange, have an effect on the second animal, including the occurrence of further social actions by that animal. By being too eager to discover causes (e.g., who started it) or functions (i.e., effects with presumed survival value for the social group being considered), we can distract attention from the patterns of events involving both participants in the exchange and occupying considerable spans of time. A "causal" study may focus too exclusively on each successive move in a prolonged social encounter and fail to see patterns in the whole, while a "functional" study may be concerned too exclusively with the functional consequences of the encounter and so miss the details of its moves. When we study the behavior of freely interacting animals, we should perhaps be less eager for quick answers in terms of causes and effects.

Much of this chapter is concerned with the methods for, and possibilities of, discovering patterns in prolonged social encounters, including the virtually life-long social "encounters" of group-living species. As a convenient shorthand in the argument, the phrases "reliably follows," "reliably precedes," "consequence," and "outcome" will be used freely. To say that an action B in a reacting animal "reliably follows" a display S by the animal's partner is to mean that, in a study of the occurrence of actions A, B, C to R in the reactor, within a specified time after the occurrence of the displays and actions S, T, U through to Z in the partner, the sequence partner gives display S and reactor does B is the most frequent one. To say that a fight, for example, was an "outcome" of an event or pattern of events is to emphasize the correlation between the occurrence of the fight and the events in question which occurred before the fight erupted. Cullen (1972) has discussed how behavioral sequences can be controlled in artificially simplified situations, and Slater (this volume) describes some methods currently available for analyzing behavioral sequences.

In this chapter, I develop a point of view which is especially concerned with those aspects of social behavior that can make it so excitingly confusing. This standpoint combines my approaches in two studies. The first was an examination of the interactions of pairs of domesticated Siamese fighting fish (*Betta splendens*) whose members were initially strange to each other. My purpose was to discover by what patterns of events in the display interaction involving the two could an observer predict the outcome, in terms of which fish ceased to display first (and was thus thereafter the subordinate one of the pair) (Simpson, 1968, 1970). The second approach was to the grooming and other interactions of free-living male chimpanzees, to discover the kinds of relationships, in terms of such activities as grooming, supplanting, and displaying, that each male had with his fellows (Simpson, 1973). The first approach was concerned with simple relationships (dominance–subordination) arising out of a complex-seeming interaction, while the second was concerned with complex relationships, as I found them. This complex pattern of chimpanzee relationships posed questions about the processes by which it arose. This chapter considers methods for studying such processes.

III. OBSERVING SOCIAL BEHAVIOR

In this section, I pretend to forget the literature, my field notes, and my event records, and I try to return to those moments when I was so absorbed in what I was watching that I forgot that I was writing notes, or pressing keys. In order to maintain some contact with my readers, I shall define a "display" as a conspicuous (to the observer) action made by an animal, whose conspicuousness led me to believe that he could, through the display, be having some effect on his companions. The merits of this definition are, first, that it leads us to ask how we may show that the display affects the animal's social companions and, second, that it reminds us how a definition of "display" can reflect the attitudes and dominant sensory modalities of the human observer, with his emphasis on what he sees and hears. I discuss the more commonly used definition of "display," with its reference to the presumed adaptive value of displays, later in this section.

My experience of social behavior leads me to make the following claims:

1. Most of the events in animal social interactions seem inconsequential, especially when consequences of presumed survival value are sought.
2. Many of the more conspicuous social actions, including those called "displays," come as surprises to the observer, and after most such actions the behavior of the displayer and his companions can also be surprising.
3. Nevertheless, one somehow becomes able to predict much of the social behavior of particular individuals in their groups, and the prediction becomes quite precise.

The relations among these three claims are somewhat paradoxical. This is especially the case for (2) and (3), which seem to contradict each other. The second claim refers, however, to what in many species are relatively rare items of social behavior. The third claim need not contradict the first, for one can become able to say what an animal will do next without being able to say how one knows or being able to describe the events in terms of function and cause.

A. Social Events Often Seem To Be Inconsequential

If events seem inconsequential, it may be because the observer is expecting particular consequences, and it may also be because he is expecting them soon. Zoologists who study free-living animals often have the possible consequences of actions by those animals in mind. They are especially concerned with consequences they believe to have survival value or to be the functions of particular actions. In the resulting descriptions of behavior, the cases where dramatic "displays" have been easily connected to consequences of obvious survival value have been overrepresented. Such series of obviously connected events leading to "consummatory" events (see discussion in Marler and Hamilton, 1966) such as mating are often contrasted with the more miscellaneous category which includes "appetitive" behavior patterns, which are less clearly connected to particular consummatory situations. Examples are described by Marler and Hamilton (1966).

Also relatively neglected in the study of social behavior have been the prolonged and apparently inconsequential play sessions of many mammal species (see Loizos, 1966, 1967; Ewer, 1968; Lumia, 1972) and the prolonged primate grooming sessions. Inconsequential-seeming sequences studied in some detail include the elaborate communal displays of certain duck species. (See Cullen, 1972; McKinney, 1961; Weidmann and Darley, 1971. Weidmann and Darley were able to tease out some of the social correlates of displays by males and females in complex social groups of mallards.) Further examples are the repeated and elaborate displays of birds already "bonded," such as the "triumph ceremonies" of greylag geese (described by Lorenz, 1966; also in Hinde, 1970), and many of the repeated displays of great crested grebes (Huxley, 1914, 1968); the elaborate duets of such birds as the laniarius shrikes (Hooker and Hooker, 1969; Thorpe and North, 1965; Thorpe, 1972) are more examples of social actions whose immediate consequences are not obvious.

To define an action as a "display" is often to imply something special about the consequences it is presumed to have: that the action is adapted to having particular social effects on fellow animals. Moynihan (1955) defined "displays" as "those peculiarly standardised and often exaggerated performances, including all vocalisations and many movements and postures, which have become specialised and modified as social signals and releasers." In the study of any species, it is an empirical matter to discover which movements have which effects on the behavior of conspecifics, as it is to discover whether all such effective social actions are "peculiarly standardized and exaggerated," and to discover how reliably they "release" behavior patterns in the animal's fellows. Work on primates (e.g., Menzel, 1971; and see Kummer, 1971, for a general discussion) suggests that to focus on peculiarly standardized movements is to miss many signals also effective in social interaction.

To classify a display as a "courtship display" or an "agonistic display" is to make rather particular implications about the display's effects. To suggest that such displays are to be found in animal interactions is to imply that, in the study of a particular species for the first time, some of its displays may eventually be fitted into "courtship" or "agonistic" categories of behavior, or even a mixture of both (see Section VI, Parts C and D). Such implications are understandable if the species' actions are being viewed in terms of their presumed survival value and if one is concerned with classifying the actions in functional terms. Thus a new animal species may have a number of conspicuous actions, and some of these actions may be peculiar and unique to it (so that they can be used as means for identifying the species; e.g., Heinroth, 1911). In contrast to these special and peculiar actions, there will be many others which will be easier to recognize, because we shall already have seen them in the many species which we have already studied. Such actions include approaches, departures, precipitate flights, chases, acts which actually cause damage or seem likely to do so to in the opinion of the human observer, copulations or spawnings, and the actions by which nests, eggs, and young ones are protected and fended for. Some of these actions are obviously of direct survival value, such as mating, while it can be all too easy to argue the survival value of some others, such as acts of violence which have as a consequence more stable social and spatial organization of individuals around limited resources (see the discussions by Eisenberg *et al.*, 1972; Crook, 1970; Crook and Gartlan, 1966; Wynne Edwards, 1972; Hinde, 1972c).

At this stage in the study of our new species, some of its peculiar displays may be seen to be quite closely linked in time and sequence to the more obviously functional actions. For example, the only time when a female *Betta splendens* nudges the male's flank with her snout is shortly before the pair rolls into the clasping spawning position, where the male's body is bent over the upside down female so that the eggs emerge close to his anal fin and urogenital aperture, there, presumably, to be fertilized (see Simpson, 1970; Kühme, 1963). An adult male chimpanzee's acts of staring at a particular female, erecting his hair and conspicuous penis, and twitching an adjacent piece of vegetation in rather a controlled and precise way often precede the female's approach to him and his mating with her (van Lawick-Goodall, 1968; McGinnis in preparation). The vegetation-twitching display is almost unique to this situation.

Very many conspicuous animal actions are not reliably and easily linked to particular outcomes. Thus male chimpanzees erect their penises in many exciting social and feeding situations, and in most primate species both sexes often mount members of their own sex in situations apparently unconnected with ejaculation (for examples, see Bertrand, 1968). In *Betta splendens,* most of the display actions performed by a male when with a female are also found when he is with a stranger of his own sex, and the same actions occur when two females, initially strange to each other, meet (Simpson, 1968). Presumably, it is the combination of the component actions which determines what the outcome of the interaction will be.

Clearly, we could list many displays which cannot easily be fitted into any simple functional scheme. In *Betta splendens,* it becomes clear whether a pair of fish is going to fight or spawn only when the participants relax their brilliant fins and cease their exaggerated performance, the more effectively to bite and evade each other's bites or the better to coordinate their clasping for spawning.

To relegate actions, not reliably linked in time and sequence to obviously functional consummatory outcomes, to a class which we call "appetitive" will not help to solve the problems of their separate causes and effects during the appetitive stages (Marler and Hamilton, 1966). Sequences of actions called "play" pose the same kinds of problems if they are approached from "functional" viewpoints, and it is also difficult to unravel the separate causes and effects of the component actions. In mammals, "playful" fights and "real" fights can be remarkably similar (e.g., Poole, 1966, 1967), and the courtship of many species, including man and domestic dogs, has many "playful" qualities, as well as some special displays (e.g., Eibl-Eibesfeldt, 1972).

Loizos' (1966, 1967) discussion of play makes it clear that play sequences raise many problems that are also raised by the display sequences of birds and fishes. In both, observers recognize a degree of ritualization (Loizos, 1966; see also Lumia, 1972): thus both show "behavior that borrows or adopts patterns that appear in other contexts where they achieve immediate and obvious ends." A rhesus monkey play bout may include some quite hard hits and also bites, mounts, and thrusts. Tinbergen (1959) has pointed to the similarities between the threat posture of a displaying herring gull and the positions adopted as the gull drives home a real attack. In both cases, the gull is raised up on its legs, so that the ankles show, the wing carpals are slightly apart from the body, and the bill is pointed downward. In both display and play sequences, the elements appear to be ordered differently than their orders when used in functional contexts. A displaying *Betta splendens,* even when not traveling through the water, moves its paired pelvic fins in ways also seen in a nondisplaying fish as it accelerates from a stationary position (pelvic fins laid back) and brakes to a halt (pelvic fins extended). When the displaying fish turns to face its opponent, it folds back both pelvic fins to lie alongside its belly, and when it turns broadside again it always erects the pelvic fin nearest to the partner, and it may also erect the other. In addition, the broadside fish may flicker the offside pelvic fin to and fro (Simpson, 1968, 1970). In both display and play, many movements are exaggerated. Thus the bouncy walk of a rhesus juvenile about to play may be compared with the fully spread fins and zigzagging course of a male *Betta splendens* soon after a ripe female has appeared.

Both play and display sequences can tempt observers to think in terms of "moves" and "games" (see also Kalmus' 1969 discussion). Such a viewpoint suggests that a move by one partner can affect the other's next action and that there is some system whereby symmetry is maintained during the encounter. Thus displaying *Betta splendens* of the same sex face other in turn, and a pair of kid goats may take turns to displace each other from a precarious perch on a fallen palm trunk. A "game" viewpoint also suggests that the series of moves leads to some outcome for the relationship between the players, or displayers, and that the outcome is somehow linked to that history of moves. Analyses by Oehlert (1958), Rasa (1969), and Simpson (1968) of the relationships between events in fish display encounters and outcomes of those encounters are examples.

This "gamelike" view of behavior is discussed further below (Section VIB). Other people's games often seem infuriatingly inconsequential to the observer. The fact that sequences of behavior can seem inconsequential should not in itself be taken as evidence for their gamelike nature. Even a successful analysis of such a sequence, using a "games" point of view, may not be adequate evidence.

B. Conspicuous Actions, and the Interactions That Follow Them, Can Surprise the Observer

In his observation of the individuals in an established social group, an observer is not always able to predict when a particular individual will perform a display. A robin may burst into song apparently spontaneously, and so may a chimpanzee perform one of his charging displays.

Displays seem especially unpredictable when the group contains rapidly developing individuals; there must be a first time for a baby primate's temper tantrum or for the full display by a late adolescent chimpanzee in the presence of several prime male fellows. The occurrence of many displays seems to depend on factors in addition to the immediate social context. Thus time lapsed since the performance of the last display may be important (e.g., Rasa, 1971a), and apparently irrelevant environmental changes may trigger displays; a chimpanzee may start a chorus of pant hoots when a large cloud covers the sun, and a gust of wind in the trees may set off a male rhesus monkey's branch-shaking display. (Of course, there are also reliably predictable displays, such as the first social events that occur when strangers of many vertebrate species meet; see Lorenz, 1966, for examples, and Lack, 1943, for an early analysis.)

Once a display has begun, the subsequent behavior of those involved can continue to offer surprises. This statement could be rephrased in a more challenging form as "The more conspicuous a social action, the less predictable are its consequences." When two individuals, initially strangers to each other, are put into the same container, it is very predictable that some form of display will ensue. But the first few minutes of display may give little hint about what the interaction is going to lead to. In the Siamese fighting fish, all the display actions seen in encounters between fishes of the *same* sex may also be seen in encounters between opposite-sexed pairs. Encounters involving opposite-sexed pairs may lead to a spawning on the same day, the next day, or not at all, and whether there is spawning or not, there may or may not be biting attacks on the female by the male and periods during which the female keeps out of the male's sight. Occasionally, encounters between opposite-sexed pairs end in fights, and very occasionally encounters between females lead to "spawning" behavior. If the fish are like-sexed, the encounter usually ends when one of the two turns pale, relaxes its fins, and begins to avoid the other, who is thereafter able to supplant it. This outcome is predictable only as the encounter approaches its end (Simpson, 1968). Until then, the encounter seems "equal and mutual" (Braddock and Braddock, 1958) or a "Frage- und Antwortspiel zweier aufeinander abgestimmter Subjeckte" (Lissmann, 1933).

The observation that many conspicuous displays lead to events which are difficult for us to predict has been accommodated by the ethological theories of conflict reviewed by Hinde (1970), and discussed in Section VI, Parts D and E. Thus, after a particular display, the behavior of the displaying individual and of his fellows in terms of the probability of attack, flight, and mating may be difficult to predict: all may be likely. To be fair to the theory, it accounts for the fact that displays may be followed by two or three of the above nondisplay possibilities. The theory also relates the occurrence of a particular kind of display to the situation the displayer finds himself in and to the situation's presumed attack-, flight-, and mating-provoking qualities. Moreover, the theory suggests that particular kinds of display may be correlated with particular subsequent relative frequencies of attacking, fleeing, and mating in the displayer. Thus a gull which is very likely both to attack and to flee may give one kind of display; one which is rather unlikely to either attack or flee, a second kind of display; and one which is very likely to attack and rather unlikely to flee, a third kind of display. Moynihan (e.g., 1956, 1958) summarizes interpretations of gull displays according to this kind of model, and Grant (1963) similarly interprets some rat displays.

An assertion that the more conspicuous social actions lead to less predictable outcomes can be fitted into a survival value (functional) explanation as follows, if:

1. In some species, individuals perform certain displays more frequently, and with less provocation, as they mature.
2. Such displays are likely to stir all individuals nearby into social action.
3. Such "stirred social action" may express, for the first time for the observer, aspects of the current social situation.

The "stirred social action" may also reflect aspects new to the social group, as when a rapidly maturing male chimpanzee shows himself able to display in the presence of several other males. In general, low-status males, in terms of the number of prime males they supplant, display *less* frequently when other prime males are present than when they are not present, while high-status males display even more frequently in the presence of fully adult males (Simpson, 1973). The individual's display may also precipitate further interaction which may have as an outcome new social relationships "surprising" to participants and observer alike. The young male chimpanzee may find that his display leads others to display in parallel with him, or he may be thoroughly beaten up for his boldness, or the others may seem to ignore him. Such different immediate outcomes may have different long-term effects on his subsequent social behavior.

In colloquial terms, a situation where individuals frequently display, with apparently little provocation, is a situation where the prevailing social relations in the group are continually being "tested." Lissmann's (1933) comment above, about the question and answer aspect to *Betta splendens* interaction, has the same implication: the strangers are "testing each other out" to discover whether their relationship shall be one of courting or rivalry and, if of rivalry, who shall be subordinate.

It is easy to write functional "explanations" of behavior and easy to propose hypotheses about interactions in colloquial terms. The balance of this chapter will be concerned with establishing my hypothesis about social relationships as outcomes of interactions. Meanwhile, it should be emphasized that the occurrence of dramatic displays by particular individuals may lead to changes in the group's social relationships, irrespective of how those displays are set off, whether by particular androgen levels, a thunderclap, or a threat from another individual. An individual need not be "using" his displays as deliberately planned social exploration. Motivational explanations of displays are discussed further in Section V, Parts D and E.

C. Individuals Become More Predictable in Their Characteristic Ways of Behaving in Their Social Groups

The predictions which an ethologist believes he can make about those individuals he knows intimately may be set against some quantitative descriptions of aspects of social interaction, which could be caricatured with such an example as "After the 36 occasions when a display of type X was given, the displayer approached his partner on 25 of the occasions, stayed still four times, and moved away seven times. On those occasions, the *partner* approached the displayer on three of them, remained where he was on 29, and moved away on four. From this, it may be concluded that the animal who gives display X is likely both to approach and to leave his partner, but more likely to approach, while in the presence of an animal who gives display X, the partner is equally unlikely either to approach or to leave." (For real examples, see Stokes, 1962; Chalmers, 1968.) Such quantitative descriptions have the value of leading to explicit classifications of actions, such as display X, in terms of what they enable an observer, and perhaps those conspecific animals who see display X, to predict about the displayer. The descriptions also provide observers with the means to classify actions in terms of their effects on the partners' subsequent behavior. But someone who knows the individuals who give and see display X might believe himself able to predict, on *each single one* of the 36 occasions when display X is given, exactly what subsequent actions are to follow.

Such an ethologist could, with reason, object to the arbitrariness involved in defining all the terms necessary before the analysis can proceed. What is to count as a leave and what as an approach? How long must neither occur after display X before the animal is classed as "staying still"? To what extent is display X really a lumped category, including several different-looking displays? Surely it matters which individual gives display X and which sees the display? For the person who knows the animals well, the quantitative approach can seem to ignore much that is relevant to his prediction, on any one actual occasion, of what will ensue in the two individuals after the display has been given.

Such objections could be made about many of the summary descriptions of behavior sequences available in the literature, which can make the most disappointing reading. Thus Simpson's (1968) account of those display activities of opposite-sexed *Betta splendens* which sometimes led to spawning used 50 verbs to describe what usually occurred in the sequences. These verbs were qualified with a total of 22 modifying phrases, including one "not necessarily," one "especially likely," four "usually"'s, and 11 "may"'s. He was attempting, in words, to give an impression of the relative frequencies with which events followed each other. Many descriptions use diagrams for this purpose. Examples are reviewed in Delius (1969), and include Morris' (1958) diagram for idealized and actual courtship sequences in the ten-spined stickleback (*Pygosteus pungitius*), Neil's (1964) summary of courtship responses in a male cichlid fish, and others shown in Marler and Hamilton (1966). Liley (1966) describes what courting guppies, including *Poecilia* =

Lebistes reticulatus, usually do, both with diagrams and in words. His ratio of qualifying phrases to verbs is 32:62. It is perhaps significant that when Simpson used words to describe interactions between like-sexed *Betta splendens,* which he understood better than the courting interactions of opposite-sexed pairs, he used one particular real case as an illustrative example. Marler and Hamilton's (1966) discussion of social interactions also includes single real examples, namely, Larsson's (1956) descriptions of sequences of male rat sexual behavior.

Disappointment arising out of such quantitative descriptions implies expectation. In this case, the expectation is that when particular individual guppies or fighting fish are interacting, each succeeding act can be predicted more precisely than a final quantitative or pseudoquantitative picture implies.

It is argued below that quantitative studies are necessary, but it must be emphasized that to end a study at the general level of description referred to above can be to foreclose study of more intricate processes. Moreover, current quantitative frameworks cannot accommodate many patterns of events that one can "see" in interacting individuals.

For example, one morning, when one group of the captive Cambridge rhesus monkeys was beginning to move about on the ground after their rest which follows the feed, the infants began approach–withdrawal play with each other, and *Bee's three-month infant,* Kerry, *ran against* the dominant adult female, *Morag.* Morag waved her arm at him and, from a sitting position, lunged her head and trunk at him rather desultorily. Kerry squeaked, and ran toward his mother, squeaking again. Bee ran toward Kerry, past him, and away from him, toward Morag. *Bee presented to Morag,* then approached Kerry, who made contact with her.

One pattern of events in this single anecdote which seems meaningful is that presented by the sentence "Bee's three-month infant ran against Morag, and Bee presented to Morag." In the anecdote, the relevant phrases have been italicized. It should be clear that they refer to events separated by other events and separated by time. Most existing quantitative methods of analysis have been applied only to adjacent events occurring within defined time intervals.

The foregoing discussion could be summarized under such a tag as "meaningful anecdote *vs.* quantitative description." Analogous conflicts are common in psychiatry, where it is often concluded that the more measurable something is, the less meaningful it seems to be when a particular patient's history is being considered (Pond, personal communication), and where "statistical" and "clinical" methods of prediction may be contrasted (Meehl, 1954). Thus one kind of psychiatrist sees an intricate pattern connecting apparently separate events which include a patient's statements about his fantasies now, a dream of the previous night, an incident 15 years ago, and the silences between his statements about these events. For another psychiatrist, the concentration of adrenocorticotropic hormones in the blood and the frequency with which the patient uses affect-laden words may be important information. The prognosis by one may be as accurate as that by the other, and in real life each may be responding to material which is the special province of the other: the first to the patient's smell, which may well correlate with his hormonal state, the second to the content of the patient's comments about the bloodletting necessary for the ACTH measurements.

Each side surely has much to offer the other: quantitative descriptions should be made within a meaningful context, and methods for discovering meaning and structure in a series of events need to be made more explicit, whether that series of events is drawn from 20 sec worth of activity in a penful of rhesus monkeys or 50 min worth of talks by a patient on an analyst's couch. In this chapter, methods by which such patterns may be discovered in animal interactions are described. In the field of analytical psychology, Seaborne Jones (1968) has reviewed the kinds of "pattern" that are the concern of analysts and the prospects of devising methods whereby hypotheses about such patterns may be made more explicit and tested.

Summarizing my claims about animal interactions being both inconsequential and potentially predictable if the individuals become known, it should be emphasized that many sequences can be both predictable and inconsequential, because there are no immediate consequences or because we fail to recognize the relevant consequences. If, in the rhesus monkey example above, we had expected Kerry's mother to run to him and gather him up when he squeaked, we would have been surprised, for first she ran past him and away from him. On other occasions, when Kerry's older sister attempted to lean against Kerry and Bee, Bee hit the older sister, then groomed her, then seemed to ignore her, and then hit her again, before leaving her.

IV. SOCIAL BEHAVIOR OF INDIVIDUAL ANIMALS

I have suggested that social behavior seems inconsequential because its consequences may be separated from it by time and by other actions. When certain conspicuous social actions lead to new social relationships involving an individual, then social behavior may appear surprising as well as inconsequential. It could, however, be the very simple case that many social interactions *really are* inconsequential and confused, in addition to being confusing to human observers. If, in certain interactions of a certain species, null hypotheses about the randomness of sequences of social events are never disproved, while comparable null hypotheses lead to orderly pictures in other species, then we have some evidence for randomness in the first species. But it would always remain possible that, had we tried some other as yet unthought of null hypothesis, we would have found order. It should also be noted that techniques based on information theory, reviewed by Cullen (1972), can be used to measure the amount of order in a sequence of events.

It must almost always be the case that the many separable components of elaborate display postures and sequences act in a merely cumulative way on certain aspects of the partner's CNS and hormonal state, so that there are general levels of behavioral and physiological description where we can talk of the partner as being aroused, or excited, or perhaps sexually aroused. It would be easy, for example, to describe the courtship of a male chimpanzee from this point of view. (For full descriptions see van Lawick-Goodall, 1968; McGinnis, in preparation.) If a glance and an erect penis in the male do not bring the female to him, then he may raise his hair and twitch vegetation, and I have seen a young male circle the female, swinging from side to side as he faces her and waving a detached leafy twig as he swings from the other arm. It is easy to regard the progressive extension in space of the male's display, from his face and penis, to his enhanced silhouette (hair erect), to adjacent branches, and perhaps ultimately through a whole tree around the female, as contributing to her "sexual arousal." It would also be easy to explain the many display components seen in courting ducks as having evolved to keep up the female's interest. Fentress (1968 and this volume) reviews the usefulness and limitations of such general arousal interpretations.

Meanwhile, I believe that the task of looking for the patterns in which social events in behavioral sequences occur is worth pursuing. At the present stage in research, my strongest grounds for this belief are esthetic and intuitive, arising out of my experience with a so-called lower vertebrate, *Betta splendens*. But, as is apparent from Slater (this volume), methods for the study of complex sequences of behavior are developing rapidly, and with the help of computers it could very soon be possible to test hypotheses about quite complex and time-extended sequences (see also Fentress, in press).

Such an assertion of belief can be constructive if it leads to explicit procedures for describing and making generalizations about social interactions. I suggest that we should focus on the social behavior of *individuals* in social groups in which the members interact freely with each other. Of our individuals, we should ask how their *relationships* with their fellows change through interaction with them. Such studies could add a "process" dimension to our generalizations about particular species' social behavior. Many of our present descriptions are rather "static": take, for example, the statements "in the species *Pan troglodytes*, an 'affinitive' expression, the 'vertical bared-teeth' display, appears most frequently in the oldest member of an interacting (often meeting) pair" (van Hooff, 1972); "in the free-living group of this species in the Gombe Stream National Park, some individual males behaved consistently in terms of whom they would supplant, and to whom they would give way" (Simpson, 1973). "Dynamic" process generalizations would be about how, in this species, individuals came to behave toward each other as if their relative ages mattered, or what social interactions led to males' becoming able to supplant more of their fellows and what social interactions led to males' coming to supplant fewer. In chimpanzees, one kind of process may be involved, perhaps involving the histories of displays in the individuals concerned; in rhesus monkeys, different kinds of processes may be involved, perhaps involving more physical violence.

Of course, there is corresponding scope for studies of social "process" from the physiological point of view. Thus Rose *et al.* (1971) show that blood testosterone concentrations of individual male rhesus monkeys are correlated with those monkeys' places in a stable dominance hierarchy, the top eight males in the hierarchy having a higher mean level than the remaining 26. Further studies following how blood testosterone levels change in individuals whose places in the hierarchy change will reveal much about the physiological processes involved.

Brockway's experiments (reviewed in Brockway, 1969) show that a particular class of male budgerigar (*Melopsittacus undulatus*) vocalization, the "soft warble," greatly stimulates ovarian activity and egg-laying in females with breeding experience caged in continual darkness without nest-boxes. "Loud warbling" in the male is stimulated by hearing others' loud warbling, and Brockway suggests (1969) that the performance of loud warbling in the male may be the source of stimuli which promote the performer's testicular activity.

To return to my proposal for studying social process: an *individual* in a social group is classified and defined in terms of his *relationships* with others in the group, and *relationships* are defined in terms of social behavior. The members of a shoal of fish have particular ways of behaving toward their fellows (e.g., Cullen *et al.*, 1965), and so do the members of a herd of domestic cattle (e.g., Schein and Fohrman, 1955). In the former case, however, all members may behave alike to all others; in the latter case, they behave differently in terms of the order in which they gain access to a limited food supply and in terms of whom they butt. Only in the latter case may we call the component members *individuals*. Thus it is by consistently different relationships that individuals in a social group are distinguished. From this definition, it should be clear that group members can have different kinds of individuality in different contexts. Thus, in terms of the suckling relationship between a particular cow and calf, all other calves and cows may be equivalent, equally to be respectively avoided and pushed away. During a violent thunderstorm, all cows in the herd may behave alike and temporarily cease to show any individuality.

This chapter is concerned with the problem of how relationships emerge in groups of individuals. Such groups include pairs of individuals, such as a male and female *Betta splendens* displaying in the first minute of their encounter, at the end of their first hour when the male has just chased and bitten the female and she is quietly hiding in his territory, when they are spawning after 8 hr, and half a minute after the last spawning when the male is violently chasing the female out of his territory.

When a pair of female domesticated *Betta splendens* meets for the first time, they begin to display at each other almost at once, and for the first few minutes of display they are not behaviorally distinguishable as individuals, as we have seen. (An observer can "distinguish" the two, because they come in a variety of colors and patterns.) After perhaps 8 min, a clear behavioral distinction emerges quite suddenly; one has become paler, relaxed her fins, and is avoiding the other. Thereafter, the two no longer display at each other for long, and their supplanting relationship remains: the first will give way to the other, who may occasionally briefly chase the first.

Have differences between the two emerged *during* the display, at some stage before its dramatic end? Examination, minute by minute, of a *population* of encounters between different pairs shows that in most respects the displays of the two members in each pair remain remarkably similar until within 2 min of the end of the encounter. Indeed, up to this point, within-pair differences are smaller than between-pair differences. Thus, within a pair, the two "match" their gill cover raising behavior in terms of the proportions of the successive 2-min periods spent with their gill covers erect. (By raising their gill covers, and by extending the black brachiostegal membranes which lie behind them, a fish dramatically increases its frontal area; see Simpson, 1968, 1970). Moreover, through the encounter, the proportion of time spent with gill covers erect increases in *both* partners. Toward the end, however, this proportion ceases to increase in the loser-to-be. The correlation between this within-display differentiation of relationship and the subsequent outcome suggests that study of this aspect of the display interaction may throw more light on the process by which the relationship emerges.[1]

In the study of animal behavior, we are seldom so fortunate as to find such clearly focused interactions leading so abruptly to such definite outcomes. For example, in established primate groups, the usual way for a

[1]I was able to observe a few encounters between wild-caught male *Betta splendens,* and these seemed very similar to the female–female encounters described here. Encounters between domesticated male *Betta splendens* are too violent for such a study.

stranger to arrive is as a newborn infant. As the infants begin to leave their mothers and interact with the others in their group, it gradually becomes possible for the observer to distinguish them as individuals, through their social behavior. Some of the distinguishing aspects of individuals' behavior may arise through their interaction. Thus one particular male baby rhesus monkey may come to end a greater proportion of his play sessions by retreating than does his peer, perhaps because his peer is heavier, perhaps because his peer's mother threatens him when he plays roughly. The individual may come to distribute his social behavior toward the others in his social group in his own way; he may discover that by sitting quietly next to the largest adult male he may avoid rough play overtures from the others, while his peer may sit close to his (dominant) mother in such circumstances.

To take an example of social differentiation proceeding over an even longer time scale, early prime male chimpanzees come to perform displays more frequently (sustained brisk runs, hair erect, hitting and slapping of ground and/or vegetation and/or resonating objects, including own chest, often with pant-hooting vocalization: full descriptions in van Lawick-Goodall, 1968, 1971, and van Hooff, 1971, and a description of the comparable gorilla display in Schaller, 1965). The males also become able to supplant more of their fellows. It is likely (van Lawick-Goodall, 1968, 1971) that their histories of display interaction and their abilities to supplant those others are connected.

But we must squarely face the problem of discovering which are the most relevant events leading to the changes in the relationships by which individuals become differentiated. If social happenings can have "meaning" to individuals, even when those individuals' behavior does not change all at once, this problem becomes very difficult. For example, a particular chimpanzee Y, who was originally regularly supplanted by another male Z, may, after a period of 2 weeks in which Y performed five charging displays in Z's presence, four of which caused Z to move aside a few centimeters, become able to supplant Z. Clearly, we cannot argue much from a single case. We need several similar histories whereby individuals became able to supplant others who had previously supplanted them. Only then can we make such generalizations as "a reversal in supplanting relationship between two individual chimpanzees is always preceded by a history where at least four of the previous five displays led to the original supplanter's flinching away from the supplanter-to-be."

It is easy to make methods of behavioral analysis seem easy, using imaginary examples. In real life, a fortnight which included five displays by one male chimpanzee Y in the presence of another particular male would also have included several displays by the other male, displays by Y when male Z was not there, several prolonged grooming sessions, many of them involving groups of at least three individuals, several foraging journeys, and various complex greeting and reassurance interactions as Z's temporary parties met other parties.

In short, Y could have been *involved in* a complex history of social interaction, and the example has suggested that, so far as Y's ability to supplant Z went, the *relevant* or *meaningful* part of that *history* was five of Y's displays and Z's response to them.

In operational terms, a "relevant" or "meaningful" history is that whose pattern is most strongly correlated with the outcome, for which the history is presumed to be relevant. Operationally, given such an endpoint as a change in two individuals' supplanting relationship, we ideally start with a kind of null hypothesis where all preceding events and combinations thereof are equally relevant to the outcome of current interest. We then set out to discover that history of pattern of events *most strongly correlated* with the outcome. This process can be repeated for any kind of outcome. Thus Y's grooming relationship with X may change, so that Y comes to initiate more sessions with X. The relevant history may include prior grooming experience, and it could also include prior experience of each other's displays (see the discussion in Reynolds and Luscombe, 1969; Simpson, 1973).

The *relevant* or *meaningful* history for a particular outcome is distinguished, by the observer, from the more *total history of events in which the participants are involved.* This relevant history, for each participant, can also be referred to as that participant's *experience* of the social events in which he was involved. It is a matter for empirical study, in any one species, to discover how many kinds of relationship are formed by its individuals (discussion in Simpson, 1973). How many aspects of the social events in which an individual is involved are "experience" relevant to each of the various relationships is then a matter for further study.

At this stage in the discussion, and at any initial stage in the study of a particular species, the term "involved in" must be used in the widest possible sense. Thus social events in which an individual is involved could include the following:

1. Events initiated by the individual, as when Y displays near Z and Z moves away.
2. The individual's behavior, however initiated, as when Y's display seems to be spontaneous.
3. Events observed by an individual but not obviously involving his immediate participation, as when, in Y's presence, W displays close to Z.

Such "latent" effects can be demonstrated. Thus the Hayes' chimpanzee

Vicki imitated the act of applying lipstick after she had observed it in her human caretaker (Hayes and Hayes, 1952; Hayes and Thompson, 1953), and Hall's patas infant came to avoid a basket which had startled its mother, even though the infant had not seen the snake in the basket (Hall and Goswell, 1964).

The distinction between *being involved in* and *having experience of*, as applied to an individual in a social group whose members are free to interact with each other, is analogous to a distinction made by von Uexküll (1934). In considering the behavior of the members of a particular species in its environment, he made a distinction between that environment, as it might be described by a broad-minded ecologist, and those aspects of the environment which actually relate to the species' behavior. These latter aspects together constitute the species' "Umwelt." An important part of the Umwelt of a hungry mated female tick waiting on the end of a grass blade for a passing mammal is butyric acid vapor; in the presence of this, she is especially likely to let go of her grass blade. An analogous distinction in a species' social sphere is made explicit in the conduct of experimental releaser studies (examples in Tinbergen, 1951). For example, a herring gull chick is involved, by the experimenters, in a series of events, some of which turn out to be more effective than others in directing the orientation of the chick's beak and thus come to be regarded as better "releasers" for this aspect of the chick's interactions with its parents (Tinbergen and Perdeck, 1950).

The difference between the "releaser" discovered in a series of experimental trials and the inference of "events experienced" from events in free-running social interactions involving individuals is a difference in complexity. Thus, for the fighting fish who emerges from a display interaction as a subordinate, the relevant "events experienced" may be a difference between the duration of his gill cover erections and those of his partner. It should be noted that an experiment which modeled this effect by using a suitable puppet would hardly be said to "manipulate" or "control" the fish's behavior in the usual sense. For to control a puppet so that it provides gill cover erections of a particular duration relative to those of the fish being studied demands that the controller of the puppet should respond suitably to the live fish's gill cover erections.[2]

The next two sections will be concerned with the relevance of the kinds of study outlined above for the study of social structure in groups of individuals and with the study of the ontogeny and differentiation of social relationships in members of particular species. The final section will discuss practical research strategies suggested by the general kind of research program proposed above.

[2]Methods for making such a puppet are described by Simpson (1968).

V. SOCIAL STRUCTURES IN GROUPS OF INDIVIDUALS

The previous section included a detailed description of an isolated process of social change involving a certain species of domesticated fish. When we study established social groups, the relationships we find between the individuals in them are presumably the outcomes of processes of social interaction and change. This section starts by considering how we, as observers, discover and classify these relationships.

Consistent individual differences in social behavior provide starting points. But how do we get any further? Surely we shall find consistent individual differences in almost any measure of social behavior we choose to take. Thus Simpson's (1973) study of the social relationships among 11 male chimpanzees revealed that individuals behaved consistently in regard to whom they supplanted, how frequently they displayed, in whose presence they displayed most frequently, how frequently they groomed, whether each individual usually groomed for a longer or shorter time than each other individual, how frequently they presented for grooming, and even how frequently they clapped their teeth during a grooming session. Where should we start?

In view of the preceding discussion about consequences, it is clearly mistaken to start with preconceived ideas about the survival value or roles of adaptive significance (e.g., Gartlan, 1968) of these individual differentiations. For our new species may be facing problems of adaptation that we have not met before. In an animal study, to say that any activity is adaptive, or of survival value, or that the characteristic activity of an individual is his role is to put forward one hypothesis about the consequences of that activity. (The concept of "role" has a special use in human studies, as Brown, 1965, explains.)

Having discovered that individual chimpanzees are consistently distinguishable from their fellows, both in the *overall* frequency with which they perform various social actions *in the group as a whole* and in how they *distribute* those social actions among *their particular fellow members*, we may suspect that some of these separately described frequencies and distributions are correlated. Thus those who display frequently in social groups may also be frequently involved in social grooming (e.g., Reynolds and Luscombe, 1969), and those who display often in the presence of particular fellow members may often become involved in grooming sessions with those fellows (e.g., Simpson, 1973).

In short, in Simpson's (1973) study displaying companions were also likely to be grooming companions at a statistical level of significance corresponding to a Spearman rank-order correlation coefficient of $+0.60$ for the 11 individuals.

A deeper study, concerned with other activities, could reveal that individual chimpanzees distributed and shared other aspects of social behavior differently among their fellows. Companions to initiate play sessions with and companions to travel with could well be different from companions with whom much or little grooming or displaying was done, and correlations between distributions of individuals' traveling (say) among their fellows could be independent of their distribution of grooming and displaying. In short, a chimpanzee could have different kinds of relationships, in terms of companions, depending on whether he was traveling or grooming. The idea that an individual can distribute his social activity differently among his fellows, according to the activity of the moment, is entertained by Lorenz' (1935) "Kumpan" concept.

Through studies which examine correlations between distributions, by the males in this case, of their different social activities among their fellows, we can discover how many kinds of social relation they could have, where we distinguish "social relations" according to the statistical independence of the distributions found. In the study where grooming, displaying, and traveling are examined, there are two distinguishable relationships: a traveling relationship and a displaying–grooming relationship.

But the evidence connecting grooming and displaying involves a not-very-high statistical correlation coefficient: the generalization about their connection is very general. In the event, although an individual does tend to groom more with higher-status fellows (in terms of displaying and supplanting), different individuals have different favorite grooming partners. Such differences suggest more detailed studies; perhaps other factors determine choice of grooming partner in addition to displaying and supplanting behavior. Thus, other things being equal, males may groom longer, and more frequently-per-opportunity, with partners whom they see less often (Simpson, 1973), and van Lawick-Goodall (1968) has suggested that playmates may become preferred grooming partners as the individuals grow up.

I have suggested that even a pair of individuals can be treated as a group. My example, taken from the interaction of opposite-sexed *Betta splendens*, suggests that some social relationships can be transitory: the male-biting and female-hiding stage in *Betta* or the male dorsal pricking and female waiting in the three-spined stickleback (*Gasterosteus aculeatus*) (Wilz, 1970) can last only minutes. Such an example reminds us that whether we treat a social situation as a static entity or as a dynamic process is a matter for convenience and level of analysis. The facts that these stages differ in whether and when they occur in different pairs of fish and that courtship sessions involving different pairs of fish have different subsequent outcomes suggest that they could be important in a study of the whole courtship interaction.

A view of a current state of affairs in a social group as an outcome of processes, including processes of social interaction, will not be new to social anthropologists. An example is Barth's (1966) discussions of the kinds of models that could be used to explain particular "forms" in social life. Barth pointed out that much effort in social anthropology had been concentrated on the necessary step of constructing static models describing "forms" in social life, where "forms" were defined as series of regularities in a large body of individual items of behavior, such as the regularities in chimpanzee behavior already referred to. The next stage, Barth suggested, should be to make models which were not static descriptions of the current state of affairs ("the observed social regularities of forms") but models designed so that they could, by specified operations, generate such regularities. In the example where a certain pair of chimpanzees' supplanting relationship reversed, the "specified operations" refer to means of correlating particular aspects of past display history with present status relationship. In a free-running social system, "specified correlations" might be a better term; several correlations are hypothesized, and analysis shows which is the strongest. In a social system open to manipulation, such as a human family which seeks psychiatric help, particular operations can be performed, such as the application of a particular reinforcement schedule. But the effectiveness of such manipulations in producing social change need not imply that corresponding processes operate when such social changes occur without intervention (see Section VI F).

The term "specified operations" has implications which may be illustrated with references to zoologists' "explanations" of behavioral states of affairs in such terms as their functions for maintaining pair bonds, or "social cohesion." Thus a description of an elaborate system of song duetting in pairs of shrikes may be explained in terms of "pair bonding" (e.g., Thorpe, 1972c) and the occurrence of social grooming in primate groups in terms of "social cohesion" (e.g., Crook, 1970). Such explanations may be regarded as prescriptions for longitudinal study: in an area containing a number of *Laniarius aethiops* shrikes, what aspects of duetting interaction lead to what kinds of social relationship? In the next section, principles likely to be revealed in, and relevant to, the study of the ontogeny and differentiation of social relationships are discussed.

VI. DIFFERENTIATION OF RELATIONSHIPS BETWEEN INDIVIDUALS

A. Introduction

Studies of the events involving two or more participants in a prolonged social exchange can reveal that some histories or patterns of events are better correlated with an outcome, such as a particular supplanting relationship between the two, than are other histories. We may infer that the best-correlated histories we observe are integrated by one or both participants' CNS, and we may suggest that correlated histories constitute the relevant experience for the participants. In this section, I return to the theme of Section III and review some methods available for studying the differentiation of social relationships between members of particular species. First I consider some of the problems arising out of a close examination of prolonged social interactions, such as those between Siamese fighting fish or human games players. Then the participants are examined in more detail, as separable actors and reactors. This separation is easy to make when encounters are short. When they are longer, and clearly involve several steps, each step may be considered separately, with one animal considered as an actor and the other as a reactor. A particularly clearly analyzed example is described by Schein and Hale (1965), and many others are reviewed by Tinbergen (1951) and Hinde (1970).

When we think of an animal as a "reactor," we may be tempted to focus on the effects of his stimulus situation, including what his partner has just done. When we think of him as an "actor," we may focus more on his internal state. In prolonged encounters, where an animal may be "reacting" not only to what his partner last did but also to his partner's penultimate move, and perhaps also to the relation between his own previous move and his partner's last move, this distinction between current stimulus situation and internal state may not be a helpful one. Have the events of the previous 10 sec, which may include the partner's responses to the animal's own previous moves and the partner's response to the move before that, become part of the animal's "internal state," or may the last 10 sec be regarded as part of the "current stimulus situation"? Can we describe such "internal states" in the motivational terms currently available? In the second part of this section, I use Smith's (1963, 1965) distinction between the *meaning* of an actor's display *for a reactor* (or *recipient*, in Smith's discussions) and the *message* of that display *about the actor* as a starting point for developing a distinction between interpretations of display in terms of meaning and interpretations in terms of motivation. Motivation approaches are briefly reviewed, and the value of the attention they draw toward temporal patterns in behavior is remarked.

Finally, the application of reinforcement theories to the differentiation of social relationships between individuals is discussed.

B. Participating in Encounters and Games

I introduce this discussion by making further reference to the 8 or 10 min during which two female domesticated *Betta splendens*, who have not seen each other before, establish a dominance–subordination relationship. It has been suggested that one fish's experience of a *difference* between the duration of her own gill cover erections and those of her partner can lead the one with the shorter durations to end the interaction by ceasing to display (above, and Simpson, 1968). How is one fish able to perceive that her gill cover erections are shorter than those of her partner? Neither has a stop-watch.

The two do, however, share a temporally regular pattern of interaction. Thus each individual alternates a bout with her gill covers erect with a bout with them down, and during the period when they are down a fish is often broadside to her partner. Moreover, when one fish is broadside, with her gill covers lowered, the other may be facing, with her gill covers erect. And the two fish tend to face each other in turn.

When her partner is facing her, with her gill covers erect, the broadside fish, in the course of her broadside bout, will often first flicker her offside pelvic fin to and fro, then beat her tail one or more times, and then briefly close and open the fin rays in her tail, so that the tail briefly closes like a fan.

The facts that the broadside fish shows an organized behavioral sequence during her broadside bout and that this sequence will be shown in the presence of a stationary puppet suggest that there is a corresponding temporally organized pattern in her CNS.

Such a pattern could provide the time base by which each fish could assess her partner's gill cover erections' durations; an erection that persists after the third tail beat may be perceived as different from one that ends after the first.

Simpson (1968) suggested that each fish could modify its own gill cover duration, when its turn to raise its gill cover came, according to the duration of its partner's bout. Through the first few minutes of the display, the durations of both fishes' gill cover erections increase in parallel, but eventually one of the fish reaches her limit. Thereafter, the gill cover duration of one of the two continues to increase, and the other may either begin a biting attack or give up after 2 or 3 min more.

The explanation of the means by which the loser-to-be perceives that it is time to give up was cast in S–R–S terms. Each fish's behavior at any stage in the display depended on events including the "stimulus" partner-raising-her-gill-covers, the fish's own "response" fish-beating-her-tail-while-partner-has-her-gill-covers-erect, and the "stimulus" partner's-response-by-lowering-her-gill-covers to the fish's tail-beating behavior.

It should now be emphasized that the terms "stimulus" and "response" have unfortunate connotations for this account; they imply that the events described might follow in a chainlike manner. But the fish may have beaten her tail anyway. What is important to her is the timing of the partner's act of gill cover raising and lowering relative to her own temporally patterned display including its tail beats. I suggest that an individual's current behavior depends on a past temporal pattern that includes the relationship between her partner's gill raising, her own broadside display including its tail beats, and her partner's gill lowering. In this explanation, the fish is seen as currently acting on a simultaneous or "atemporal" representation of the temporal sequence of the preceding events. Such arguments have been elaborated by Lashley (1951) and Bruner *et al.* (1966).

The fishes' shared ritual of "taking turns" to face and raise their gill covers provides a kind of "grammar" or "game" which allows differences between them to emerge, to be perceived by the participants as well as the observer with his stopwatch. The connotations of the terms "grammar" and "game" are worth examining (see also Kalmus, 1969), although it is all too easy to overinterpret the behavioral regularities described in such terms. In this context, the terms are being used to direct attention to the interaction as a series of actions between two participants which is such that differences between them can emerge, to be appreciated by at least one of the participants—the-loser-to-be in the present example. This distinction between a ritual series of actions and differences that arise within the ritual is analogous to the distinction made by Watzlawick *et al.* (1967) when they describe every communication as having a "relationship" and a "content" aspect. In many animal interactions, the "relationship" aspect may be sufficiently defined by the fact that there are only the two individuals together (see also Gregory Bateson, 1969). But those individuals can also define their relationship further, as for instance when a monkey prefaces the act of hitting another with a "play-face" (see Blurton-Jones, 1967, and Altmann, 1967*b*, for further examples of acts which have been classified as meta-communication).

Human grammar differs from the gamelike fish interaction in many ways. The number of potential moves to be made within the temporal structures of grammar is much greater and perhaps unlimited (see McNeill, 1966). Thus there are more verbs which a human being can insert into the blank space in "The fish ——— her partner" than there are distinguishable actions that one fish can do to her partner in the first 3 sec after her partner raised her gill covers. The point of this particular comparison is to emphasize that whether the second fish does not beat her tail at all, does so once, does so three times, or bites the first fish is *not* determined *only* by the preceding actions of the two: it is an aspect of the second fish, which is to be revealed in the context of the display. A bee's "honey dance" as a context within which different messages, although with a limited number of referents, can be transmitted is an analogous case to the fish example, and different from the grammar example, because those referents are limited to information about direction and quantity of nectar source (see also Hockett, 1960; Altmann, 1967c).[3]

In this context of "games," we should remember the similarities between some prolonged animal displays and some mammal play sessions. Perhaps primate play sessions in which individuals establish relative statuses (if they do in the course of such sessions) differ only from *Betta splendens* display sessions in the number of possible moves available to each participant at each turn.

The point of view so far offered, that an individual's relationship with his fellows can be the outcome of his experience of social events involving him and them, is a very general one. Thus the term "involving" has the widest possible implications: the individual can merely be present when another member of his social group does something. In the study of animal behavior, hypotheses which include reference to the effects of experience have been less general. Some of these hypotheses will be considered with reference to the "meaning" of a social action to an individual who is present when it occurs, drive theories of display, goal theories of display, and reinforcement theories of display. The obvious role of imprinting is finally discussed briefly.

C. Meaning, Reacting, and Acting

Cognitive words and phrases, including "experience," "perception of a difference between own and partner's display," "atemporal representation of a temporal sequence," and "meaning," appear frequently in this chapter. This section discusses a definition of "meaning."

Observers of primate social interaction have long been aware (e.g., Marler, 1965; Altmann, 1967b; Bertrand, 1968) of the way in which the effects of a particular social action can depend on its context, which includes the identities of the performer and the receiver of that social action. Indeed,

[3]Attempts have been made by human beings to communicate with their pet chimpanzees, notably Premack's (1971) with Sarah and the Gardners' (Gardner and Gardner, 1969) with Washoe. Complex systems (especially Premack's token system) and large vocabularies have been mastered by the chimpanzees, but neither appears to have told her owners much that would be new or surprising to an observer who knew the situations in which the animals were working and who was able to interpret the "nonverbal" behavior of chimpanzees. It will be interesting to see whether chimpanzees can use their new systems of signs to tell stories about past, future, or imaginary experiences. Perhaps the Hayes' chimpanzee Vicki approached this point: she could "run" herself on picture series and make the movements cued by the pictures (Hayes and Hayes, 1952; Hayes and Thompson, 1953). Such a situation could be used to discover whether particular "meaningful" orders of pictures were preferred and whether a chimpanzee also taught sign language could tell herself similar stories without the pictures. Washoe sometimes indulged in "monologs" where she made her signs when she was alone.

whether one monkey responds to a threat by another may often depend not only on the identities of those two but also on the availability of a third for the support of one of the two and where the third is looking at the time (e.g., discussion by Kummer, 1967, of protected threat).

To take another primate example, many apparently arousing situations can lead a rhesus monkey to raise its tail vertically (see also Andrew, 1972), and a 5-week-old baby rhesus who has begun to make his first excursions away from his mother is likely to raise his tail when he begins moving, whether he is away from or near his mother when he does so. When he is near her, tail raising could lead us, and perhaps his mother, to expect him to leave her, and when he is already away from her we might expect him to approach her or to move still further away. In all three cases, it is of obvious functional advantage if the act of tail raising leads his mother to attend to him: the moment may not be right for him to leave her, and once he is out of her immediate reach, he may be about to approach her because he has been startled, or he may be about to wander still further away. The corresponding appropriate actions by the mother may be, respectively, to hold onto his tail, to run to meet him, or to move to a place where she can keep him in sight.

This way of looking at the effects of a signal is developed by Smith (1963, 1965; and see 1970), and Smith's work is also discussed by Hinde (1972b) and Jolly (1972). Smith discusses the effects of a signal in terms of its *meaning to a recipient*. "Meaning" is defined as "the response elicited by the signal in the recipient from all the responses available to it." If an observer has information about the recipient's repertoire of responses, he should then be able to discover the meanings of signals in particular contexts.

With reference to the rhesus baby described above, so far as the *mother's* response goes the raised tail "means" different things in the different contexts: grab erect tail if baby is within reach and, if conditions are not safe away from her side, watch him especially carefully if he is out of reach.

Smith derives his method for defining "meaning" from Cherry (1957), and the same method is implied in Altmann's (1967b) definition of communication as occurring when "an act by one animal changes the probability distribution of the subsequent acts by another." To describe possible and observed responses in terms of their "probability distributions" is merely to take account of the fact that most responses occur sometimes in most situations; rather few lie dormant until elicited by some key signal. This situation does not alter the principles implied by the definition of "meaning"; it only recognizes that the process of discovering this meaning may be difficult in real life.

Smith distinguishes "meaning" from "message" which is in some way descriptive of some aspect(s) of the state of the CNS of the individual giving the signal. Thus the observer watching the baby rhesus monkey may infer the "message" "likely to move" or "excitement" from the erect tail on one particular occasion in an individual baby, because such characteristics are common to all rhesus monkeys with erect tails.

Smith's approach thus makes a distinction not always made in the classical "ethological" approaches to displays, whose interpretations of animal displays have emphasized motivational descriptions of the CNS states of the animals involved in display episodes. In the rhesus baby example above, the distinction is easy to make, because the baby's excitement at being about to move away from his mother's side is so clearly different in affective value from the mother's prompt and cool tail grabbing. The distinction is more difficult, and more important, when two animals are involved in some encounter that is more prolonged and more mutual, as when they are fighting or courting each other. Then it is easy to slip into a "message" or "motivational" way of thinking: the increase in the "aggressiveness" of one animal's display is increasing the other's "aggression and/or fear." Such language can detract attention from what the animals are actually doing with each other.

In the discussion of the display of two female *Betta splendens*, which introduced this section, it was suggested that, for one fish, her partner's act of lowering her gill covers after the fish's own first tail beat could "mean" something different from her partner's act of lowering her gill covers after her own third tail beat. In this example, context, so far as the partner's act of gill cover lowering goes, depends on temporal and sequential relationship to the fish's own tail-beating behavior, which provides a reference point against which certain acts by the partner may be late or early. If (1) further analysis showed that the partners of losers-to-be always persisted in their gill cover erections until after the losers-to-be had beaten their tails for the third time, while the losers-to-be always lowered their own gill covers after their partners' first tail beat, and (2) if this difference (like the already observed difference in gill cover erection duration) appeared sometime before the loser-to-be finally ceased to display, then we would have a case where an interaction at one time had "meaning" for a social relationship emerging at another time.

In short, patterns of events which lead to succeeding correlated social changes after an intervening delay can be assigned meaning operationally in the same way as can events which lead immediately to correlated social changes. The technical difficulties involved in working with delayed effects are obviously greater, and are considered further.

D. Being Driven and Striving: Drive Theories of Display

In the preceding section, and also when the disappointing nature of published behavioral descriptions for those interested in the behavior of *individuals* was pointed out, I was concerned with processes leading to unique responses in particular situations and processes leading to unique social outcomes in particular histories of interaction.

In this section, I discuss descriptions and explanations which refer to those behavioral changes that are reversible (see also Fentress, in press and this volume; McFarland, 1971). Thus a rhesus monkey's "excitement" could be said to increase and decrease, and this statement could, and should, summarize a whole group of correlated behavioral and physiological changes. The same is implied for a rat as it gets more thirsty, and is then allowed to drink, and is no longer thirsty for a while (e.g., Miller, 1957). Ideally, motivational hypotheses should (in my opinion) refer to measurements and manipulations that are actually or potentially possible to make on the animal's CNS.

In the study of animal behavior, statements true for many occasions in the lives of many different members of the species are often likely to seem rather general. For example, study of a number of occasions when an animal gives a particular kind of display may reveal that, in the succeeding 5 sec, the displayer is both likely to approach and likely to flee from its partner, but more likely to flee. This finding can be "explained": the display characterizes any member of the species who "has" the particular balance of approach and flight tendencies suggested by the observed approach and flight frequencies. Such descriptions and explanations are reviewed by Hinde (1970).

However, before we can accept this "drive" story, we must know more about the study of our displaying individual. If approach and leaving frequencies were the *only* correlates of the display that were looked for, we should perhaps not be very interested in the result. For in space as we know it, individuals can only approach and leave their fellows (if for the moment we exclude the more complex temporal patterns of behavior we call chasing, following, leading, and avoiding). Animals being integrated systems, there will be *some* measurable correlation between every pair of measures of their activity that we choose (see discussion in Meehl, 1967; Lykken, 1968), and we have only to take large samples and control our conditions to make those correlations statistically significant.

Obviously, we must examine *many* possible correlates of the display and discover which are the strongest. Wiepkema's (1961) factor analysis of the courtship behavior of the bitterling (*Rhodeus amarus*) reveals that the acts of approaching and fleeing fall into separate clusters of social gestures; groups of display action thus reflect approaching and fleeing tendencies in this species. Van Hooff's (1971) cluster analyses of chimpanzee social behavior, however, show that approaching and leaving do not clearly fall among separate clusters of social gestures. This should not be surprising when it is remembered that chimpanzees are rather socially cohesive primates. There

are indeed occasions, as in male courtship, when the act most likely to precipitate attack on the female by the male is her refusal to approach him (van
Lawick-Goodall, 1968; McGinnis, personal communication).

It can obviously be a useful first step in the analysis of a species' display
repertoire to discover which displays are best correlated with approaches in
its members, which with flight, and which with staying; examples of analyses
along such lines include those of Stokes (1962), Chalmers (1968), Dunham
(1966), and Dunham et al. (1968). But it could be a mistake to end analysis
at such a point, for displays could affect the displayers' fellows in other ways.
Very often, a display will be followed by another *display* in the partner, so
that the interaction will be inconsequential as far as approach and leaving
behavior goes. Moreover, in some interactions, such as approach–withdrawal
play sessions in mammals and the prolonged courtships of *Betta splendens,*
both participants may make *repeated* approaches and withdrawals, so that
the moment-by-moment changes in their approach and fleeing behavior
become too complex to be accommodated by simple hypotheses in terms of
the individuals' prevailing readiness to approach and leave.

Drive hypotheses of displays as responses to the integration in the
individual of the approach-provoking and the flight-provoking aspects of
the situation in which he finds himself suggest that, so long as different kinds
of approach-provoking and flight-provoking situations lead to the same
provocation to approach and flee, however measured, the displays should be
the same. Blurton-Jones (1968) has put this hypothesis to rigorous test and
found that, in fact, certain great tit displays are affected by the nature of the
situation provoking the approach and preventing that approach (results also
summarized in Hinde, 1970).

Andrew (1972) points out that the ethologists' classical emphasis on
aspects of internal state presumed to reflect individuals' readinesses to attack,
flee, and mate has distracted attention from many other aspects, including
some which may be physiologically measurable, such as autonomic states
corresponding to a mammal's or bird's being cold or being excited. Other
aspects of the internal state are less measurable by physiological methods:
noticing "stimulus contrast" (Andrew, 1964) or engaging in a protective response (Andrew, 1963). The last interpretation of display can be strikingly
confirmed by the form of the movements: a young rhesus monkey wiping a
sunflower seed husk off its face makes a very similar face to a threatening
monkey. We have also seen how similar display movements can be to those
of overt attack. The moral to be drawn from this situation is perhaps that,
given any new species, we should be as imaginative as we can about the
forms of its display movements (see also Section VII C): they may give us
interesting clues about motivation, or they may be totally misleading.

Motivational theories of display can be valuable when they lead to parallel physiological studies of CNS correlates, although the temptation to describe behavior in terms of response frequencies and thus to "scale" behavior and make it comparable to, say, a hormone concentration could distract attention from sequential and temporal patterns.

Of course, there are motivational theories which account for (e.g., Rasa, 1971a,b; Wilz, 1970) and classify (e.g., McFarland, 1969, 1971) sequential patterns. The study of such patterns often demands that the situation of the animal be kept as constant as possible, and the influence of moment-by-moment social changes on them thus may not be considered. Indeed, it is possible that it is the function of such patterns to be independent of moment-by-moment social change; I have suggested that the alternation of facing and broadside bouts in displaying *Betta splendens* could be independent of moment-by-moment changes in the individuals' behavior so as to provide a relatively stable context within which differences between the two may emerge (Simpson, 1968).

Temporal patterns of behavior, on the 10-sec-by-10-sec and minute-by-minute time scales, have been rather neglected (but cf. Fentress, 1968; McFarland, 1971; Sevenster, 1968; and, of course, those concerned with schedules of reinforcement, such as Ferster and Skinner, 1957; Kelleher *et al.*, 1959). Whether we can at this stage suggest biological functions for them or not, they may provide important constraints on patterns of social interaction and learning (Sevenster, 1968).

For example, a 4-month-old rhesus infant made a series of excursions from his mother's side, and the durations of these excursions as they occurred were 14.3, 14.9, 30.4, 58.0, 27.8, and 8.5 sec. To take a simpler case, the excursions of five free-living chimpanzee children, between 3 and 7 years old, from their mothers' sides were examined (Anne Simpson, personal communication). Confining attention to those excursions both started and ended by the children, over the successive minutes an individual spent away from its mother's side the probability of a return first declined. If an individual reached his fourth minute without returning to his mother, he was extremely unlikely to return in that minute. Subsequently, the individuals became progressively more likely to return.

Whatever CNS and motivational explanations we may find for such patterns, they provide helpful guidelines for further study of social processes. In a chimpanzee study, we might now pay special attention to the first 6 min of the young ones' excursions away from their mothers and discover what happens to them in the successive minutes. Individual differences could arise out of differences in timing of events relative to the individual's time of departure; a child who was often hit by another in the second minute of his excursion might well be differently affected than one who was not often hit until the fourth minute, when he was very unlikely to return.

Drive theories may be invoked when we have to explain reversible changes in behavior (e.g., Fentress, this volume; Hinde, 1970; McFarland, 1971), and so may goal theories. The latter can be more satisfying when applied to prolonged sequences of behavior with unique aspects, such as the display sequences of *Betta splendens*. Thus all display episodes are similar in that they end in a dominance–subordination relationship, although they do so by slightly different behavioral routes, and the outcomes are, of course, unique so far as the individuals are concerned.

E. Being Driven and Striving: Goal Theories of Display

When goal theories are used to explain reversible sequences of behavior, they often seem able to account for the unique courses of events that lead to the common goals. The relationship may be discerned especially easily when it is expressed spatially, as when a troop of baboons visits its waterhole by different routes on different days (Altmann and Altmann, 1970) or a dog follows a different route in searching for its ball every time the ball is thrown, with the particular route clearly adjusted to cues of scent, discovered moment-by-moment as the search progresses. They have been applied extensively by Miller *et al.* (1960) and recently discussed with reference to social behavior by Blurton-Jones (1972) and MacKay (1972).

Such theories may be useful when the course of behavior has a clear, but variable, relation to the endpoint (Hinde and Stevenson, 1969; Thorpe, 1963).

In this section, both a goal theory and a drive theory of conflict will be used to "explain" a behavioral example and to illustrate the similarity between drive and goal theories (see also McFarland, 1971).

If an observer stands near a pen full of rhesus monkeys, he will find that some individuals stay where they are, whether they were near his part of the pen or far away when he arrived. These will probably keep silent and may even look at him. Others will move away, maybe rapidly and silently. The large male may charge toward him once, possibly with a roar, probably silently. The most noisy monkeys are those between 5 months and 2 or 3 years old. They will also be the ones who approach and leave the observer's part of the wire quite frequently during the 5 min he spends there. The place where they make most noise (mainly pant-threats; see Rowell, 1962) will be their nearest point of approach to the observer, even when they do not come as near as the cage allows.

The fact that monkeys may also stay at their point of nearest approach to the observer suggests that monkeys at that point are in a situation where factors promoting both approach and leaving are simultaneously present, and balanced. In common with many other species in such situations, the frequent approaches and leavings characterize individuals most likely to threaten. Often, such measurable frequencies of approaching and leaving are described as measures of "tendencies" to approach and leave (Hinde, 1970). In this case, the tendencies are the same and balanced. In a generalization which connects threat behavior to balanced tendencies to approach and to leave, we have a redescription of the state of affairs, with the tendencies to approach and to leave being used as intervening variables. When, however, we suggest that the tendencies are in conflict, we risk classing the tendencies as entities or things, which can conflict, and our redescription can become a hypothesis about drive states.

A goal-type model could be used to bring the description to a similar point: a human observer standing close to the wire provides the younger monkeys with an interesting social object to be explored by interacting with it. The obvious way for a monkey to initiate and maintain interaction with a live object, human, monkey, or passing lamb, is to approach as close as he dares, threaten, perhaps "display" (shake wire), and then leave quickly and see what happens. The adults do not threaten, or even move, because they are more used to human observers looking into the pen. Even an adult, however, will threaten if the observer does something unusual, such as persistently attempt to engage in eye contact with it. Moreover, those who threaten in the presence of a human observer usually threaten more when the observer stares at them (Richards, 1972): an observer doing something so unusual invites further exploration.[4]

In this goal model, the two entities postulated are as follows:

1. The monkey's state of knowledge of the human observer to date, which extends to nonstaring human observers in the adults but not in the young ones.
2. What the monkey observes of the current behavior of the human observer.

If the process of comparison between (1) and (2) in the monkey reveals a discrepancy, a strategy or "plan" (Miller et al., 1960) is set into operation; in this example, the plan includes attempts at initiating interaction. This model is analogous to Hinde's (1970) model of what causes an individual to

[4]Note that a simple-minded ethological drive theory of display might predict the opposite result: the human stare, conveying the motivational message "aggression," could increase the "fear" of the monkey and decrease its threatening behavior. A more sophisticated theory would suggest that, in the threatening monkey, the aggressive tendency was at first stronger than the fear tendency, but when the human stared at the monkey the fear tendency of the threatening monkey was increased, so that it came to balance the monkey's aggressive tendency more exactly and thus to exacerbate the monkey's conflict and intensify its threat behavior.

explore more in a new environment than a familiar one and to "notice" and explore changes in a familiar environment. It is also analogous to Sokolov's (1960) "neuronal model." This latter model has been discussed by Hinde (1970), Horn (1967), and McFarland (1971). Horn (1967) proposes an alternative system which is drivelike in its lack of feedback loops and comparators, and McFarland (1971) suggests that behavioral description, which cannot yet be related to experimental manipulations of anatomically and/or pharmacologically separable parts of the CNS whose separate activities are presumed to represent (1) and (2) above, cannot let us distinguish drive-type and goal-type theories.

At a behavioral level, either or both models can be used if they seem to help the organization of research. To both, it can be objected that they bring behavior descriptions too easily to premature and relatively uninteresting ends, for both easily lead to postulations about unitary internal states to explain complex behavioral interactions.

Blurton-Jones (1972) describes the effect on calling in a nursery child of asking the teacher to ignore his calls. His impression is that the alternative strategies of gaining attention "are basically arranged on a scale related to increasing time and probably to increasing strength of whatever stimulus initiated the call." Simpson (1968) found similar effects on tail beating in one fish, when that fish was not allowed to influence a stimulus fish's gill-lowering behavior (because the stimulus fish, displaying at its reflection on the bright side of a one-way vision screen, could not see and therefore could not be affected by the tail-beating fish). Clearly, a goal-model approach directs attention to one aspect of the temporal relationship between the activities of the participants.

Its disadvantage lies in the way it restricts attention. Thus we have suggested (Section IV and Section VIB) that, *after* one bout of gill-raising and tail-beating behavior, the fish's *subsequent* behavior may vary according to the exact pattern of events: e.g., whether partner's gill lowering follows one tail beat or three tail beats. The goal-theory approach confines attention to the tail-beating *during* the partner's gill-raising bout.

F. Reinforcement Theories of Social Differentiation

Traditional reinforcement studies have been concerned with responses that are easy to record automatically, such as lever presses, and consequences that are easy to deliver automatically, such as dry pellets of food for rats (see Skinner, 1966).

In common with other approaches to learning, and in contrast to studies made within the frameworks of goal and drive theories, operant approaches are especially concerned with the ways in which an individual's own past history may be correlated with, or manipulated to produce, his present behavior. Such approaches direct attention to certain temporal contingencies of past histories: a present response frequency in a certain defined situation is correlated with the recurrence of past patterns of events, where certain consequences follow occurrences, in the defined situation, of the response in question. They usually follow soon afterward, if they follow. If they do not follow invariably, they may do so according to particular schedules (see Ferster and Skinner, 1957).

Experimental reinforcement approaches to social behavior have been reviewed by Dimond (1970). Social responses and the social consequences thereof are not easily delivered automatically, nor is it easy to devise an apparatus which "responds" only to restricted kinds of social responses. Examples of successful attempts to do this include those by Thompson (1966), Thompson and Sturm (1965), Hogan (1967), and Sevenster (1968). An automated study must perforce work at rather a "global" level: the "response" may be the act of swimming through a hoop (Thompson, 1963) or a maze (Thompson and Sturm, 1965) or to a special place (Rasa, 1971a), and the "reinforcement" may be the opportunity to see and/or to perform a display. Clearly, the "reinforcement" could contain hosts of variable further responses and consequences.

In the interpretation of any particular experimental reinforcement study, it should be remembered that it remains likely that all deliverable consequences could have *some* effect on all responses that could be allowed to produce those consequences (e.g., discussions by Premack, 1959, 1965). It thus becomes empirically interesting to discover, for each kind of consequence selected for study, what its reinforcing effect on *a range of different response* might be (cf. Meehl's 1950 discussion of "weak" and "strong" laws of effect, and see Perkins, 1968). For each response selected, does a particular consequence have an especially strong reinforcing effect, or a particularly weak effect? Thus Sevenster (1968) found that he could train a stickleback to bite a rod if the act of biting led to the presentation of another male to display to but not if the same act led to the presentation of another female. It might further be asked: if the consequence B of a response A in situation C affects the subsequent frequency of occurrence of A in situation C, does it also affect the subsequent frequency of other actions, such as R, S, and T?

Until more comprehensive studies of reinforcement have been made, the suggestion that reinforcement interpretations merely reflect available techniques for manipulation through reinforcement has some force.

For students of free-running interactions, reinforcement interpretations suggest one point of view, which can be illustrated with reference to Jensen and Bobbit's (1965) account of how, during its development, a baby pigtail macaque (*Macaca nemestrina*) learns to approach its mother on those occasions when she makes a particular face at it (the LEN face) and to approach her less often on other occasions. It was observed that the consequences of approach for a baby to a LENning mother were likely to include being allowed access to the nipple, while the consequences of approach to the mother at other times were quite likely to include being hit and bitten by her. Through such naturally occurring schedules of reinforcement, baby pigtails may learn the "meaning," in the sense developed above, of a LEN expression in their mothers. Observable differences in such free-running schedules could be correlated with subsequent differences in mother–infant relationship. To take a hypothetical example, some mothers might hit their babies sometimes even after LENning at them, and their babies might never develop the ability to discriminate the LEN face or they might lose an initial ability (cf. Sackett's 1970 and 1966 studies). Some babies might approach their mothers regardless of the mothers' hitting and LENning behavior, with subsequent consequences for these two aspects of the mothers' behavior.

Liley's (1966) interpretation of how male guppies, of the species *Poecilia = Lebistes reticulatus, P. picta, P. parae,* and *P. vivipara,* become selective in directing their courtship behavior to females of their own species was made in terms of reinforcement concepts. While females (even virgin females) are selectively responsive to males of their own species, males are more catholic, especially when naive. In a natural situation, they presumably learn to respond only to their own species, because female members of that species are most responsive to their advances. If, however, they are placed in a situation where heterospecific females are more responsive than their own species, then they come to direct more courtship to the former. If males are reared with males only, males come to prefer to direct courtship to males only; possibly even a noncourtship display in response to a male's courtship advances reinforces such courtship behavior.

The pigtail monkey and guppy fish examples are presented as descriptions of correlations between particular outcomes, such as directing courtship toward certain kinds of partner, and foregoing histories, such as the responsiveness of those classes of partner to courtship overtures by developing males. If such histories, occurring in free-running situations, comprise the events most strongly correlated with the outcome we are concerned with, then we can suggest that there is a causal relationship between history and outcome. We may then proceed to manipulate the life history of the fish, by deliberately imposing schedules of reinforcement. If we thereby succeed in controlling the adult courtship preferences of our fish, we have confirmed our hypothesis, so far as the experimental situation is concerned. However, our success at achieving such control in deliberately impoverished situations does not exclude the possibility that, in the natural situation, guppy responsiveness develops in other ways. We only know that we *can* exert such control.

This chapter has been concerned about discerning processes from correlations connecting social events at different stages in individuals' social histories. Causal language has been kept in the background or used as a shorthand. At a descriptive level, causal language may not be very helpful. For example, Skinner (1972) concludes that it is a mistake to expect to explain a social outcome, such as an indolent child, by supposing that somewhere a causal sequence has been initiated. Yet we may readily describe a rat in a Skinner box as having "come under the control of a particular schedule of reinforcement," and we may suggest that the schedule somehow causes the rat's behavior. Note, however, that the rat must press the lever before there can be any schedule. In reinforcement interpretations of *social* situations, the place of the scheduled Skinner box is taken by the animal's social companions, and he also takes the place of such a "Skinner box" for them. A reinforcement interpretation encourages us to look for temporal and sequential contingencies in freely interacting individuals' social transactions, but it leaves open the question of the internal processes contributing to the kind of "schedule" which each animal generates in interaction with his partner.

G. Social Differentiation and Imprinting

This chapter has been concerned with individual histories of interaction, where the outcomes are relationships involving particular pairs or groups of *individuals*. The question of whether, within the species, certain relationships are more readily formed with some kinds of partners than with other kinds arises. In those species for which this turns out to be the case, further developmental questions about how this situation comes about arise. Such questions are especially the concern of the imprinting literature but, of course, have also been approached in other ways. Sackett's (1970) review of rhesus monkey social development provides an example, from Pratt and Sackett's (1967) work. They gave 2-year-old monkeys the choice of three kinds of strangers to sit next to, where the strangers were classified according to their rearing conditions: having had free access to peers, or having been in separate cages but able to see and hear peers, or having been totally isolated for their first 9 months of life but shown a variety of pictures of monkeys and subsequently tested in various social conditions. Each of the 2-year-old monkeys being tested came from one of the same three kinds of rearing condition. A monkey preferred to sit near a strange partner who had been reared in the same way as he had. This example illustrates the potential subtlety of processes in social development.

VII. METHODS AVAILABLE FOR STUDYING SOCIAL DIFFERENTIATION

A. Introduction

The foregoing discussion raises certain practical problems of behavioral analysis. This section briefly discusses those problems whose solutions are becoming available. Thus computing systems help us with the forbidding tedium of most analyses. No mechanized systems, however, can yet help us to draw up our initial list of the social actions of a species new to us, when those actions must be defined according to the movement patterns involved and according to their immediate consequences. Such systems inevitably beg questions about the behavioral route by which the consequences were achieved and about the behavioral definition of the consequences.[5] Having obtained records of behavior, in terms of our initial list of actions, methods are available for discovering action sequences characteristic of the species. Once we know the usual sequences with which actions follow within and between individuals, we can discover which sequences are variable and which of those sequences allow us to characterize different individuals in their social groups. We then face the difficulty that there are many ways for characterizing an individual, and we must discover how many of these apparently separate classifications are correlated.

B. Computer-Linked Event Recording and Computer Aid

Without computer aid and the computer-linked event-recording systems that are becoming available (e.g., White, 1971; Dawkins, 1971), the program of work involved in longitudinal studies of groups with named members would be forbidding. Fortunately, systems and components are evolving so rapidly that it should soon be possible to take computer-compatible recording systems into the field. Perhaps even more important is the increase in the rate at which data can be analyzed; comparison between the rate at which I analyzed ink-and-paper event records of *Betta splendens* interactions and the rate at which I can now analyze comparably complex rhesus monkey records suggests that what was 80 hr of work by hand now takes about one-hundredth of the time by machine, and this time could easily be reduced a further hundred fold. One very important consequence of access to such help is that a particular problem for analysis can be embarked on much more lightly, for one's intuitions may be tested on the day they appear rather than a fortnight later.

[5]Hinde (1970) has spelled out the distinction between descriptions of behavior by movement pattern and descriptions by consequence.

Now that analyses can be conducted more quickly, new problems of data assimilation arise. Meehl (1967) has described those which follow from the ease with which statistically significant results may be obtained from biological data. This happens because organisms are integrated systems: many measures of an animal's behavior will be correlated to some degree with any other measure one chooses, and that correlation will be statistically significant if the sample is large enough; moreover, any "treatment" in an experimental situation will have a "significant" effect on any dependent variable chosen, provided only that the sample is large enough and that the situation is sufficiently well "controlled." Obviously, statistically significant effects and correlations are not by themselves very interesting, and they may not even become very interesting in relation to hypotheses concerned only with the existence or direction of "effects." At our present stage, I believe that we should be looking for patterns in our results, and for this to be possible we need as our starting point the results of a range of experimental treatments (cf. the suggestions in the previous section about experimental reinforcement studies) and from correlational analyses between a wide range of behavioral elements.

C. Distinguishing Social Actions in the Stream of Behavior

It is worth attempting to state explicit procedures by which social actions in individuals of a new species may be distinguished, when we do not allow ourselves to classify those actions according to their presumed consequences. The reasons for not starting from presumed consequences can be illustrated with an example. In chimpanzee social action, the act of "presenting for grooming" (see van Lawick-Goodall, 1968; van Hooff, 1971) could be defined in terms of its consequence: the partner starts grooming. However, partners start grooming males who approach without presenting, and 21 % of the acts of presenting by adult males are followed by the *presenter,* rather than the partner, being the one to make the first act of grooming (Simpson, 1973). It is possible that individuals could *differ* in the readiness with which their partners responded to their acts of presenting. To define presenting only in terms of its effects on the partner's grooming behavior could be to prevent the discovery of this possibility.

Clearly, the process of defining separate "actions" places a great burden on the observer's ability to recognize and define actions according to the patterns of movement made by the actor. An observer's success at distinguishing, for the first time, a "new" action in a species new to him requires the kind of connoisseurship which a histologist develops as he becomes able to recognize different kinds of tissue in species new to him. The development of connoisseurship implies an apprentice type of training, in parallel with more experienced observers.

It should at once be objected that if our recognition of actions depends on such ill-defined processes, then any further description of social interactions, and social situations, in terms of frequencies of those actions will be correspondingly insecure. To make this objection, however, is to confuse the following:

1. The ill-defined process by which an observer comes to "see" an action in a species new to science.
2. His subsequent description of that action, which can be made very explicit and public.

It could still be objected that our lists of actions will remain dependent on the initial ability of observers to perform the first task. At any one stage in the study of a species, we will not be discriminating as many actions as we could. It is certainly the case that, with hindsight, we can see that we have not been making all the discriminations we could have. For example, guppies (*Poecilia reticulatus*) and Siamese fighting fish (*Betta splendens*) are common enough domestic and classroom animals. Yet the act of "wheeling" in the former was "recognized" only recently (Liley, 1966), and the act of "pendeln" or "pelvic fin flickering" which occurs in almost every display has not often been described in the latter (Laudien, 1965; Simpson, 1968). In the domestic cat, the act of rubbing the lip on other creatures and objects has only recently been distinguished from the act of so rubbing the tail (Prescott, personal communication). Clearly, we must be humble about the *completeness* of the list of actions that we have at any one stage.

But although lists of actions are incomplete, observers usually agree where their lists overlap: Simpson described "pelvic fin flickering" before he saw Laudien's paper, and the occasions when observers have started their studies of the same primate species in different places have led to encouragingly similar descriptions; for example, compare Altmann's (1962) with Hinde and Rowell's (1962) and Rowell and Hinde's (1962) descriptions of rhesus monkey social repertoires.

Those setting out to distinguish "actions" in the behavior of species new to them should bear in mind such "principles" or "generalizations" about displays as are already available in the literature, including Darwin's "principle of antithesis" (Darwin, 1872; also in Hinde, 1970), where one kind of exaggerated posture leads an observer to expect an exaggeration in the opposite direction; Tinbergen's (1959) and Andrew's (1963) suggestions about the forms of display movements; the concept of ritualization (e.g., review by Blest, 1961), including its emphasis on the ways in which conspicuous structures can function to enhance signal movements; and Morris' (1957) discussion of typical intensity. As a corrective to expectations that all social actions should be clear-cut and easily distinguishable, the graded nature of mammal sounds (e.g., Rowell, 1962; see also Marler and Hamilton, 1966) and postures (e.g., illustrations in Ewer, 1968, and Hinde, 1970) should be remembered. Some of the mammalian postures and gestures may be extremely subtle: chimpanzees seem able to communicate to their fellows whether they have food, and how much, and where, before they come within sight of the food object (Menzel, 1971), and human observers have not been able to discern obvious corresponding signals in the chimpanzees.

D. Discovering Action Sequences Within and Between Individuals

At a descriptive level, the limiting case of a within-individual action sequence is the "fixed action pattern" (discussions by Barlow, 1968; Hinde, 1970): two or more distinguishable actions follow so reliably that they can be classed as one pattern. A descriptive list of actions which begins at too fine a level of detail will reveal such patterns. For example, in my initial study of *Betta splendens* display, I recorded both the fish's orientation with reference to its partner and whether or not both pelvic fins were laid back flush with the belly. However, whenever the fish turned to face its partner, it always laid its pelvic fins back, and when it turned broadside, it always raised at least the nearside one, and sometimes both. To record *both* the act of turning to face and that of laying the fins back was thus to waste a key on the event recorder. (See van Hooff, 1971, for further discussion of "splitting" and "lumping.") When the fish is broadside, it sometimes flickers one of the pelvic fins to and fro and sometimes does not. Thus it remained possible that individuals could differ in the proportion of broadside bouts which included pelvic fin flickering and in their responsiveness, by pelvic fin flickering, to the partner's different actions, and even to particular actions performed by different partners, so that it was worth continuing to record this movement.

At a descriptive level, the limiting case of a between-individual sequence is the releaser–released–response relationship. Thus in the displaying Siamese fighting fish or in the grooming chimpanzee, the respective act of flickering the pelvic fin or being first to start grooming could have been reliably "released" by the corresponding act of turning to face in the fish or presenting for grooming in the chimpanzees. In the event, neither was released reliably; a displaying fish was more likely to flicker its pelvic fin in the first few seconds after its partner had turned to face it, and a chimpanzee was first of the two to start grooming in nearly 80% of the occasions when he was presented to.

If, as seems to be the rule, within- and between-individual sequences are not invariant, we must start, at least implicitly, with the null hypothesis that, given any starting action, all possible succeeding actions are equally frequent. Studies with such starting points include those of Altmann (1965), van Hooff (1970, 1971, 1972), Wiepkema (1961), Grant (1963), and Delius (1969).

Such studies can lead to the production of enormous matrices describing what follows what. The problems they pose for statistical testing and human understanding are discussed by Slater (this volume) and referred to again below.

It is difficult for those who know their animals well to accept as worthwhile any enterprise which looks only from one social action to its successor. It is, of course, possible to work by treating groups of more than one action as "the preceding action," as Altmann (1965) has shown. But the problem with groups and sequences of actions is that they take finite periods of time to occur. Thus the action that follows a series of four in one animal may depend not only on what those four are, and their order, but also on the time lapsed since the first of the four and on any other actions that have meanwhile occurred in the animals' neighbors, as Nelson (1964), Simpson (1968), and Delius (1969) have discussed.

Simpson's (1968) methods, for showing how, through successive time intervals after some reference event, the frequencies of other events varied could provide one starting point. Against an initial null hypothesis that the occurrence of all actions is temporally independent of all reference events, it soon becomes apparent that some reference events are followed predictably by a series of actions through a series of time intervals succeeding the reference event. Thus if we know when an individual *Betta splendens* has turned broadside, we know when pelvic fin flickering may be the most frequent event (in the first $1\frac{1}{2}$-sec time interval), when (in the second interval) pelvic flickering may be less likely, when (in the third) tail beating may be most likely, and so to the fifth, when the act of turning to face may be most likely. If we know when a young chimpanzee has left his mother, we may predict when he is least likely to return to her side.

In trying out a series of reference events, we may find that some enable us to predict more events and some to predict events further ahead than may others. This is the beginning of one possible taxonomy by which groups or clusters of actions can be recognized. In the fish and chimpanzee examples mentioned above, acts of turning to face or of leaving mother's side provide obvious-seeming reference events. But it remains possible that other events will prove to be better, in terms of the economy with which they enable us to describe the regularities in the behavior. Moreover, an emphasis on single "reference" events may be misplaced: *combinations* of events occurring close together in sequence, regardless of their exact order of occurrence, may prove to be higher-order units. Fentress (in press) is actively exploring methods for discovering such higher-order units, comprised of groups of actions.

The resultant "patterns" are not fixed action patterns, but they are suggestive of patterns of neural organization (Section VIC), and they provide us with null hypotheses for further study. For both the case of the broadside

Betta splendens and the chimpanzee child away from its mother, the null hypothesis is that there is only one kind of pattern, which is relatively independent of the moment-by-moment changes in the behaving animal's situation. Against this hypothesis we may compare real cases; perhaps broadside fish differ in the order of beating their tails and flickering their pelvic fins, or perhaps they would flicker their pelvic fins less if their partner never faced them when they were broadside, as is the case when the "partner" is a mirror image. Perhaps chimpanzee young would return to their mothers in the fourth minute more often if there were no others to play with, or perhaps they would spend longer away if there were no others to chase them.

Simpson's (1968) methods for analyzing the relations between events separated by different intervals are rough and ready compared to those suggested by Delius (1969). Delius advocates the application of "power spectra" and "cross spectra" analyses, for which mathematical techniques are already available, and shows how these methods may be applied to behavioral data. The advantages of these methods include the following:

1. The fact that they take time into account, as do those of Simpson, Fentress (in press), and Nelson (1964).
2. Some neurophysiologists concerned with the behavior of neuronal networks, in terms of sequences of neuronal spikes, already use these methods, so that biologists' experience is available.
3. The methods provide frameworks for restricted predictions about behavior in free-running situations.

Delius' (1969) presentation includes a succinct review of methods available for making behavioral analyses, while Slater (this volume) covers the more recent literature on methods for analyzing sequences of behavior.

E. Variable Action Sequences and the Characterization of Relationships Between Individuals

To take a simple example of variable action sequences, chimpanzee males start grooming sessions without presenting for grooming more often than with presenting. It could thus be the case that all males present equally infrequently (and this is a kind of null hypothesis for this stage in the study) or some males may consistently present more frequently than others. In the event, there were considerable between-male differences, with Mike presenting once in every 18 grooming sessions at which he was present, and Goliath presenting once in every two of his sessions. Once a chimpanzee has presented, the partner is first to start the grooming session on nearly 80% of the occasions. We can work against the null hypothesis that all partners are equally eager to start grooming all individuals who actually present for grooming. Inspection of the scant figures (Simpson, 1973) suggests that the null hypothesis would be supported: the act of presenting among males may be equally effective, whoever the participants are.

Thus the frequency of presenting by individuals in grooming situations distinguishes these individuals, but the events that follow immediately after their acts of presenting do not. We can work through a number of action sequences and sift out those sequences which reveal consistent differences between individuals.

From this stage, analysis of individual differences will depend very much on the kind of groups that are being considered: it is relatively easy if the individuals can be compared pair by pair, or if the behavior of individuals in groups can be ranked in terms of their social behavior, regardless of whom it is directed to. Thus eleven male chimpanzees may be compared for the frequencies with which they display, or supplant other males, by a simple ranking procedure. However, more detailed study usually reveals that individuals show characteristic *distributions* of behavior among their fellows: each may, for example, have a particular male in whose presence he displays least often.

Moreover, for many measures of social behavior, it becomes necessary to compare the individuals in a complex social group pair by pair, and the data must be represented in matrix form (e.g., Simpson, 1973; Sade[6] 1973). Matrices pose two kinds of problems: that of testing for the statistical significance of the pattern of results contained therein and that of extracting a representation of the state of affairs that is both humanly assimilable and biologically meaningful. The former problem is considered by Slater (this volume).

Out of data in matrices, very evocative and "meaningful" biological pictures can be produced. Examples are Wiepkema's (1961) and van Hooff's (1971) analyses of social behavior. Thus van Hooff's cluster analysis presents a two-dimensional picture of the relationships between more than 50 behavior patterns in chimpanzees. Morgan (1973) has used three cluster analysis techniques to illustrate matrices representing the frequencies with which human subjects in noisy environments respond to spoken letters with other letters, and he shows how some groups of letters are more readily confused with each other than others. His methods are nonparametric multidimensional scaling (MDSCAL), also described by Kruskal (1964a,b); single-link cluster analysis (SLINK), also described by Johnson (1967); and the B (2) cluster analysis of Jardine and Sibson (1971). Morgan (personal communication) has applied his methods to data about the numbers of hours individual chimpanzees spend together in a particular area in the Gombe Stream

---

[6]Sade reviews the relevant sociological literature, and presents elegant approaches to information contained in matrices.

National Park, and in this case representations of the groupings of the individual chimpanzees with their fellows emerge.

VIII. CONCLUSIONS

This chapter is an attempt to share my thinking about social behavior rather than to present conclusive findings and arguments. I start with the dissatisfaction I feel when I compare social behavior as I see it with the descriptions remaining after current methods of analysis have been applied. This dissatisfaction stems partly from an inability to treat social interactions adequately as processes occupying time during which one or more participants may change progressively. This inability may partly be the result of too ready a reliance on causal and functional interpretations of interactions. I introduce this point by comparing social episodes in animal interactions with human accounts of their dreams. In so doing, I give myself the license to describe prolonged interactions as if they were inconsequential; that is, as if recognizable causes and effects, including effects presumed to have survival value, could not be assigned to every distinguishable social action when the actions were taken one by one. I emphasize the patterns that can be discerned when they are considered in groups.

In Section III, I work from my own limited experience with animals, rather than from the literature, to suggest how classifications of social actions, in terms of their presumed causes and effects, can foreclose study of the patterns connecting them. Examples are to be found in the studies of Siamese fighting fish interaction, referred to in this chapter; examples of within-individual patterns in bird song are described by Hall-Craggs (1969), while Thorpe and North (1965) and Thorpe (1972) describe patterns involving pairs and trios of birds.

Section IV discusses how an animal's current social behavior in his group could be correlated with discoverable patterns of events which had involved him and his fellows at earlier stages.

Section V pays special attention to the methods for distinguishing the different kinds of social relationship that can involve an individual with the others in his group.

In Section VI, I consider in more detail how an animal's social behavior can be correlated with patterns of events which occurred at an earlier stage, and which included an animal's own, and his fellows', actions. The first part of Section VI develops a "games" view of social interaction to suggest an alternative to the "causal" and "functional" descriptions. The "games" view distinguishes two kinds of events in an interaction: those contributing to the relationship aspect and those contributing to the content aspect (Watzlawick et al., 1967). The relationship aspect refers to the respects in which both partners remain similar in what they do. Two fighting fish may face each other in turn (Simpson, 1968), and this may insure that both have repeated opportunities to erect their gill covers and to see those of their partners. Those features common to a particular human language which provide speakers with a common framework for saying "new" things (see McNeill, 1966; and cf. Lyons, 1972) may be analogous. In the fish interaction, "content" or "news" aspects include the actual gill cover erection durations occurring within the facing-in-turn relationship of the display. In Part C of Section VI, a distinction between a "meaning" and a "message" interpretation of a social action is made. An animal's most probable social response to an action by its partner, in a particular context, may be used to define the "meaning" of the partner's action. The "meaning" interpretation refers to the relation between what the animal does and what the partner has just done in the particular context. This relationship could be extended over several previous moves by both. A "message" interpretation refers to aspects of the CNS presumed to be common to all individuals who perform a particular action. "Message" interpretations can overlap with "motivational" interpretations, and the relations between various motivational interpretations and their accompanying behavioral descriptions are discussed. Reinforcement techniques and interpretations are also discussed in this section, because they too can limit our approaches to social interaction.

Animal interactions are obviously complicated. It is all too easy to write science fiction about the patterns that might be discerned in the stream of events involving a particular animal. Section VII refers to methods which could be used for discovering whether such patterns exist, starting from viewpoints which avoid specially committed attitudes about the causes and functions of the social actions studied.

IX. ACKNOWLEDGMENTS

Any constructive suggestions emerging from this chapter stem from my interaction with those at the Sub-department of Animal Behavior, Cambridge, since 1963; at the Department of Psychiatry, The London Hospital Medical College in 1967 and 1968; and at the Gombe Stream National Park, Tanzania, in 1969 and 1970. I am grateful for my Fellowship at St. John's College, Cambridge, which enabled me to work in London and Tanzania. I wish to thank Dr. P. P. G. Bateson and Professor R. A. Hinde for their criticisms and editorial help. The responsibility for persisting errors and obscurities is mine. This chapter was written while I was on the staff of the Medical Research Council.

X. REFERENCES

Altmann, S. A. (1962). A field study of the sociobiology of rhesus monkeys, *Macaca mulatta. Ann. N.Y. Acad. Sci.* **102**:338–435.

Altmann, S. A. (1965). Sociobiology of rhesus monkeys. II. Stochastics of social communication. *J. Theoret. Biol.* **8**:490–522.

Altmann, S. A. (1967a). Sociobiology of rhesus monkeys. IV. Testing Mason's hypothesis of sex differences in affective behavior. *Behaviour* **32**:50–69.

Altmann, S. A. (1967b). The structure of primate social communication. In Altmann, S. A. (ed.), *Social Communication Among Primates*, University of Chicago Press, Chicago, 392 pp.

Altmann, S. A., and Altmann, J. (1970). *Baboon Ecology. African Field Research*, University of Chicago Press, Chicago and London, 220 pp.

Andrew, R. J. (1963). The origin and evolution of the calls and facial expressions of the primates. *Behaviour* **20**:1–109.

Andrew, R. J. (1964). Vocalization in chicks, and the concept of "stimulus contrast." *Anim. Behav.* **12**:64–76.

Andrew, R. J. (1972). The information potentially available in mammalian displays. In Hinde, R. A. (ed.), *Non-verbal Communication*, Cambridge University Press, Cambridge, 443 pp.

Barlow, G. W. (1968). Ethological units of behavior. In Ingle, D. (ed.), *The Central Nervous System and Fish Behavior*, University of Chicago Press, Chicago and London, 272 pp.

Barth, F. (1966). Models of social organisation. Royal Anthropological Institute, Occasional Paper No. 23.

Bastock, M. (1967). *Courtship: A Zoological Study*, Heinemann, London.

Bateson, G. (1969). Metalogue: What is an instinct? In Sebeok, T. A., and Ramsey, A. (eds.), *Approaches to Animal Communication*, Mouton, The Hague and Paris, 261 pp.

Bernstein, I. S. (1970). Primate status hierarchies. In Rosenblum, L. A. (ed.), *Primate Behavior: Developments in Field and Laboratory Research*, Vol. 1, Academic Press, New York, and London, 400 pp.

Bertrand, M. (1968). The behavioural repertoire of the stumptail macaque. *Bibl. Primatol.* **11** (Karger, Basel).

Blest, A. D. (1961). The concept of ritualisation. In Thorpe, W. H., and Zangwill, O. L. (eds.), *Current Problems in Animal Behaviour*, Cambridge University Press, Cambridge, England, 424 pp.

Blurton-Jones, N. G. (1967). An ethological study of some aspects of social behaviour of children in nursery school. In Morris, D. (ed), *Primate Ethology*, Weidenfeld and Nicolson, London, 374 pp.

Blurton-Jones, N. G. (1968). Observations and experiments on causation of threat displays in the great tit (*Parus major*). *Anim. Behav. Monogr.* **1**(2):75–158.

Blurton-Jones, N. G. (1972). Non-verbal communication in children. In Hinde, R. A. (ed.), *Non-verbal Communication*, Cambridge University Press, Cambridge, England, 443 pp.

Braddock, J. C., and Braddock, Z. I. (1958). Aggressive behavior among females of the Siamese fighting fish, *Betta splendens. Physiol. Zool.* **28**:152–172.

Brockway, B. F. (1969). Roles of budgerigar vocalization in the integration of breeding behaviour. In Hinde, R. A. (ed.), *Bird Vocalizations. Their Relations to Current Problems in Biology and Psychology*, Cambridge University Press, Cambridge, England, 394 pp.

Brown, R. (1965). *Social Psychology*, Free Press, New York, 785 pp.

Bruner, J. S., Olver, R. R., and Greenfield, P. M. (1966). *Studies in Cognitive Growth*, Wiley, New York and London, 343 pp.

Chalmers, N. R. (1968). The social behaviour of free-living mangabeys in Uganda. *Folia Primatol.* **8**:263–281.

Cherry, C. (1957). *On Human Communication*, Science Editions, New York, 333 pp.

Crook, J. H. (1970). Social organisation and the environment: Aspects of contemporary social ethology. *Anim. Behav.* **18**:197–209.

Crook, J. H., and Gartlan, J. S. (1966). Evolution of primate societies. *Nature* **210**:1200–1203.

Cullen, J. M. (1972). Some principles of animal communication. In Hinde, R. A. (ed.), *Non-verbal Communication,* Cambridge University Press, Cambridge, England, 443 pp.

Cullen, J. M., Shaw, E., and Baldwin, H. A. (1965). Methods for measuring the three-dimensional structure of fish schools. *Anim. Behav.* **13**:534–543.

Darwin, C. (1872). *The Expression of the Emotions in Man and the Animals,* John Murray, London.

Dawkins, R. (1971). A cheap method of recording behavioural events, for direct computer access. *Behaviour* **40**:162–173.

Delius, J. D. (1969). A stochastic analysis of the maintenance behaviour of skylarks. *Behaviour* **33**:137–178.

Dimond, S. J. (1970). *The Social Behaviour of Animals,* Batsford, London, 256 pp.

Dunham, D. W. (1966). Agonistic behaviour of captive rose-breasted grosbeaks *Pheuctitus ludovicianus* (L). *Behaviour* **27**:160–173.

Dunham, D. W., Kortmulder, K., and van Iersel, J. J. A. (1968). Threat and appeasement in *Barbus stoliczkanus* (Cyprinidae). *Behaviour* **30**:15–26.

Eibl-Eibesfeldt, I. (1972). Similarities and differences between cultures in expressive movements. In Hinde, R. A. (ed.), *Non-verbal Communication,* Cambridge University Press, Cambridge, England, 443 pp.

Eisenberg, J. F., Muckenhirn, N. A., and Rudran, R. (1972). The relation between ecology and social structure in primates. *Science* **176**:863–874.

Ewer, R. F. (1968). *Ethology of Mammals,* Logos Press, 418 pp.

Fentress, J. C. (1968). Interrupted ongoing behaviour in two species of vole (*Microtus agrestis* and *Clethrionomys britannicus*). II. Extended analysis of motivational variables underlying fleeing and grooming behaviour. *Anim. Behav.* **16**:154–167.

Fentress, J. C. Development and patterning of movement sequences in inbred mice. In Kiger, J. (ed.), *The Biology of Behavior,* Oregon State University Press, Corvallis, in press.

Ferster, C. B., and Skinner, B. F. (1957). *Schedules of Reinforcement,* Appleton-Century-Crofts, New York, 741 pp.

Fillenbaum, S., and Rapoport, A. (1971). *Structures in the Subjective Lexicon,* Academic Press, New York.

Freud, S. (1900). *The Interpretation of Dreams,* Trans. and ed. by J. Strachey, 1967, George Allen and Unwin, London, 692 pp.

Gardner, R. A., and Gardner, B. T. (1969). Teaching sign language to a chimpanzee. *Science* **165**:664–672.

Gartlan, J. S. (1968). Structure and function in primate society. *Folia Primatol.* **8**:89–120.

Grant, E. C. (1963). An analysis of the social behaviour of the male laboratory rat. *Behaviour* **21**:260–281.

Grant, E. C. (1972). Non-verbal communication in the mentally ill. In Hinde, R. A. (ed.), *Non-verbal Communication,* Cambridge University Press, Cambridge, England, 443 pp.

Hall, K. R. L., and Goswell, M. J. (1964). Aspects of social learning in captive patas monkeys. *Primates* **5**:59–70.

Hall-Craggs, J. (1969). The aesthetic content of bird song. In Hinde, R. A. (ed.), *Bird Vocalizations. Their Relation to Current Problems in Biology and Psychology,* Cambridge University Press, Cambridge, England, 394 pp.

Hayes, K. J., and Hayes, C. (1952). Imitation in a home-raised chimpanzee. *J. Comp. Physiol. Psychol.* **45**:450–459.

Hayes, K. J., and Thompson, R. (1953). Nonspatial delayed response to trial-unique stimuli in sophisticated chimpanzees. *J. Comp. Physiol. Psychol.* **46**:498–500.

Heinroth, O. (1911). Beiträge zur Biologie, namentlich Ethologie und Psychologie der Anatiden. *Verh. 5th Internat. Ornithol. Congr.,* pp. 589–702.

Hinde, R. A. (1970). *Animal Behaviour: A Synthesis of Ethology and Comparative Psychology*, McGraw-Hill, New York, 876 pp.

Hinde, R. A. (ed.) (1972a). *Non-verbal Communication*, Cambridge University Press, Cambridge, England, 443 pp.

Hinde, R. A. (1972b). Social behavior and its development in subhuman primates. Condon Lectures, Eugene, Oregon.

Hinde, R. A. (1972c). Aggression. In Pringle, J. W. S. (ed.), *Biology and the Human Sciences. The Herbert Spencer Lectures, 1970,* Clarendon Press, Oxford, 139 pp.

Hinde, R. A., and Rowell, T. E. (1962). Communication by postures and facial expressions in the rhesus monkey (*Macaca mulatta*). *Proc. Soc. Zool. Lond.* **138**:1–21.

Hinde, R. A., and Stevenson, J. G. (1969). Integration of response sequences. In Lehrman, D. S., Hinde, R. A., and Shaw, E. (eds.), *Advances in the Study of Behavior,* Vol. 2, Academic Press, New York.

Hockett, C. F. (1960). Logical considerations in the study of animal communication. In Lanyon, W. E., and Tavolga, W. N. (eds.), *Animal Sounds and Communication,* Intelligencer Printing Co., Washington D.C., 443 pp.

Hogan, J. A. (1967). Fighting and reinforcement in the Siamese fighting fish (*Betta splendens*). *J. Comp. Physiol. Psychol.* **64**:356–359.

Hooker, T., and Hooker, B. I. (1969). Duetting. In Hinde, R. A. (ed.), *Bird Vocalizations. Their Relation to Current Problems in Biology and Psychology,* Cambridge University Press, Cambridge, England, 394 pp.

Horn, G. (1967). Neuronal mechanisms of habituation. *Nature* **215**:707–711.

Huxley, J. (1914 and 1968). The courtship habits of the great crested grebe (*Podiceps cristatus*); with an addition to the theory of sexual selection. *Proc. Zool. Soc. Lond.* **35**; and (1968) *The Courtship Habits of the Great Crested Grebe,* Jonathan Cape, London, 98 pp.

Jardine, N., and Sibson, R. (1971). *Mathematical Taxonomy,* Wiley, New York.

Jensen, G. D. J., and Bobbitt, R. A. (1965). On observational methodology and preliminary studies of mother–infant interaction in monkeys. In Foss, B. M. (ed.), *Determinants of Infant Behaviour,* Vol. III, Methuen, London, and Wiley, New York, 264 pp.

Johnson, S. C. (1967). Hierarchical clustering schemes. *Psychometrika* **32**:241–254.

Jolly, A. (1972). *The Evolution of Primate Behavior,* Macmillan, New York, 397 pp.

Kalmus, H. (1969). Animal behaviour and theories of games and language. *Anim. Behav.* **17**:607–617.

Kelleher, R. T., Fry, W., and Cook, L. (1959). Inter-response time distribution as a function of differential reinforcement of temporally spaced responses. *J. Exptl. Anal. Behav.* **2**:91–106.

Kruskal, J. B. (1964a). Multidimensional scaling by optimising goodness of fit to a non metric hypothesis. *Psychometrika* **29**:1–27.

Kruskal, J. B. (1964b). Nonmetric multidimensional scaling: A numerical method. *Psychometrika* **29**:115–129.

Kühme, W. (1963). Verhaltensstudien am maulbrütenden (*Betta anabantoides* Bleeker) und am nestbauenden Kampffisch (*Betta splendens* Regan). *Z. Tierpsychol.* **18**:33–35.

Kummer, H. (1967). Tripartite relations in hamadryas baboons. In Altmann, S. A. (ed.), *Social Communication Among Primates,* University of Chicago Press, Chicago, 392 pp.

Kummer, H. (1971). *Primate Societies: Group Techniques of Ecological Adaptation.* Aldine-Atherton, Chicago, 160 pp.

Lack, D. (1939). The behaviour of the robin: I and II. *Proc. Zool. Soc. Lond.* **A109**:169–178.

Lack, D. (1943). *The Life of the Robin,* Witherby, London.

Lack, D. (1966). *Population Studies of Birds,* Oxford University Press, Oxford, 341 pp.

Larsson, K. (1956). Conditioning and sexual behaviour in the male albino rat. *Acta Psychol. Gothoburg* **1**:1–269.

Lashsley, K. S. (1951). The problem of serial order in behavior. In Jeffress, L. A. (ed.), *Cerebral Mechanisms in Behavior: The Hixon Symposium,* Wiley, New York.

Laudien, H. (1965). Untersuchungen über das Kampfverhalten der Männchen von *Betta splendens,* Regan (Anabantidae, Pisces). *Z. Wiss. Zool.* **172**:134–178.

Liley, N. R. (1966). Ethological isolating mechanisms in four sympatric species of poeciliid fishes. *Behaviour*, Suppl. 13.

Lissmann, H. W. (1933). Die Umwelt des Kampffisches (*Betta splendens* Regan). *Z. Vergl. Physiol.* **18**:65–111.

Loizos, C. (1966). Play in mammals. *Symp. Soc. Zool. Lond.* **18**:1–10.

Loizos, C. (1967). Play behaviour in higher primates: a review. In Morris, D. (ed.), *Primate Ethology*, Weidenfeld and Nicolson, London.

Lorenz, K. (1935). Der Kumpan in der Umwelt des Vogels. *J. Ornithol.* **80**(2); and Companions as factors in the bird's environment. The conspecific as the eliciting factor for social behaviour patterns. In Lorenz, K., and Martin, R. D. (trans.), *Studies in Animal and Human Behaviour*, Vol. 1, Methuen, London, 403 pp.

Lorenz, K. (1966). *On Aggression*, Trans. by M. Latzke, Methuen, London, 273 pp.

Lumia, A. R. (1972). The relationship between dominance and play behavior in the American buffalo, *Bison bison. Z. Tierpsychol.* **30**:416–419.

Lykken, D. T. (1968). Statistical significance in psychological research. *Psychol. Bull.* **70**: 151–159.

Lyons, J. (1972). Human language. In Hinde, R. A. (ed.), *Non-verbal Communication*, Cambridge University Press, Cambridge, England, 443 pp.

MacKay, D. M. (1972). Formal analysis of communicative processes. In Hinde, R. A. (ed.), *Non-verbal Communication*, Cambridge University Press, Cambridge, England, 443 pp.

Marler, P. (1965). Communication in monkeys and apes. In DeVore, I. (ed.), *Primate Behavior. Field Studies of Monkeys and Apes*, Holt, Rinehart and Winston, New York and London, 654 pp.

Marler, P., and Hamilton, W. J., III (1966). *Mechanisms of Animal Behavior*, Wiley, New York.

McFarland, D. J. (1969). Mechanisms of behavioural disinhibition. *Anim. Behav.* **17**:238–242.

McFarland, D. J. (1971). *Feedback Mechanisms in Animal Behaviour*, Academic Press, London and New York, 279 pp.

McKinney, F. (1961). An analysis of the displays of the European eider *Somateria mollissima v. nigra* Bonaparte. *Behaviour*, Suppl. 7.

McNeill, D. (1966). The creation of language. In Lyons, J., and Wales, R. J. (eds.), *Psycholinguistic Papers*, Edinburgh University Press, Edinburgh, 243 pp.

Meehl, P. E. (1950). On the circularity of the law of effect. *Psychol. Bull.* **47**:52–75.

Meehl, P. E. (1954). *Clinical Versus Statistical Prediction; a Theoretical Analysis and a Review of the Evidence*, University of Minnesota Press, Minneapolis.

Meehl, P. E. (1967). Theory-testing in psychology and physics: A methodological paradox. *Philos. Sci.* **34**:103–115.

Menzel, E. W., Jr. (1971). Communication about the environment in a group of young chimpanzees. *Folia Primatol.* **15**:220–232.

Miller, G. A., Galanter, E., and Pribram, K. H. (1960). *Plans and the Structure of Behavior*, Holt, Rinehart and Winston, New York, 226 pp.

Miller, N. E. (1957). Experiments on motivation. *Science* **126**:1271–1278.

Morgan, B. J. T. (1973). Cluster analysis of two acoustic confusion matrices. Perception and Psychophysics **13**:13–24.

Morris, D. (1957). "Typical intensity" and its relation to the problem of ritualisation. *Behaviour* **11**:156–201.

Morris, D. (1958). The reproductive behaviour of the ten-spined stickleback (*Pygosteus pungitius* L.). *Behaviour*, Suppl. 6.

Moynihan, M. (1955). Remarks on the original sources of displays. *Auk* **72**:240–246.

Moynihan, M. (1956). Notes on the behavior of some North American gulls. I. Aerial hostile behavior. *Behaviour* **10**:126–178.

Moynihan, M. (1958). Notes on the behavior of some North American gulls. II. Non-aerial hostile behavior of adults. *Behaviour* **12**:95–182.

Neil, E. H. (1964). An analysis of the colour changes and social behavior of *Tilapia mossambica. Univ. Calif. Publ. Zool.* **75**:1–58.

Nelson, K. (1964). The temporal patterning of courtship behaviour in the glandulocaudine fish (Ostariophysi, Characidae). *Behaviour* **24**:90–146.

Oehlert, B. (1958). Kampf und Paarbildung einiger Cichliden. *Z. Tierpsychol.* **15**:141–174.

Perkins, C. C. (1968). An analysis of the concept of reinforcement. *Psychol. Rev.* **75**:155–172.

Poole, T. B. (1966). Aggressive play in polecats. *Symp. Zool. Soc. Lond.* **18**:23–44.

Poole, T. B. (1967). Aspects of aggressive behaviour in polecats. *Z. Tierpsychol.* **24**:351–369.

Pratt, C. L., and Sackett, G. P. (1967). Selection of social partners as a function of peer contact during rearing. *Science* **155**:1133–1135.

Premack, D. (1959). Toward empirical behaviour laws: I. Positive reinforcement. *Psychol. Rev.* **66**:219–233.

Premack, D. (1965). Reinforcement theory. In Levine, D. (ed.), *Nebraska Symposium on Motivation,* University of Nebraska Press, Lincoln.

Premack, D. (1971). Language in chimpanzee? *Science* **172**:808–822.

Ransom, T. W., and Ransom, B. S. (1971). Adult male–infant relations among baboons (*Papio ursinus*). *Folia Primatol.* **16**:179–195.

Reynolds, V., and Luscombe, G. (1969). Chimpanzee rank order and the function of displays. In Carpenter, C. R. (ed.), *Second Conference of the International Primatological Society,* Vol. 1: *Behaviour,* Karger, Basel and New York.

Rasa, A. O. E. (1969). Territoriality and the establishment of dominance by means of visual cues in *Pomacentrus jenkinsii. Z. Tierpsychol.* **27**:825–845.

Rasa, A. O. E. (1971a). Appetence for aggression in juvenile damsel fish. *Z. Tierpsychol.,* Suppl. 7.

Rasa, A. O. E. (1971b). The causal factors and function of "yawning" in *Microspathodon chrysurus* (Pisces; Pomacentridae). *Behaviour* **39**:39–57.

Richards, S. M. (1972). Tests for behavioural characteristics in rhesus monkeys. Ph.D. thesis, Cambridge University.

Rose, R. M., Holaday, J. W., and Bernstein, I. S. (1971). Plasma testosterone, dominance rank and aggressive behaviour in male rhesus monkeys. *Nature* **231**:366–368.

Rowell, T. E. (1962). Agonistic noises of the rhesus monkey (*Macaca mulatta*). *Symp. Zool. Soc. Lond.* **8**:91–96.

Rowell, T. E., and Hinde, R. A. (1962). Vocal communication by the rhesus monkey (*Macaca mulatta*). *Proc. Zool. Soc. Lond.* **138**:279–294.

Sackett, G. P. (1966). Monkeys reared in visual isolation with pictures as visual input: Evidence of an innate releasing mechanism. *Science* **154**:1468–1472.

Sackett, G. P. (1970). Unlearned responses, differential rearing experiences, and the development of social attachments by rhesus monkeys. In Rosenblum, L. A. (ed.), *Primate Behavior. Developments in Field and Laboratory Research,* Vol. 1, Academic Press, New York and London, 400 pp.

Sade, D. S. (1972). Sociometrics of *Macaca mulatta.* I. Linkages and cliques in grooming matrices. *Folia Primatol.,* **18**:196–223.

Schaller, G. B. (1965). The behavior of the mountain gorilla. In DeVore, I. (ed.), *Primate Behavior: Field Studies of Monkeys and Apes,* Holt, Rinehart and Winston, New York and London, 654 pp.

Schein, M. W., and Fohrman, M. H. (1955). Social dominance relationships in a herd of dairy cattle. *Anim. Behav.* **3**:45–55.

Schein, M. W., and Hale, E. B. (1965). Stimuli eliciting sexual behavior. In Beach, F. A. (ed.), *Sex and Behavior,* Wiley, New York, London, Sydney, 592 pp.

Seaborne Jones, G. (1968). *Treatment or Torture. The Philosophy, Techniques and Future of Psychodynamics,* Tavistock, London, 324 pp.

Sevenster, P. (1968). Motivation and learning in sticklebacks. In Ingle, D. (ed.), *The Central Nervous System and Fish Behavior,* University of Chicago Press, Chicago and London, pp. 233–245.

Simpson, M. J. A. (1968). The display of the Siamese fighting fish, *Betta splendens. Anim. Behav. Monogr.* **1**:1–73.

Simpson, M. J. A. (1970). The Siamese fighting fish. In Boswall, J. (ed.), *Private Lives,* B. B. C., London, 160 pp.

Simpson, M. J. A. (1973). The social grooming of male chimpanzees. In Crook, J. H., and Michael, R. P. (eds.), *The Comparative Ecology and Behaviour of Primates,* Academic Press, London and New York.

Skinner, B. F. (1966). Operant behavior. In Honig, W. K. (ed.), *Operant Behavior,* Appleton-Century-Crofts, New York.

Skinner, B. F. (1972). *Beyond Freedom and Dignity,* Knopf, New York, 215 pp.

Smith, W. J. (1963). Vocal communication of information in birds. *Am. Naturalist* **97**:117–125.

Smith, W. J. (1965). Message, meaning and context in ethology. *Am. Naturalist* **99**:405–409.

Smith, W. J. (1970). Displays and message assortment in *Sayornis* species. *Behaviour* **137**:85–112.

Sokolov, E. N. (1960). Neuronal models and the orienting reflex. In Brazier, M. A. B. (ed.), *The Central Nervous System and Behavior,* Josiah Macey Junior Foundation, New York, 475 pp.

Stokes, A. W. (1962). Agonistic behaviour among blue tits at a winter feeding station. *Behaviour* **19**:118–138.

Thompson, T. I. (1963). Visual reinforcement in Siamese fighting fish. *Science* **141**:55–57.

Thompson, T. I. (1966). Operant and classically conditioned aggressive behavior in Siamese fighting fish. *Am. Zoologist* **6**:629–641.

Thompson, T. I., and Sturm, T. (1965). Visual-reinforcer color, and operant behavior in the Siamese fighting fish. *J. Exptl. Anal. Behav.* **8**:341–344.

Thorpe, W. H. (1961). *Bird-Song. The Biology of Vocal Communication and Expression in Birds,* Cambridge University Press, Cambridge, England, 143 pp.

Thorpe, W. A. (1963). *Learning and Instinct in Animals,* Methuen, London, 493 pp.

Thorpe, W. H. (1972*a*). The comparison of vocal communication in animals and man. In Hinde, R. A. (ed.), *Non-verbal Communication,* Cambridge University Press, Cambridge, England, 443 pp.

Thorpe, W. H. (1972*b*). Vocal communication in birds. In Hinde, R. A. (ed.), *Non-verbal Communication,* Cambridge University Press, Cambridge, England, 443 pp.

Thorpe, W. H. (1972*c*). Duetting and antiphonal song in birds: Its extent and significance. *Behaviour,* Suppl. 18.

Tinbergen, N. (1951). *The Study of Instinct,* Clarendon Press, Oxford, 228 pp.

Tinbergen, N. (1959). Comparative studies of the behaviour of gulls (Laridae); a progress report. *Behaviour* **15**:1–70.

Tinbergen, N. (1969). *The Study of Instinct,* 1969 reprint, Clarendon Press, Oxford, 228 pp.

Tinbergen, N., and Perdeck, A. C. (1950). On the stimulus situation releasing the begging response in the newly hatched herring gull chick (*Larus argentatus argentatus* Pont.). *Behaviour* **3**:1–39.

Tinbergen, N., Broekhuysen, G. J., Feekes, F., Houghton, J. C. W., Kruuk, H., and Szulc, E. (1962). Egg shell removal by the black-headed gull, *Larus ridibundus* L.; a behaviour component of camouflage. *Behaviour* **19**:74–117.

van Hooff, J. A. R. A. M. (1970). A component analysis of the structure of the social behaviour of a semi-captive group of chimpanzees. *Experientia* **26**:549–550.

van Hooff, J. A. R. A. M. (1971) *(Aspecten van het sociale Degrag ende Communicatie bij humane en hogere niet-humane Primaten (A Structural Analysis of the Social Behavior of a Semi-captive Group of Chimpanzees),* Bronder-Offset, Rotterdam, 188 pp.

van Hooff, J. A. R. A. M. (1972). A comparative approach to the phylogeny of laughter and smiling. In Hinde, R. A. (ed.), *Non-verbal Communication,* Cambridge University Press Cambridge, England, 443 pp.

van Lawick-Goodall, J. (1968). The behaviour of free-living chimpanzees in the Gombe Stream Reserve. *Anim. Behav. Monogr.* **1**:161–301.

van Lawick-Goodall, J. (1971). *In the Shadow of Man,* Collins, London.

von Uexküll, J. (1934 and 1957). *Streifzüge durch die Umwelten von Tieren und Menschen (A Stroll Through the Worlds of Animals and Men: A Picture Book of Invisible Worlds),* Springer-Verlag, Berlin; and (1957) *Instinctive Behavior. The Development of a Modern Concept,* Trans. and ed. by C. H. Schiller, International Universities Press, New York, 328 pp.

Watzlawick, P., Beavin, J. H., and Jackson, D. D. (1967). *Pragmatics of Human Communication. A Study of Interactional Patterns, Pathologies and Paradoxes,* Norton, New York.

Weidmann, U., and Darley, J. (1971). The role of the female in the social display of mallards. *Anim. Behav.* **19**:287–298.

White, R. E. C. (1971). Wrats: Computer compatible system for automatically recording and transcribing behavioural data. *Behaviour* **40**:135–161.

Wiepkema, P. R. (1961). An ethological analysis of the reproductive behaviour of the bitterling. *Arch. Néerl. Zool.* **14**:103–199.

Wilz, K. J. (1970). Causal and functional analysis of dorsal pricking and nest activity in the courtship of the three-spined stickleback *Gasterosteus aculeatus. Anim. Behav.* **18**:115–124.

Wynne-Edwards, V. C. (1972). Ecology and the evolution of social ethics. In Pringle, J. W. S. (ed.), *Biology and the Human Sciences. The Herbert Spencer Lectures, 1970,* Clarendon Press, Oxford, 139 pp.

Chapter 8

DOES THE HOLISTIC STUDY OF BEHAVIOR HAVE A FUTURE?[1]

Keith Nelson

California State College
Sonoma
Rohnert Park, California

What I understand by "holism" is what the Greeks called ἁρμονία. This is something exhibited not only by a lyre in tune, but by all the handiwork of craftsmen, and by all that is "put together" by art or nature. It is the "compositeness of any composite whole," and, like the cognate terms κράσις or σύνθεσις, implies a balance or attunement.

—D'ARCY THOMPSON, *On Growth and Form*

I. INTRODUCTION

D'arcy Thompson did not consider "what the Greeks called ἁρμονία" to be a proper subject for scientific investigation: "it is on another plane of thought from the physicist's that we contemplate their intrinsic harmony and perfection, and 'see that they are good'" (Thompson, 1961, p. 7). But recently the question has been reopened by the theoretical physicist David Bohm (1968*a,b*). His "remarks on the notion of order," at one of those posh conferences on theoretical biology, created a major disturbance, and may signify an incipient revolution in biological thinking. I cannot summarize what he had to say; his eloquence should be allowed to speak for itself. But

This paper is dedicated to the memory of Richard Few and Donald M. Wilson.
[1]Reprints not available.

I must state that I count myself as a Bohmist, whether or not he chooses to acknowledge his disciples.

My subject is the nature of behavioral organization. My hope is that by examining some quite complex behavior, the singing of certain thrushes, as well as the behavior of simple model systems, we may begin to understand how that may arise which we unscientifically refer to as "organization."

A. "Dimensional Complexity"

I just spoke of "complex behavior" and this raises two initial questions. The first is "How do you tell the simple from the complex?" and the second is "Assuming that the distinction can be made, why study complex behavior patterns when there appear to be so many simple ones around?" Well, how does one know that apparently complex patterns will not turn out to be simple when appropriately studied—and perhaps the apparently simple ones complex—unless one does subject them, after all, to study?

For the most part, behavioral patterns are at least "dimensionally complex"; that is, their attributes occupy too many dimensions to make their description and recording convenient. Nevertheless, we seem to have little difficulty in finding natural units among the patterns displayed by an animal. A train of behavior usually seems to divide itself into units, and we decide that these units fall into a reasonably small set of categories. In fact, most studies of "animal behavior" with which I am familiar take the categories as givens, unanalyzed, and look only to classify them according to stimulus situations which alter their development or frequency of occurrence —they are dealt with thereafter as *names* and in an atomistic fashion. Such questions as whether the patterns are indeed separable units in any objective sense and whether they have gestalt-like features are rarely asked, because asking them leads directly into the mare's nests of "dimensional complexity" and holism.

B. The Advantages of Studying Acoustic Behavior

When we record acoustic behavior such as bird song, all of the multi-dimensional complexity of muscle and membrane vibration reduces to a one-channel, two-dimensional sampling, dimensionally the absolute minimum: amplitude with respect to time of a wave train within the cone sectioned by our microphone and parabolic reflector. Compared with the visual observer of behavior, we are in a fortunate position. We need not take it upon ourselves to reduce or confound the dimensionality of the behavior, as we must ordinarily with a motion-picture record, for example. All of the confounding has been done by the time the signal leaves the bird, and we can assume, reasonably, that a tape recording so made contains everything about the signal (physically conceived) that a similarly placed conspecific could possibly be interested in.[2]

We also assume that the recording would not be too different were it made from a different angle or distance with respect to the performer and that the recording we have made, if it is a good one, is representative of the signal that reaches the ear of the performer itself, differing mainly in amplitude of different frequency components but not at all in their patterning. Acoustic signals may be nearly unique in this respect: they allow the performer to "see himself as others see him" and this incidentally provides a physical basis for imitation (see Davis, this volume).

Thus the signal with which one works in studying bird song is one of already minimal dimensionality which has undergone relatively little simplification and distortion in the process of transduction onto the tape. All of its complexity which is within a band likely to be of meaning seems to be there on the tape for leisurely and repeated study. The auditory signal is dimensionally so impoverished that we must "invent" a new dimension to appreciate it properly; various instruments such as the sound spectrograph transform the amplitude–time pattern of the tape into a three-dimensional one of frequency–amplitude–time, but at the cost of a certain amount of confounding. Budgetary matters force us to lose information as to *precise* frequency, amplitude, and time, including all information as to phase relationships among different frequency components, in exchange for the three-dimensionality of the picture.[3]

II. THE PHENOMENA: THE MORNING SONG OF THE SWAINSON'S THRUSH[4]

A. Introduction

In the early hours during the spring and summer, males of such birds as the thrushes perform in a manner almost ideal for my purposes. For a period of up to an hour, such a male will remain on the same perch, perhaps

[2]This is not strictly true. The frequency response of our tape recorder might have to be wide enough to encompass sufficiently high overtones to establish the timbre, if such is important in species recognition or in distance estimation based on differential attenuation.

[3]Other than evidence of tonotopic central representation of frequency (Potash, 1970), there is little *a priori* evidence that birds perform a frequency analysis similar to that of the sound spectrograph; the same *kind* of information could be extracted in other ways. But as I will show, frequency or something similar ("pitch"?) is an important variable in the organization of thrush song.

[4]I am indebted to D. H. Morse for recordings of two males of this species, made at Thieves' Island, Maine, in June 1968.

occasionally turning or preening, but mainly doing little other than singing. Further, such a male does not seem to be influenced much by phasic or "triggering" environmental stimuli, other perhaps than that auditory feedback which results from his own performance. To a far greater extent than we are used to when dealing with "communicatory" behavior, the male together perhaps with his auditory feedback forms a system which we may regard as informationally closed.

The olive-backed or Swainson's thrush, *Hylocichla*[5] *ustulata swainsoni,* is a North American bird most commonly found in undisturbed coniferous forests (Dilger, 1956). Males utter *songs*[6] lasting from $1\frac{1}{2}$ to $2\frac{1}{2}$ sec, separated by silent intervals of at least 5-sec duration.

B. The Song Types

By superimposing tracings of sound spectrograms of such songs on spectrograms of other songs, one may soon establish that the songs of a particular record may be classified into a discrete number of "song types"; the spectrograms of two "song tokens" of the same song type are almost exactly superimposable, that is, nearly alike in every detail of frequency–time patterning. Figure 1 illustrates the six song types[7] from one record of a male, hereinafter designated Swainson's thrush I. The exactness of the superimposability of song tokens of the same type was and continues to be astonishing to me; superficial inspection reveals no obvious difference between types 1, 3, and 4, for example, yet by the criterion of superimposability a song token belonged clearly to one and only one of the three. You must accept my word that there were no intermediates. Differences between type 1 and types 3 and 4 are indicated by the solid arrows, a difference between types 3 and 4 by the hollow arrow.[8]

C. The Primary Pattern Types

We find it easy to divide each song into about six or seven "primary patterns," each separated from its neighbor by about a 50-msec interval, and if for the moment minor differences such as those designated by the arrows are ignored, it is easy to see that in this bird's repertoire there are just

[5]Or *Catharus* (Dilger, 1956). I prefer *Hylocichla.*
[6]In German, the preferred term is *Strophe,* which from its Greek antecedents seems an equally unfortunate designation. I have considered other terms such as "phrase," "sequence," "sentence"; all such terms seem to have objectionable overtones. Perhaps a completely new terminology will be justified when we know a little more: for example, LIBU, for Long-Interval-Bounded-Unit, would be a pronounceable designation for "song."
[7]Omission of a terminal group of notes, pattern *L,* was occasionally found with types 3 and 5.
[8]Such differences were first discovered by the late Richard Few.

12 such primary pattern types, labeled *A–L* in Fig. 1. Primes (e.g., *E*, *E'*, *E"*) indicate variations in a pattern type in different song types; in other cases, only slight differences in frequency (kilohertz) exist between instances of the same pattern type in different song types (pattern types *B* and *C*, for example).

D. The Scales

In fact, all of the primary pattern types within a song type seem to be different in frequency (kilohertz) from their representations in other song types. And, in fact, each song seems to be sung to a *musical scale* characteristic of its type: the frequency spectrum for a song shows sharp peaks at roughly equal (logarithmic) intervals approximating those of a pentatonic scale (five intervals per octave), and there are just six such scales seen in the record of this bird (Fig. 2). That is, each song is sung to a different *key*. The corresponding notes of a primary pattern type in different song types fall on corresponding *degrees* of the respective scales; that is, the corresponding instances of the primary pattern are musical *transpositions*.

With such small differences between corresponding degrees of the scales, you may be assured that I measured and measured. I am pretty confident in the relative position of the scales in Fig. 2, although I still have an occasional doubt about types 3 and 4. The scales are all similarly close to pentatonic, with some tendency toward larger intervals at the lower and perhaps upper extremes. In other birds of this and related species, the tendency was for greater deviation toward larger intervals at the lower end of the scale (below 2 kHz).

E. The Ordering of Primary Patterns in a Song

I hope that I have convinced you that there are two more or less alternative ways of classifying this bird's output, by what I have called "primary pattern type" and by song type; the latter breaks down into a particular ordering of primary pattern types and a musical scale or *key*. First, I wish to examine the ordering of the primary patterns within the song (Fig. 1).

I note first that the frequency band within which a primary pattern (less its obvious overtones) lies becomes progressively wider, with both maxima and minima in general rising during the course of a song. Furthermore, it appears to me that a primary pattern tends to be a variant or elaboration of the preceding one; there is progressive "pairwise similarity" and "pairwise difference" between the successive patterns of a song. This is most evident in such pairs as (*D, E*), (*D", I*), and (*G, H*); it is less apparent in other cases. For example, to me at least, pattern *B* is an elaboration of pattern *A*, and *C* of *B*.

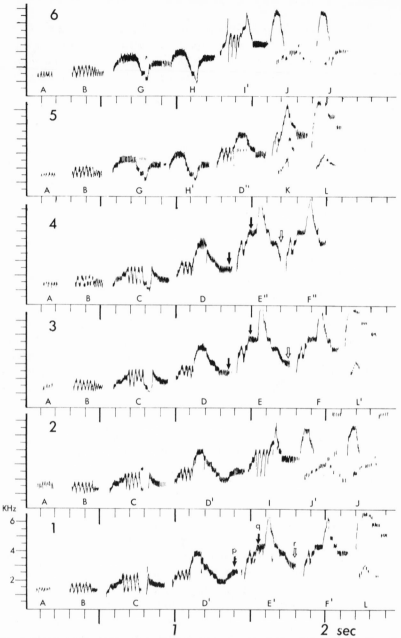

Fig. 1. Tracings of spectrograms of song types in repertoire of Swainson's thrush I. Ordinate, frequency in kilohertz; abscissa, time in seconds. Capital letters and primes designate primary patterns and variants. Arrows point to differentiae of song types 1, 3, and 4: *p* and *q* distinguish 1 from 3 and 4; *r* distinguishes 4 from 1 and 3.

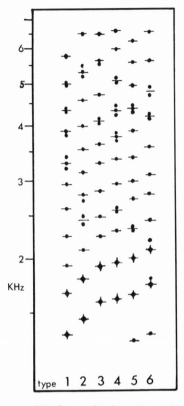

Fig. 2. Swainson's thrush I. Somewhat idealized "spectra" of the six song types in Fig. 1. Ordinate, frequency (on log scale) in kilohertz; abscissa, song types ranked according to keynote frequency (i.e., the lower diamond in each scale is the arbitrarily designated keynote, the lower frequency in primary pattern *B*, Fig. 1). Horizontal strokes mark most likely positions for true scale degree frequency. Diamonds designate the pair of notes in the trill of *B*. Consistent variations judged to be significant in the patterning are indicated; other minor appogiaturas are left out.

Further pairwise comparisons may be made between patterns and their variants in corresponding positions in different song types. Some of these have already been mentioned: *A* and *B* found in all song types, the rest of the patterns in song types 1, 3, and 4. In addition, my eye and to a certain degree my much less well-trained ear finds such pairwise similarities between *C* and *G* and between *I*, *E*, and the variant *D″* in song type 5. I am confident that if the rewards were greater, I could work hard enough to produce a perfectly precise, even quantitative terminology with which to express these pairwise similarities (and differences)[9] to the satisfaction of all. For now, I will content myself by noting that in the trill in the first half of primary patterns *B*, *C*, *D*, *E*, *F*, and *I*, as well as in their variants, the modulation period appears to be the same, and in all of those cases but *B*, the trill begins after a short ascending sequence.

[9]As well as those between the corresponding patterns of different individuals. The overall primary pattern shapes in different Swainson's thrushes are quite distinctive, except for patterns *A* and *B*, which are similar in all examples at my disposal. Differentiation begins with the third pattern in the song.

F. Evidence for an Overall Pattern to the Song

I have already noted that each succeeding primary pattern in a Swainson's thrush song appears to be a (usually more elaborate) variant on the preceding pattern. In addition, looking at the song as a whole there seem to be several alternating ascending and descending *movements*. Furthermore, each upward drift in frequency covers a wider frequency band than either the preceding or the following descending movement, hence the characteristic appearance of an overall rise in frequency during the song. Finally, in this individual the descending movement is steeper and "smoother" (i.e., a glide or a simple arpeggio) than the preceding upward drift—although in other individuals the reverse may be the case. Either way, the overall effect on the spectrogram is of two series of roughly parallel ascending and descending diagonal bands.

Change from ascending to descending motion, or the reverse, may but need not coincide with primary pattern boundaries. I contend that this also is evidence for an overall temporal organization to the song perhaps transcending the ordering of neighboring primary patterns. To overcome any resistance you may have on this point, I wish to turn briefly to an examination of the song of another thrush. This will also enable me to discuss a complication not well shown in Swainson's thrush, namely polyphony.

G. Polyphonic Organization in the Veery

The Swainson's thrush seems not to make much use of a "two-voice" capability, evidently conferred on at least most passerines by the possibility of independent manipulation of two sound sources, the internal tympaniform membranes of the syrinx (Greenewalt, 1968).[10] We may say that of the primary elements of music, the Swainson's thrush makes use of melody (movement and rhythm) but hardly any use of the potentials of harmony.

The veery (*Hylocichla fuscescens*), however, is decidedly polyphonic, although one "voice," usually the upper, has the greater amplitude (Fig. 3). In this species, primary patterns are again evident, by virtue of their contribution to different song types, but in contrast to Swainson's thrush songs there need not be an interval between primary patterns, and an interval may be present in the middle of the primary pattern in one voice but not in the other. This raises the possibility that primary patterns in the two voices may not always be temporally coincident—I have no clear case of this, however.

The opening pattern in the veery is a blurred upward note, the buzziness

[10]However, as indicated by the temporal overlap of two nonharmonically related notes, some use is made of a second voice, in primary pattern *J*, song type 2, Fig. 1, for example.

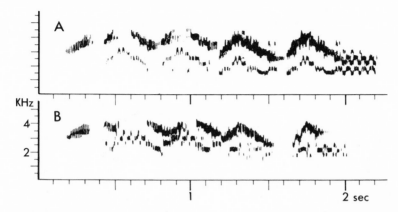

Fig. 3. Tracings of spectrogram of veery song types, from Federation record (Borror and Gunn, undated). A: Bird 2, song 3. B: Bird 1, song 8. Ordinate, frequency in kilohertz; abscissa, time in seconds.

probably created by the participation of both voices at slightly different frequencies. There follows in each voice a pattern of ascending and descending movements as in the songs of the Swainson's thrush; in contrast to the Swainson's thrush example, the *descending* movement in the veery is characteristically slower, less "smooth," and covers a wider range of frequency than the ascending, and hence there is an overall downward drift in frequency. At the end of most songs, the two voices come together to cooperate in a characteristic extended trill of *overlapping arpeggios* (song *A*); sometimes this "cadence" appears to be left to the lower voice alone (song *B*).

The ascending–descending motion in the two voices may be in phase (song *A*) or out of phase (song *B*); in either case, in the samples at my disposal from the Federation of Ontario Naturalists record (Borror and Gunn, undated), there is a very prominent and characteristic feature which is the point of this digression: *the movement, whether upward or downward, is usually carried from one voice to the other*. That is, descending motion of the upper voice is continued by the succeeding descending movement of the lower voice, and an ascending motion beginning in the lower voice sometimes continues into the next ascending movement of the upper. This happens sufficiently often and with sufficient regularity to give sometimes the overall appearance of a pantograph, in which neither "nodes" nor extremes need fall at primary pattern boundaries. Thus there seems to be cooperation between the voices, not just to produce the terminal series of overlapping arpeggios, but to continue also the slower ascending and descending motion beyond the range of either voice alone, in true polyphonic style.

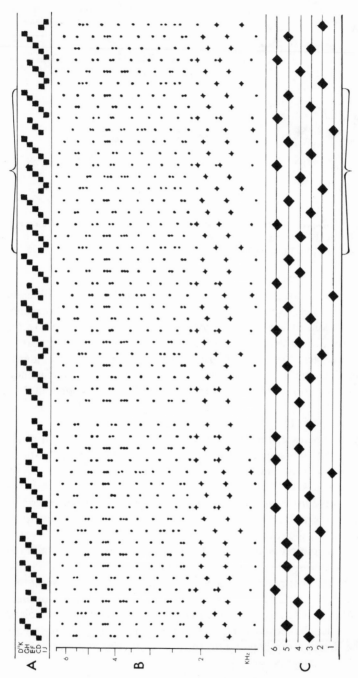

Fig. 4. Sequencing of songs, Swainson's thrush I. The sample is composed of two sequences separated by a flyaway interval. A: Sequencing of recombination units. Ordinate, recombination unit type; abscissa, sequence of recombination units (i.e., each successive pair of squares represents one song token). Bracketed, a repeated series of 28 recombination units. B: The corresponding sequence of scales. The symbols are as in Fig. 2. C: The corresponding sequence of keynotes ranked according to their frequency (in kilohertz) on the ordinate. Bracketed, a repeated series of 14 songs.

H. Overall Primary Pattern Sequence

Returning to Swainson's thrush I, I wish to describe the way in which songs are strung together in a sample. In most of my long samples of *Hylocichla* song, most "primary patterns" are unique to a particular song type, as near as I can tell—that is, as far as recombination goes, the "primary pattern type" is coextensive with the "song type." In Swainson's thrush I, there is rather free recombination of patterns in the different song types: the "unit of recombination" is typically composed of two of what I have referred to as "primary patterns." Thus, perhaps I really should speak of these *pairs* as primary patterns, but their temporal structure suggests that at one time the members of a "recombination unit" were separable. Be that as it may, the recombination units in Fig. 1 are (A, B), (C, D), (E, F), (G, H), (I, J), (D'', K), and L. I will eliminate from the analysis consideration of pattern pair (A, B), present in all songs (A sometimes may appear to be missing), and pattern L, always of low amplitude and variable in its appearance, sometimes absent where expected.

The sequencing of the remaining five pairs—disregarding intersong intervals entirely—is given in Fig. 4A. The vertical ordering of the pairs was arbitrarily chosen to minimize the number of "discontinuities" in the graph, as between (G, H) and (I, J), and (D, K) and (C, D). It has not necessarily any significance, but it does serve to emphasize the precision of ordering. The bracket encloses a sequence of 28 recombination units or 56 primary patterns (or about 90 counting (A, B)s and Ls) which is repeated exactly. In the earlier sample, prior to the "flyaway interval" during which the bird was not recorded, the ordering was nearly the same if not quite so precise. Essentially, the record may be considered (arbitrarily) to consist of two kinds of sequence leading up to pattern pair (D'', K): one beginning with (C, D), (I, J), or song type 2 and the other with (C, D), (G, F), or song type 1.

Similarly precise sequencing, but not of this "complexity," was first noted by Wallace Craig (1943) in the song of the wood pewee, *Myiochanes* (*Contopus*) *virens* (Tyrannidae), which is not an oscine or "songbird." The three primary patterns in the *species* (not just individual) repertoire were typically uttered in repeated strings such as (3,1,3,2) or (3,1,3,1,2).

Other passerine species have since been shown to have rather regular sequencing, but as is so often the case in behavior studies, such investigations have tended to be dead ends, for want of any evident basis for the order. As I mentioned, the vertical ordering of the recombination units is arbitrary and was chosen mainly for clarity. Craig's pioneering analysis assessed the wood pewee's song as *music*: he chose pattern 2 as the string terminator on the basis of its musical "finality," but many of us are not yet prepared to accept such notions as fully scientific. Todt (1968) has found that in the

European blackbird (*Turdus merula*) neighboring songs (or strophes) often contain one or more primary patterns (or elements) that are the same or similar, and he suggests this as a principle of ordering. But the evidence is meager and vague.

The most general principle which has emerged concerns the *recurrence time* or *recurrence number*, the duration or number of patterns of other types which intervenes between two successive occurrences of the same type. Nice (1943) reports that the song sparrow tends to go through its repertoire without repeating (although each song type when it does recur is given in a long bout); the principle in the wood pewee seems to be, in my interpretation, a minimization of the variance in recurrence number (Craig, 1943). Todt (1968) finds that in the blackbird the distribution of recurrence number for a particular song type (or strophe class) shows several peaks, the later ones not necessarily at integral recurrence-number multiples of the first. This is obviously the case with Swainson's thrush I: the "allowable" recurrence numbers for (*I, J*) are 4, 6, 8, and 10, and for (*D, K*) they are 4, 8, and 10. It seems to be generally the case in passerine song that whether a particular pattern will occur or not is related in part to when it last occurred, although not necessarily in any simple way.

I. The Keynote Sequence and Modulation Order

Figure 5A gives a first-order kinematic graph (Ashby, 1956) of the transitions between Swainson's thrush I's recombination units. It can be seen that there are two branch points, at (*C, D*) and at (*G, H*); were the sequence of primary patterns the only information available, one might conclude that the only "choice" was exercised in the middle of the song, not between songs. This is not the case.

In Fig. 4B, the sequence of song tokens is represented according to the scales (or keys) of the corresponding song types. We may (somewhat arbitrarily) designate the lower tone of the trill in pattern *B* in each song type as the "tonic" or "keynote" for that scale and rank them according to frequency: song type 1 has the lowest keynote and song type 6 the highest. We may then abstract from Fig. 4B just the sequence of keynotes, ranked according to frequency, as given in Fig. 4C. There is a very striking order: the keynote of a song is most often either two higher or three lower than the preceding keynote. To convince you of this I will further abstract from Fig. 4C a function of *relative* key which I will call the *modulation order*.

I obtain from the sequence of Fig. 4C new sequences of arbitrary length, in this case 14, in which the course of the keynotes is given *relative* to the keynote of the particular starting song token; thus the sequence

$$(2, 4, 6, 3, 5, 1, \ldots)$$

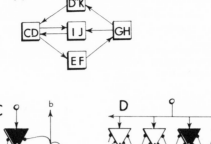

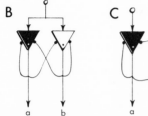

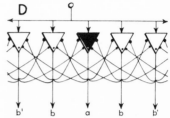

Fig. 5. A: Kinematic graph, showing the first-order transitions for the recombination unit types of Fig. 4A. Note that only at (*C, D*) and (*G, H*) does branching occur. B: Harmon's (1964*b*) pair of self- and reciprocally inhibiting neuromimes with a common input. Arrows are excitatory connections; terminals with dots are inhibitory. C: Another representation of the same pair, emphasizing that from the point of view of neuromime *a*, neuromime *b* is part of a positive feedback loop to neuromime *a*, so long as *b*'s excitation remains high enough to keep it firing at all. D: Part of a linear array of neuromimes such as are portrayed in B, each of which has inhibitory connections to its four nearest neighbors, two on either side, and in turn may be inhibited by four neighbors, two on either side. Self-inhibitory connections are not shown.

becomes

$$(0 + 2 + 4 + 1 + 3 - 1 \ldots)$$

where the sign indicates whether the keynote for the token in that position is above or below the keynote (0) with which the sequence begins. This is done for each token in the sample,[11] and all such sequences are tallied in a graph of the form of Fig. 6A. Such a graph[12] gives the expected course of relative keys ± 0,1,2, . . . irrespective of the key of the origin song token.

We see immediately from Fig. 6A that "up two, down three" patterning is dominant for the first few positions, fades, and becomes prominent again as a recurrence number of 14 song tokens (corresponding to the bracketed 28 recombination units in Fig. 4A) is approached. The pattern has a symmetry of 180° rotation about the origin: to find the relative key sequence or modulation order *preceding* the origin token, turn the graph upside down.

[11]Theoretically, I should stop the procedure 14 tokens before the end of the sample; as my samples were often fairly short and I wanted to take advantage of all the available information, I didn't.

[12]An elaboration of the "expectation density function" representation of Huggins (1957), related to auto- and cross-correlograms.

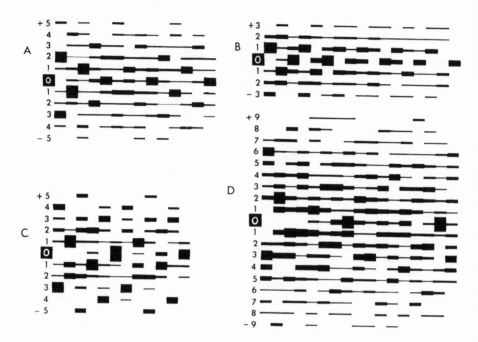

Fig. 6. A: Graph of the expected course of "relative key," or *modulation order*, obtained from Fig. 4C. Overall ordinate, position of rank of key of song token relative to rank of key of reference token (origin), that is, the rank of the song token minus the rank of the reference token (origin). *Height* of rectangle (small ordinates) at each sequence position–relative key locus, other than that of the origin, is proportional to the corresponding frequency in the sample. The *row* containing the origin is an approximation to the expectation density function of Huggins (1957); it is related to the right half of an autocorrelogram. The other rows are related to the cross-correlogram. B: The same, for a short sequence from another Maine Swainson's thrush with a repertoire of four song types. C: The same, for a short sequence of a California hermit thrush (shown in Fig. 7D). D: The same, for a Maine hermit thrush (Fig. 7C).

The "up two, down three" ordering is a special case of a pattern of upward- and downward-going diagonals: *rhomboidal* or *diagonal patterning* again. Figure 6B shows the expected relative key function or modulation order for a shorter sample from another Swainson's thrush; here the repertoire consists of just four song types. Figure 6C was obtained from a hermit thrush (*Hylocichla guttata*) with six song types; here, the pattern is clearly "up four, down three." In other examples, especially when there is a larger repertoire size, regular patterning of song types is less evident, but diagonal or rhomboidal structure is still apparent in the graph of modulation order (Fig. 6D).

J. The Emergence of "Song Cycles": Interaction Between Primary Pattern and Key Sequencing

It is not clear whether just the keynote alone of a song type, or its entire scale, is involved in the sequencing, for the rhomboidal patterning of the successive scales in Fig. 4b is equally striking. What is clear is that in this genus the ordering of successive patterns is related to their component frequencies, in such a way that the order of modulation follows very specific rules. But in Swainson's thrush I at least, the primary pattern or recombination unit sequence seems to follow equally specific rules, and there are indications that these rules of pattern order and key order interact to produce the actual sequence of songs. For example, only in the case of the transition (2,4) is one (C, D) of the five recombination pairs of Fig. 4A seen to follow itself in successive songs. Other cases where the "up two, down three" rule would produce immediate repetition of a recombination unit, as in the transitions (1,3) and (4,1), are not seen at all.

In general, as in other passerines, there seems to be an avoidance of low recurrence numbers, both for recombination units [excepting, of course, (A, B)] and for keynotes (there are only four cases of a recurrence number less than four; see the "0" line of Fig. 6A). Transition (5,2) follows the "down three" rule but results in a disturbance of the short-term regularity of the pattern sequence. Transition (5,1) preserves the regularity of the pattern sequence but does not follow the "down three" rule; furthermore, it continues to upset the regularity of the sequencing, for the following reason. If, according to the "up two" rule, song type 1 is now followed by type 3, the same pattern sequence (C,D,E,F) will be immediately repeated. The transition (1,4) is no better, and disturbs the key sequencing as well. The transition (1,2) would be adequate in terms of recombination unit patterning, perhaps [although it would mean a repetition of (C,D)], but only in three cases is a key followed by one immediately above or below (Fig. 6A). Transition (1,5) is out of the question; the immediately preceding transition was (5,1). Evidently, the transition (1,6) is the least objectionable. In some way, these situations are resolved to produce the observed sequence.

Admittedly, such after-the-fact rationalizations are a bit laughable, and they remind even me of the centipede who became a functional paraplegic when asked to explain how he walked. However, it seems clear that the rules governing primary pattern or recombination unit sequencing are on some occasions in conflict with those governing key sequencing and that it is the resolution of these conflicts that produces the richness of patterning which is seen. What one sees is an overall sequencing which is not a simple regular repetition of a few patterns but something entirely more elaborate yet "just as determinate"; and there appear to emerge larger, regular patterns or song "cycles"—e.g., (2,4,6,3,5) or (1,6,3,5,)—which are in turn organized in an even larger repeated sequence of 14 songs—a "supercycle"—something, in fact, in form very much like a sonnet.

K. Summary: The Patterning of Song in *Hylocichla*

I hope that I have demonstrated to you the following points. Singing is organized into temporally defined groups of notes which we may call *song tokens*. Each song token is classifiable into one or another *song type* on the basis of *superimposability* of the *scale* to which it is sung and of its *pattern* of notes or *melody*. When a melody is decomposable into *units of recombination* in different song types or on the basis of their temporal pattern, we may speak of these smaller units as *primary patterns*. Thus a song token is a series of one or more *pattern tokens* selected from a discrete number of *pattern types* and given in a particular order and according to a particular scale or *key*.

The melody or order of the string of pattern types within a song type is such that an overall pattern appears, of repeated, approximately equally spaced ascending and descending *movements,* or frequency trends, the relationships and spacing of which may be characteristic of an individual or a species. On a finer temporal scale, within a primary pattern, similar ascending and descending series of notes are found, most clearly in the trills of overlapping arpeggios that characterize the "cadence" of the veery and that are found in the songs of other species.

The ordering of songs was found to be closely related to their key, in such a way that the expected course of the keynotes of the sequence tended, again, to conform to a pattern of approximately equally spaced ascending and descending motions, designated as *diagonal* or *rhomboidal patterning*.

Thus similar tendencies to conform to ascending and descending diagonal movement were found at three or perhaps four *levels of organization* of the singing: within a primary pattern; in the alternating upward and downward movement "transcending" the primary pattern; perhaps in the overall upward or downward movement of the song (not necessarily found in the hermit thrush or in the perhaps-more-distantly-related wood thrush, *Hylocichla mustelina*); and finally at the level of the sequencing of songs and groups of songs. The seeming interaction of these tendencies with others based on the sequencing within and between primary patterns (where these can be seen independently of the key sequencing) imparts an orderliness to larger and larger units of behavior, until in some cases even higher levels of organization, *song cycles* and *supercycles* may become recognizable.

III. THE PHENOMENA: DIAGONAL OR RHOMBOIDAL PATTERNING IN OTHER TAXA

To be able to demonstrate in a simple fashion the existence of this sort of patterning in the behavior of a species, it is necessary first to be able to order the repertoire of behavioral units along a single dimension, such as frequency in the case of bird song. With most behavior, perhaps including most bird song, the patterning may be too complex for such a simple ranking to be very meaningful.

A. The Western Meadowlark

I have had the opportunity to study the song types in the repertoire of a western meadowlark (*Sturnella neglecta*), belonging to the family Icteridae. In this individual, each song type shows some indication of diagonal patterning in its infrastructure; each song type is given in a series or *bout* of variable length before switching to another type occurs (Fish *et al.*, 1962). The sequencing of bouts of song types ranked according to first sustained note in the song is shown in Fig. 7A, and the modulation order or course of relative key is given in Fig. 7B. I see again clear ascending and perhaps also descending diagonals, much as in Fig. 6D. On the basis of this meager evidence from two families, I am willing to make the prediction that wherever bird song types or other units can be ranked according to frequency in an unequivocal manner, rhomboidal patterning or diagonal movement will be found in the sequencing.

B. Human Music

Dare I generalize further? The hymn which begins "Oh God, our help in ages past . . ." shows extensive rhomboidal patterning; I am sure you will think of many other examples. Both Western and much Oriental music are based on the relationships of the musical interval of the fifth (or fourth; the fifth, ratio 2:3, and its complement the fourth, ratio 3:4, together "add" to make the octave and "subtract" to form the interval of the major tone, 8:9). The series of ascending fifths (descending fourths) is used to generate the musical scales not only in Western music (the major "circle of fifths" beginning with C and the minor series beginning with A) but in Chinese musical theory as well (Crossley-Holland, 1960). The ordering of sharps in the key signature of a score is that of the ascending series, and the ordering of the flats is that of the descending series of fifths; the key signatures of C-sharp (all seven degrees sharped) and A-flat minor (all seven degrees flatted) display the whole system and the resultant rhomboidal patterning.

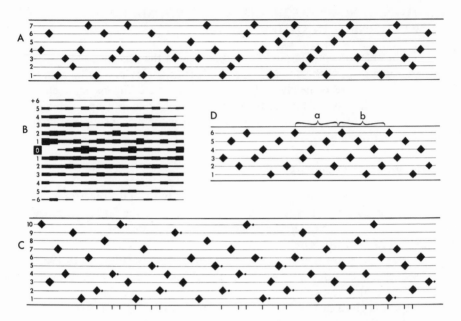

Fig. 7. A: The last 50 of a sequence of over 80 bouts of song types of a western meadowlark (Fish *et al.*, 1962), each bout consisting of from one to ten or so songs of the same type. Ordinate, song types ranked according to first major note; abscissa, position of bout in sequence. B: The graph for the same meadowlark sequence of expected relative "key" or modulation order (for "key" substitute "first major note"). Explanation is as for Fig. 6A. C: Keynote sequence for a Maine hermit thrush, part of a long record. The song tokens followed by dots are repeated 16 tokens later (see Fig. 6D); the "disconnected" pattern formed by the songs indicated by the ticks along the abscissa is repeated twice with identical relations. Other explanation is as for Fig. 4C. D: Keynote sequence for a short sample from a California hermit thrush. A repeated pattern (*a*) of all six song types (6,3,4,1,5,2) changes suddenly to the repeated pattern *b* (6,3,5,1,4,2), resulting in (or from) a reversal of the direction of the most apparent diagonal motion. Other explanations are as for Fig. 4C.

C. Plant Phyllotaxy

That a tape recording is a pattern in a spirally wound strip of matrix should remind us that the study of behavior can be regarded as part of a branch of developmental biology, specifically that which deals with accretionary growth. Therefore, we may search for analogous patterning in other kinds of accretionary growth, for example, among snail shells and plant phyllotaxy patterns. Rhomboidal patterning occurs extensively in the pigmentation of the shells of many marine gastropods; here, the "unit of behavior" is a spot or line of pigment, and the dimension in question is position along the edge of the mantle. It is the basic phyllotaxy system among the Pteropsida, the ferns, gymnosperms, and flowering plants (Snow, 1955). Here, the "unit of behavior" is the leaf primordium and the dimension, angular position around the apical meristem; the leaf primordia are almost always found at the intersections of two oppositely directed systems of one or more equally spaced helices or spirals. The numbers of such equidistant rows (parastichies) in the two systems form two neighboring terms in the Fibonacci series 1,1,2,3,5,8,13, . . . or sometimes in a related series 1,3,4,7, 11, . . . or 2,5,7,12, ... In this type of series, each term is the sum of the two previous terms. In the case of the Fibonacci series, the average angular distance around the meristem between each two successive primordia approximates the "Fibonacci angle," 137.5°, obtained by dividing the circumference in two on the ratio of the golden mean. The corresponding angles of the second and third series are 99.5° and 151.8°; in all three cases, the ratio between successive terms of the series converges, however, on the ratio of the golden mean. The actual pair of terms to which the numbers of spirals or helices conform is a function of the ratio of the rate of production of primordia to the rate of elongation of the stem relative to its diameter (Richards, 1948); where this is large, as in the flowerhead of the sunflower, the pair of terms will be high in the series, for example, 55,89.

D. A Comparison of Thrush, Human, and Plant

The series beginning 2,5,7,12, . . . is rare among plants but is interesting in that the corresponding phyllotaxy pattern resembles that produced by repeated application of the musical interval of the fifth. We first note that the numbers of black and white keys on the piano are in the ratio 5:7; together, they add up to the 12 degrees of the equal-tempered scale. If we think of a musical scale as a circular arrangement of intervals from 1 to 2, then from the mantissa of the logarithm to the base 2 of an interval we may derive a measure of "distance" which translates into angular terms for comparison with Fibonacci-like series. The angles for the fifth and its complementary fourth are 210.7° and 149.3°, respectively. The corresponding angles for the 2, 5, 7, . . . series are 208.2° and 151.8°. In Fig. 8, the series which result from repeated application of these angles are compared with a trill of overlapping arpeggios on a pentatonic scale which was found in several hermit thrushes from California and Maine.

One further point: in each case the pattern generation is evidently reasonably isolated from external disturbance. The apical meristem grows in an unresistant medium; in the soil, root primordia show no such precision of pattern. During the day, when a thrush is freely interacting with its neighbors, precision of ordering disappears—Todt (1970) has shown, however, that the order of singing in the blackbird may be influenced by the *nature* of utterances of other birds. And, presumably, the patterning of an evening's music among us is different, perhaps "less orderly," when requests from the audience are acted upon by the performers.

IV. THE INTERPRETATION OF ORDERLY PATTERNING

A. Complex Determinate Sequences

In the cases we have been concerned with, the potential exists for ongoing function of a system to serve as a constraint on the pattern of its future function. The path of constraint may be quite direct (for example, the activity of the nervous system in generating behavior no doubt directly affects its future activity) or relatively direct (as in the case of auditory feedback from the song itself), relatively indirect (as in the case of conspecifics influencing one another by their behavior) or very indirect (if community energetics or population size is modified by performance and in turn modifies the succeeding behavior).

There are several extreme kinds of patterning which bear mentioning.

1. First, I assume that there is no case to be found of "completely random" behavior; simple physical constraints within and without the animal prohibit it. However, in some cases apparently independent-trials *sequencing* of behavioral *units* of different classes has been described (as in successive throws of a die), and in others apparently times of occurrence of particular events are independent of the times of preceding occurrences (Poisson processes; such as the intervals between successive radioactive decays). In these cases, there are intraunit constraints without a doubt, but apparently such constraints as may operate between units are not involved in differentiation of the sequence.
2. Second, the sequencing of members of several classes of behavioral activity may be found to be exceedingly regular and predictable, or the intervals between events may be found to have negligible variance, as with certain circadian rhythms, for example. In these cases, the interunit constraints seem to be nearly absolute.
3. There exists the possibility, not often recognized, of still another

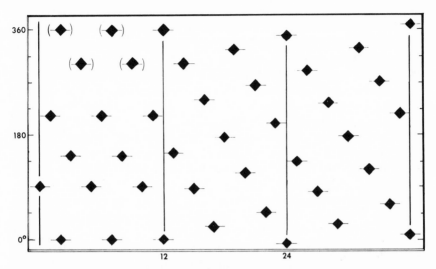

Fig. 8. Rhomboidal patterning. Ordinate, position on the octave, expressed as an angle in degrees with respect to an "arbitrary" origin and direction. The graph is really an un-rolled cylinder, with its length (abscissa) representing serial order of events. In the left-hand panel is shown a sequence of 12 notes from a trill of descending arpeggios found in several hermit thrushes (the notes in parentheses are often omitted). At position 12, the series changes to a complete sequence of 12 descending fifths (or ascending fourths). At position 24, the series changes again, to one based on the 2,5,7, . . . Fibonacci-like series, with an angle of ascent of 151.8° (see text).

"extreme case," actually a special case of the second, in which a determinate sequence is so complex that it "mimics" the independent-trials case. That is, units of behavior might be appearing in a perfect order exactly predictable from knowledge of a string of previous out-comes which was very long relative to the sample duration one was studying, so that its orderliness could not in principle be discovered. In the case of (1), the constraints in question are "invisible": "outside of," unrelated to, the ongoing sequence. In case (3), some of the constraints at least are similarly invisible, "outside of" the ongoing sequence, but now in a temporal sense.

In practice, we may rarely have to deal with such extreme cases, but I think it important to consider the difference between examples lying between cases (1) and (2) on the one hand and between (2) and (3) on the other. In both kinds of examples, we are dealing with irregular (nonrepeating) se-quences showing some lawfulness. In the first case, the system may often with justice be described as a Markov process or something similar (Cox and Lewis, 1966),[13] and the orderliness of the sequence may be expressed in a number of ways by statistics based on probability and information theory. But in the second case such a description would be at best misleading and at worst absolutely false—for example, if description in terms of a Markov process were meant as a *model* of the system for the purpose of guiding further research into its functioning.[14]

In some of my samples of *Hylocichla* song, the near-precise repetition of a very long sequence of behavior suggests that irrespective of whether the system producing it is composed of *parts* with indeterminate behavior, the *interaction* of the parts is such that the ordering which results is determinate or nearly so. In some cases, there is clear evidence of a determinate, regular sequence in one part of the sample passing over into a different but equally determinate sequence by means of a more or less complicated transient (Fig. 7D). In these cases, it is clear that description in terms drawn from probability theory is inappropriate.

But there are other examples in which an overall rather indeterminate-appearing sequence contains short groups of songs repeated exactly at precise intervals, separated by irregular, nonrepeating series. Figure 7C displays the course of a short sample from a hermit thrush, for which the graph of relative key progression was given in Fig. 6D. In other samples from the same bird, the sequencing may be yet more irregular, so that no groupings are repeated at regular intervals. Are such samples to be described as sequences degraded by "noise" toward case (1), or as one or more near-determinate transients of indefinite length? I think the latter alternative the more likely in this case, however unattractive it may appear from the point of view of simplicity of analysis. Possibly such a sequence should be regarded as a mixture, but on the face of it one in which it is impossible to sort out the determinate and indeterminate components.

B. Simplistic Models of Behavioral Organization

In the past, I believe, probability theory and its derivatives have con-

---

[13]In the simplest Markov-process type of model, the probability of the system entering into a state of one particular class among several alternatives is determined by the class of the immediately preceding state. In other words, each class is characterized by a set of numbers (which add to 1) each of which is a transition probability for a particular following class. More elaborate stochastic models utilize information about the classes of states prior to the immediately preceding one.

[14]An example of a perfectly determinate, cyclically repeating sequence whose law of construction is resistant to discovery by statistical means is . . . 2247440451710112362022474-40451.

spired with Ockham's Razor ("entities are not to be multiplied beyond necessity") and its derivatives to discourage the discovery and acknowledgment of complex modes of biological function. We are enjoined to consider the simplest hypothesis which apparently could account for the phenomena under study. It seems to me that this usually results in simplifying the data to fit the hypothesis. Or in our ignorance, alas, we may be unable to recognize whether or not the simpler of two hypotheses ignores an essential property of the system.

What often seems to happen is that we tentatively accept the apparently simplest hypothesis; then as the disagreeable facts begin to accumulate, we modify it; then we modify the modifications, etc., until we are faced with a structure which is much more complex than the more appropriate "complex" concept that eventually replaces it, once it is recognized that the simpler scheme ignored an essential property. I imagine science as proceeding upon an "adaptive landscape" similar to those once fashionable in population genetics: once a discipline is committed to a particular "adaptive peak," it may prove very difficult to cross over to a neighboring but "higher" one.

The influence of Ockham's Razor–type canons may have been largely salutory in some fields in which the "entities" at least seem to be relatively few. I doubt whether psychology has benefited much by their application. It seems to me that it was the complexity implied by gestalt theory which led more than anything else to its initial rejection. When, with the explosive development of computer techniques, the complex, holistic field concepts of gestalt theory might have become testable, paradoxically attention was instead directed (perhaps by the atomistic nature of digital processing itself) toward more elaborate atomistic models: the "Markov process" and the linear control system.

These models may become quite complex, and their detailed processing may tax the memories of large digital computers. Although analytical solutions to a control system model (via the roots of characteristic functions, usually) are often easy enough to obtain with digital techniques, plotting an actual trajectory of some state variable may be very time-consuming. And ironically, because the digital computer is perfectly determinate in its operation, clever and often time-consuming procedures are used to generate in perfectly determinate fashion sequences of "pseudorandom numbers" which model the probabilities of the Markov process.

In general, the number of transition probabilities necessary to characterize such a process is a function of the number of state classes raised to a power equal to the length of the chain of preceding states needed to specify the process. Matters soon get out of hand, and in practice the models are useful even as summary descriptions of sequences of behavior only if the number of classes is below about ten and the number of preceding states to be considered is below three. Even in such cases, a sample of several thousand behavioral events may be necessary to provide even a rough estimate of the transition probabilities; as it must in principle be approximately statistically stationary, such a sequence may be impossible to obtain.

Still, much behavior may actually be found to conform to such a model, particularly if its sequencing is subject to much disturbance from unrelated environmental factors. Even in cases where behavioral sequencing is known to deviate extensively from the closest Markovian model, it may be a useful avenue to explore. Such models have very interesting properties and, as Cane (1961) has shown, may set a kind of lower limit on what one can expect from a behavioral system. I do not suppose, however, that description in terms of a Markov process model will in itself lead to very great understanding of the behavioral organization, and it may in fact—I speak from personal experience—blind us to a different, possibly more complex but ultimately more realistic model.

The development of linear control theory was in a sense more revolutionary, for at last it was possible to admit that machines could behave purposively: even *simple* machines. Any biological system in which the components could be assumed to interact determinately and linearly (that is, in a way describable by linear differential equations) could be modeled, no matter how many components there were and no matter how complex their interconnections. That is, if one knew the input/output relations of each component and how they were connected, one could study the response of the whole system or of any subset of its components, always provided that their interactions were linear (or approximately so within some more or less limited range). The solutions to the equations for the subsets could be combined with ease; the main procedures and an inkling of the power of the method may be gained from a reading of Milsum (1966).

This power diminishes rapidly as the need to acknowledge nonlinear interactions between components becomes more imperative. Probably only a small fraction of biological systems approach linearity of interaction; this is certainly the case with behavioral systems of any interest. I suspect that in some cases which do show linearity of interaction, linearity itself has been selected for during evolution. Just as a digital computer generates "pseudorandom" numbers by determinate means, a system may be constrained to produce linear behavior by nonlinear means. In other cases, linearity may be an effect of the mass action of many nonlinear components. But I suspect that the very power of linear control theory may blind one to more realistic models. As in the case of the Markov process, its attractive simplicity gently persuades us to overlook complexity.[15]

C. Simple Nonlinear Models and the Emergence of Emergent Properties

Now, I maintain that there is a world of difference between the approach of beginning with a linear model and sacrificing linearity in a particular interaction only when its absence becomes too obvious to ignore, and one which begins with the full biological complexity and reduces interactions to linearity only reluctantly. The former is the way of the mathematician, perhaps. As a biologist, I presume Nature to be hierarchically organized, and I assume that there may be emergent properties at each level which result from but are not entirely predictable from nonlinear interactions at levels lower down in the hierarchy. It is barely possible that I am wrong, but it seems to me that such properties will only occasionally have a chance to engage our interest if we stick with the former approach; it is a bed of Procrustes which amputates emergent properties.

I believe it is better to acknowledge that one often simply cannot obtain mathematical rigor without discarding the life of the system. One may still deal with models, however. Many analog computer types are general-purpose in nature; they consist of collections of resistors, capacitors, and operational amplifiers which may be connected in various ways, mainly to model systems of linear differential equations (although nothing prohibits the introduction of certain nonlinearities). They do this perhaps none too accurately, but in a direct, dynamic way, as all the forces in the system seek an equilibrium. The digital computer, solving the same equations, has the strained quality of a politician's smile: it is sincere, no doubt, but you know it took a lot of effort.

The most interesting results in the present context have been obtained with analog computers in which the basic elements are "neuromimes" rather than the linear summers or summer-integrators of the general-purpose device; that is, their functioning is modeled on that of neurons, whose all-or-none performance is the epitome of nonlinearity of a certain sort. Numerous "system properties" of simple networks of neuromimes have been found, and in some cases this has led directly to the discovery of the same property in a real biological system.

For example, Harmon (1961) found that a simple two-neuromime chain could exhibit the property of temporal summation and in so doing could

[15]In fairness, I should note that more and more workers have assayed the analytical study of nonlinear models, some with remarkable sophistication and success.

produce rather complex patterns of output behavior. If the amplitude of the input impulses (spikes) from the first to the second neuromime was decreased, then the second would no longer follow every spike but perhaps only every second or every third one ("integral frequency division"; 2:1, 3:1 input/ output ratio). However, it might also deliver *two* output pulses for every three inputs, or even three for every five, in a regularly repeated way. As the input impulse amplitude was decreased *smoothly,* the firing pattern of the second neuromime *jumped* through a regular series of changes:

$$
\begin{array}{lll}
\text{(a)} & \ldots 1\ 1\ 1\ 1\ 1\ 1\ 1\ 1\ 1\ 1\ 1\ 1\ 1 \ldots & 1{:}1 \\
\text{(b)} & \ldots \overline{1\ 1\ 1\ 1\ 0}\ 1\ 1\ 1\ 1\ 0\ 1\ 1\ 1\ 1 \ldots & 5{:}4 \\
\text{(c)} & \ldots \overline{1\ 1\ 1\ 0}\ 1\ 1\ 1\ 0\ 1\ 1\ 1\ 0\ 1\ 1 \ldots & 4{:}3 \\
\text{(d)} & \ldots \overline{1\ 1\ 1\ 0\ 1\ 1\ 0}\ 1\ 1\ 1\ 0\ 1\ 1\ 0 \ldots & 7{:}5 \\
\text{(e)} & \ldots \overline{1\ 1\ 0\ 1\ 1\ 0}\ 1\ 1\ 0\ 1\ 1\ 0\ 1\ 1 \ldots & 3{:}2 \\
\text{(f)} & \ldots \overline{1\ 1\ 0\ 1\ 0}\ 1\ 1\ 0\ 1\ 0\ 1\ 1\ 0\ 1 \ldots & 5{:}3 \\
\text{(g)} & \ldots \overline{1\ 0}\ 1\ 0\ 1\ 0\ 1\ 0\ 1\ 0\ 1\ 0\ 1\ 0 \ldots & 2{:}1
\end{array}
$$

Each pattern was maintained over a small domain of input voltages, in range more or less proportional to the simplicity of the firing ratio. Such *nonintegral frequency division* is an "emergent property" of a single neuromime really; the first stage serves only to drive it. But although the property *could* have been predicted from the details of the model, it wasn't.

Now, however, the late Don Wilson (1964) could look for similar non-integral frequency division by the motor neurons in the thoracic ganglion of the locust, and of course he found it. Wilson demonstrated also that a modified model more closely resembling the neurons he was studying could also produce the property, using different assumptions from Harmon's.

This is the first example I will give of the synthesis of a "complex" behavior pattern. It illustrates in a beautiful manner how an understandable dynamic system may "extract" complexity from simple and unpromising antecedents. More than that, it is also the simplest example I know of which illustrates how production of a number of different temporal patterns might be controlled very economically, by change in the amplitude of the input impulses in this case. The repeated units of pattern are underlined in the tabulation above; note that they do not change in duration monotonically with a decrease in input impulse amplitude, but rather in a sort of cyclical fashion.

Input impulse *amplitude* is the kind of variable that I think of when I recall "what we formerly called drive"; that is, it is a motivational sort of variable operating to modulate behavioral output rather more slowly than the rate of changes which are occurring within and between temporally adjacent patterns. Input impulse *rate* of occurrence is another. They may be expected to interact in complicated ways. As long as input impulse amplitude is *high* enough, and input impulse frequency *low* enough, the neuromime will respond one-for-one. Beyond that, such variables as firing ratio, duration of pattern, and frequency of pattern will appear as complex functions over the domain of the two "drive" variables.

A little discouraging, isn't it? But the point is that just a slight increase in complexity of such a system might render its outputs sufficiently diverse to produce in a stable manner *all* of the diverse patterns of stridulation in the short-horned grasshoppers, merely by changes in a few parameters without extensive structural modification.

Clearly, the model is insufficient as given, however. For one thing, evolution from one to another pattern must involve not just one or two parameter modifications in the nervous system of the male locusts but must also involve a change in the pattern-sensing propensities of the females of the species, so that they become more receptive to the new pattern and less receptive to the old.[16]

For another, the model as outlined is unrealistic in that it has only a single output neuromime, sufficient perhaps to govern the pattern of contraction of the elevators, say, but not sufficient to return the leg to a resting position. At least two output neuromimes are needed. Several authors (references in Harmon, 1964*b*) have suggested that such a set of antagonists, involved in vertebrate respiration, locomotion, etc., might be governed by a reciprocally inhibitory pair (or pair of groups) of neurons. Reiss (1962) and Harmon (1964*b*) undertook to consider what would happen if two neuromimes, driven by a single input impulse stream from a third unit, were connected by both reciprocal inhibition and self-inhibition (Fig. 5B).[17] They both chose to study asymmetrical pairs, that is, pairs in which coefficients of inhibition differed for the two units and thus one unit was more or less dominant over the other.

Both found that under a fairly wide range of input impulse frequency, the pair would deliver alternating bursts, with the end of a burst from one under certain conditions overlapping the beginning of a burst from the other. Reiss found that with increasing input frequency the duration of a burst decreased for both units, but the relative shrinkage might be greater for the dominant member of the pair. Although Harmon's model was very similar to Reiss', he found more or less the opposite, that with increasing input frequency the subordinate member participated less and less. In other words,

---

[16]But I see no compelling reason for making the perceptual "filters" of the females much more complicated than the generating "filters" of the males.

[17]Harmon's study resulted directly from a suggestion by Wilson that flight in some insects might be so governed. Wilson was inspired earlier, you recall, by Harmon's earlier study. Their interaction is a good example of reciprocal excitation.

Reiss found that the disparity between their performance decreased with increasing input impulse frequency, and Harmon found that the disparity grew. Harmon does not comment on this, but the difference seems to lie in the choice of parameter values rather than in a more "structural" feature. In Harmon's model, the controlling parameters seem to be those involved in reciprocal inhibition; Reiss' model seems more to be governed by self-inhibition.

In short, the "structure" of a dynamic system determines the modes of behavior it may show, which may be many; choice of a particular set of parameter values further restricts the system to a particular mode. The modes of behavior determined by different sets of parameter values may be radically disparate.

D. Further Consequences of Nonlinearity

This raises the general question of whether it is possible or even useful to distinguish too sharply between "structure" and "function" in such a system. Whether an axonal terminal is excitatory or inhibitory on a post-synaptic neuron depends on synaptic structure, yet equally it may be said that as far as the spike-initiating process is concerned the difference is one of choice of a particular range of parameter value. It may be useful again to think very generally in terms of kinds of constraints. There is in any system a set of constraints which are relatively permanent, including most of those we generally lump under the term "structure." Then there are those constraints which fluctuate, but in general in a way which is not dependent on the dynamic behavior one is interested in, that is, in some way extrinsic to the system under study. The parameters of input impulse amplitude and frequency are of this nature relative to the neuromime models discussed above. Finally, there are those constraints whose values at particular moments are dependent on the detailed dynamics of the system as a whole; in realizable systems they exhibit "memory," i.e., state dependence on features of the past performance of the system. These dynamic constraints may be further subdivided in various ways: in models, into those such as "excitation," regarded as resulting from additive interaction of (filtered) input pulses, and those such as "output pulse" whose origin and impact are quite nonlinear.

There are several kinds of nonlinear interactions which are likely to be of importance in systems governing behavior. Two input variables, each of which by itself produces a linear effect on the output, may interact in such a way that their combined effect is nonlinear. For example, if one of the inputs to a model acts to increase the amplitude of pulses at another input, the combined effect on output may be more or less multiplicative. An important example of interactive nonlinearity is given by cases in which the threshold of a unit is driven up by an inhibitory input. If the output cannot take on negative values, once it reaches zero no further increase in the inhibitory input has any effect (Hartline and Ratliff, 1958). Reichardt and MacGinitie (1962) were able to treat analytically in a linear manner an array of units connected by reciprocal inhibition in spite of such nonlinearities, by partitioning off the "silenced" units which, as long as they were silent, had no further effect on the operation of the system.

When the effect of one variable on another is itself radically nonlinear over an interesting part of its range, treatment by analytical methods becomes more difficult. For example, the input/output relationship may be sigmoid; that is, the output may respond almost linearly over intermediate values but show no further response at both extremes of input range. In the limit, as the slope of the linear range is increased, the output may come to exist in one of two discontinuous states, according as the input state is one or another part of its (continuous) range.[18]

Systems involving this kind of function typically show *hysteresis:* the shape of the curve relating output to input varies according to whether the input variable is increasing or decreasing. In general, this takes the form of the output tending to be "captured" and "locked" in the state it is already in. Harmon (1964*b*) found that his pair of reciprocally inhibiting neuromimes exhibited hysteretic behavior of this sort when input impulse frequency was changed. The output pattern of the *pair* would jump through a particular series of changes as the input frequency was increased and through a quite different series as it was decreased. Furthermore, Harmon discovered that different values of a second parameter (time constants of reciprocal inhibition) produced somewhat different hysteresis "loops."

In such a system, a number of modes of behavior will be available, possibly as alternatives, and as in the case of the simple neuromime chain model are accessible through change in one of several parameters. Harmon observed that a transient change in the input impulse frequency—for example, even the intercalation of a single extra pulse—could be sufficient to "flip" the neuromime pair from one to another stable mode of behavior.

If you push a linear system, it adjusts to the new conditions smoothly. Push such a nonlinear one and it resists for a while—thus within limits it may be stable under minor disturbances, "noise"; push a little harder and it jumps to a new stable mode of behavior. In such polystable systems, the

---

[18]Such simple "flip-flop" behavior forms the basis of operation in digital computers.

output pattern for a particular choice of parameter values is not uniquely specified. Several relationships may be available, depending on the history of the system.

E. Complexity and Research Strategy

Even with such simple systems of neuromimes, we are evidently entering realms of complexity taxing our abilities of comprehension. Harmon (1964a) asserts that with a system composed of just three neuromimes connected by just four "synapses," all of known circuitry and individual properties, "it is extremely unlikely that anyone will be able to specify its output function." Contrast this with the linear system: here we can tackle each component separately and combine them readily no matter how many of them there are. If we are dealing with a system whose behavior is constrained by important nonlinearities, we should try to understand its components, of course, but if we are to understand the *system* we must consider it as a totality or not at all.

We are "dragged kicking and screaming" into the age of holistic analysis.

Consider the problem from another angle. Think of how many different "things" one such reciprocally inhibiting pair of neuromimes can do—that is, how many different stable patterns of output it has in its repertoire. It seems that such a model is "too good"—the theory that explains everything explains nothing.

Instead of the usual highly specialized model put forward to explain a very specific function, in collections of neuromimes connected in simple highly specific ways we seem to be confronted with "Jacks-of-áll-trádes," at least until the very last parameter value is strapped down tight. Harmon (1964a) again: "The enormous number of degrees of freedom possible even in a small assembly of neurons makes it possible to find almost any set of stimulus–response characteristics one can imagine." If this is so, it is doubtful whether single-unit studies, for example, will tell us much unless they are guided by behaviorally derived hypotheses; these will have to be detailed and sophisticated.

Harmon therefore has this advice for experimental neurophysiologists: "Rather than acquiring data where and how it can be found, cataloging complexity after complexity, why not go into a system with an hypothesis, a prejudice? . . . The point is that for the time being we may find it most profitable to seek an understanding of nervous system operation in terms of highly specific *a priori* conjectures."

Doesn't that make you uneasy? It is probably excellent advice, but it means that an enormous amount of care is going to be needed, both to arrive at the specific hypothetical mode of functioning and to determine just how it is to be sought in the physiology of the neuromuscular system. One of the implications, I believe, is that the "controlled experiment" will be of limited usefulness in elucidating neural function, and it is going to play a much reduced role in the behavioral study of the near future. This need not be entirely a bad thing. Controlled experiments are costly, usually, in time, money, animals, and effort which might at this stage of our knowledge be applied more productively in finding out just what it is that is worth doing experiments on. I hesitate to use the examples of astronomy and geophysics, but they may prove to provide better research paradigms for us than physics has. The difficulty inherent in subjecting the earth and stars to controlled experiments has perhaps made astronomy a different sort of science, but not necessarily one lacking in rigor or success.

We may, if we like, subject animals to controlled experiments, but there is no law that they must divulge their secrets to us thereby. My personal opinion is that the greater priority is on the development of new modes of behavioral description, based on rigorous notions (where such are possible) of harmony and conflict, part and whole, simple and complex, containment and change (see Bohm, 1968a,b).

I expect considerable dissent from some quarters over these points. It seems not to be generally realized that the subjects of most controlled behavioral experiments are analog models of "real animals in real situations" just as a collection of neuromimes is. As a minimal assumption, the experimenter counts on Nature's having no intentions of resisting his invasion of her privacy, and we really have no means of determining how often he mistakes her intentions. Kavanau (1967) has catalogued as a cautionary tale for us some instances of the contrariness of well-bred mice.

As I write, I am aware of my own debt to the controlled experiments of others, especially in the area of the role of auditory feedback in the ontogeny of bird song. I mean only to suggest that there is room in our field for much more careful observation and description, and checking of hypotheses against further observation, to establish the critical junctures at which the controlled experiment becomes necessary and decisive.

V. ON THE HIERARCHICAL ORGANIZATION OF BEHAVIOR

A. Different Kinds of Hierarchies

In his influential *The Study of Instinct* (1951), Tinbergen suggested a hierarchical model, in which centers governing successively smaller units of behavior are supposed to be arranged in a cascade. Energy ("impulses") is shunted down along a particular trajectory through the cascade by the gating action of specific releasing stimuli on perceptual templates, "Innate Releasing Mechanisms," resulting in the appearance of behavior of a particular sort. A more workable model of hierarchical control, rather freer of particular structural implications than Tinbergen's, was proposed in 1960 by Miller *et al.* In their model, the unit of organization is the so-called TOTE unit, composed to a "test" and an "operate" phase. In the "test" phase, the unit samples and compares some environmental feature (that is, a state external to itself) with a standard; if there exists an incongruity, *control* (not just energy) is shifted to the "operate," or effector phase. Upon *action* by the operate phase, control is now returned to "test" and the results of that action are compared with the standard. If the incongruity no longer exists, control is now shifted to the "test" phase of another TOTE unit—thus, "Test–Operate–Test–Exit."

Miller *et al.* suppose that these TOTE units are hierarchically organized: that the TOTE units governing elementary patterns are themselves embedded in the operational phase of TOTE units governing larger patterns, etc. In contrast to Tinbergen, they avoid further structural implications: the TOTE hierarchy, or *Plan,* is one of function. There is a residual implication, that "test" and "operate" occur in different locations in the brain: else why need "control" be *transferred*?

What we mean by "hierarchical organization" requires some examination. Evidently, the term carries with it the connotation of the pyramidal structure found in the military and ecclesiastical chains-of-command. But the term is also often used to describe the organization of the physical universe, of particles into atoms, atoms into molecules, and so forth. And Tinbergen and Miller *et al.* support their hypotheses with the evidence of a *temporal* hierarchical organization of behavior: related activities, e.g., those involved in nest building, are grouped in time. Presumably, this must involve a judgment as to what constitute "related" activities. These judgments are built up during the investigation of the behavior, as units of behavior are gradually defined out of the accumulating mass of observations. Often, the sole source for such judgments seems to be the temporal relations of the behavioral events, and among these relations their grouping together in time stands preeminent. It is to these temporal groupings that names are given, and temporal hierarchical organization of behavior really means temporal hierarchical organization of the elements subsumed under these names.

I think that a distinction must be made between hierarchies of *embedment,* such as the temporal hierarchy we have been discussing or the structural hierarchy of matter, and hierarchies of *connection,* such as the ecclesiastical chain-of-command. It is clear that the existence of the one does not necessarily imply and may in certain cases rule out the coexistence of the other: the Curia is not in any ordinary sense a partitioning of either the substance or the functioning of the Pope.

B. Hierarchies in the Nervous System

If we turn now to the organization of structure and function in a terrestrial vertebrate, we may glimpse a number of different senses in which it may be said to be hierarchical. There is the organization into cells, tissues, organs, and organ systems; within the nervous system, there is organization into nuclei, groups of nuclei, and organs. These units are *connected* by tracts, groups of parallel axons, and in some cases these tracts seem to bind different nuclei into a more or less clear hierarchy of connection and presumably control: the motor nuclei of the spinal cord are subordinated to centers in the brain stem and these in turn to higher centers. But in mammals, for example, the pyramidal tracts bypass the brain stem and connect the motor cortex directly with the spinal nuclei; one could name many other examples in which there is not even the appearance of connective hierarchical organization.

At first glance, the motor cortex–spinal nuclei connection might seem to be hierarchical, but we note that stimulation at a specific point on the cortex results in contraction in a specific muscle, or at least movement in a particular limb. It is found that in this way the body of a mammal (and hence its spinal motor nuclei) may be mapped onto the motor cortex, actually onto several (extrapyramidal) supplementary areas as well (Woolsey, 1958, gives the relevant references); thus the overall pattern of connection does not appear at all hierarchical. The organizational scheme appears to be top-heavy with command officers. Does convergence appear at levels "higher" than the motor cortex? My guess would be that even there control remains distributed.

The mammalian motor cortex is said to function in control of "voluntary" muscular activity; specific centers for specific activities cannot be found. Yet a number of centers for other activities have been described: a respiratory control center in the brain stem, divided into inspiratory and expiratory groupings; hypothalamic centers controlling eating, etc. There seem to be factors favoring isolation of function which, given enough time, may result in the evolution of specific centers controlling particular important functions; there is little evidence that *most* animal behavior is organized by such hierarchies of connection. Even the epochal experiments of von Holst and von St. Paul (1963) do not exclude the possibility that although they were able to elicit different patterns of behavior in the chicken by stimulating in different locations in the brain, they did so by biasing different parts of a single distributed "center" so that as a whole it responded in different ways. I believe that the evidence they present concerning interactions between the processes underlying different behavior patterns is consistent with this interpretation.

Perhaps there is a continuum between hierarchical control of old, long-established patterns of behavior and distributed control of phylogenetically recent patterns, particularly those developing during an animal's lifetime through learning and similar processes. I also imagine that control of "whole-animal" patterns such as locomotion and many courtship activities is most likely to be distributed.

In a strict hierarchy of connection, units at the same level may communicate only through their superiors; there are no "official" horizontal connections. In practice, especially when the system must cope with rapidly changing situations, such isolation must break down if the system is to survive; Toffler (1970) describes the evolution under modern pressures of the static hierarchical bureaucracies of business and government into what he refers to as "Ad-hocracy." Similarly, in the motor control systems of animals there are rich interconnections between elements at each level, with the possible exception of certain "relay" nuclei. These functional interconnections were first described by Sherrington for the level of motor nuclei. The picture of motor control which emerges, then, is of a stack of levels of interconnections of similar order of complexity, with vertical connections between levels largely organized on the lines of the topography of the body. To a particular behavioral pattern there corresponds a particular spatiotemporal distribution of activity at each level; to the extent that the vertical interconnections are parallel, these spatiotemporal distributions at different levels might be expected to be topographically similar. To complicate the picture, there may be levels arranged in parallel as well as in series: those of the cerebellum, for example.

The principle of similar topographical organization of activity at different levels clearly applies to the sensory side of the nervous system as well: the spatial relationships of the cochlea, the eye, and the somatosensory receptors are preserved in the mapping onto successively higher levels. Upon closer inspection, however, hierarchical organization of connection reappears: as one progresses "up" the visual pathway, for example, *convergence* of several units on one unit at the next level is the rule, so that for a particular cortical unit a (retrograde or antidromic) branching structure of connections exists with many terminals at the rods or cones. But *divergence* is also the rule, with the result that the visual system, neglecting the no doubt intricate interconnections within levels, may be said to consist of two sets of innumerable directed hierarchies of connection, one set (the convergent) with its branch tips at the retina, the other with its branch tips in the cortex. Only the latter hierarchies at all resemble the ecclesiastical model. Both kinds, I would guess, may be found in abundance on the motor side as well.[19]

Nevertheless, there exist various nuclei which do not appear to be organized in such topographical ways: parts of the basal ganglia and hypothalamus, for example, and the ascending reticular formation. Can these be said to sit astride the motor system in such a manner as to form a hierarchy of control? They often seem to mediate "motivational" functions, to "bias" the more topographically organized parts of the system in a nonspecific way. But their interconnections and activity also seem to be such that "command" is exercised by the brain as a whole, not by individual parts in any clear hierarchical way.

C. Hierarchies and Plans

Forgive the length of this seeming digression; the horse I have been kicking has been moribund for some time, but I find still has a surprising sway over ethological thought. I wish to return now, briefly I hope, to the more strictly functional notion of TOTE hierarchies or Plans of Miller *et al.* (1960). Although most of their concern is with the elaboration of the complex schemes of human behavior, they do postulate the existence of "innate Plans," based in large measure on their reading of Tinbergen (1951):

> In comparing Tinbergen's hierarchical description with the hierarchical organization that we refer to here as a Plan, one is struck by the fact that the higher levels in his description are not sequential . . . . Not until Tinbergen's description reaches the level of the consummatory acts does it take on the hierarchically organized sequence characteristic of a Plan. Thus we are led to think of relatively discrete, stereotyped, innate Plans for organizing actions into a consummatory act, but those acts are themselves ordered in time by some other mechanism. (pp. 77–78)

The alternative "mechanism," they feel, may be either rigid *chaining* or flexible *concatenation*, depending on the degree of environmental control of sequencing.

> It is always difficult, of course, to distinguish behavior based on a Plan from behavior based on a chain, but a simple concatenation is usually easy to distinguish. However,

---

[19]There seems to be evidence for the view that motor cortical activity is somatotopically organized in terms of individual muscles, for the view that it is organized by body region, and for the view that it is organized in terms of movements, i.e., joints (Rugh, 1965). Perhaps all three kinds of organization are present, if they are really different.

organisms are often so well adapted . . . that a concatenation may give the appearance of purposive, intentional behavior—as if the environment itself could serve as part of the animal's memory. It is almost as if the Plan were not in the organism alone, but in the total constellation of organism and environment together. (p. 78)

They do not pursue this latter point much further, unfortunately. I wish to emphasize it, and to emphasize as well the difficulty of deciding, when faced with a temporal hierarchy of behavior, whether the underlying organization is a hierarchy, a kind of chaining, a concatenation, or something entirely different. Miller *et al.* propose the TOTE unit as the fundamental functional unit of behavior, replacing the reflex; they are at pains to demonstrate the universality of Plans. At several points, they make statements such as "*complete* planlessness must be equivalent to death," and it is clear throughout that, having forgotten about chains and concatenations, they assume that an organism is hardly ever without a Plan.

D. Hierarchies and Holism

I wish to return now to Harmon's and Reiss' models of reciprocally inhibiting neuromimes. If input impulse frequency to the pair is made to undergo slow, wide excursions, it will be possible to "parse" the output pattern in a hierarchical manner: the successive levels are single output pulses from one unit, bursts from one unit, repeated patterns of alternating bursts from both units, groupings in time of related patterns. Yet the only "testing" which can be identified is the continuous comparison of "excitation" and "threshold" at the impulse-initiating site (the "monostable flip-flop" or multivibrator producing the output impulse) in each unit—either that, or "testing" is proceeding continuously at every point in the system. The "test" of whether to "transfer control" from one to another unit, for example, is distributed between the two impulse-initiating sites and not really localizable in either; it is a function of the performance of the system as a whole. Furthermore, there is no clear spatial separation between "test" and "action"; the process is all-of-a-piece, and differentiating "test" and "action" depends merely on whether the observer's attention is focused antidromically or orthodromically. Finally, it is difficult to consider the output pattern to be a function of a Plan (there are no identifiable higher-order "tests"), a chain (although what the system does is clearly a function of what it has just been doing), or a concatenation, nor is it easy to regard it as a function of some combination of the three.

Miller *et al.* consider the Plan to be a holistic notion, replacing the atomistic reflex as a unit of behavior. Yet we now can see that as long as "tests" are assumed to remain identifiable as such, and separable in some way from the total performance of the system, the concept retains a vestige of atomism. The notion of the TOTE hierarchy does represent an advance over that of the reflex. It is a concept derived in part from a heuristic programming language developed by Newell *et al.* (1959). Such heuristic "list" languages allow a much more holistic performance by the computer than do the more usual rigid algorithmic languages; during the operation of the program, there is continual reorganization, restructuring, of the material stored in memory.

It is often considered that vitalism, holism, etc., are in constant retreat before the onslaught of mechanistic science. But man models his notions of the operation of Nature on his machines. As machines become more complex, so do his notions of organism. The TOTE hierarchy is still a notion based on the machines of man; as such, it represents another retreat from an atomistic conception of Nature. We are thus faced with the paradoxical situation that as mechanism's progress is the more triumphant, the retreat of its atomistic foundation becomes the more precipitate. How much further retreat will be necessary before our notions approach consonance with Nature's complexities?

VI. PERCEPTUAL GATING ARRAYS AND THE CONCEPT OF DISTRIBUTED CONTROL

A. The Perceptual Template in Song Development and Production

I must now review briefly the state of our knowledge of the development of passerine song. All signs point to the existence of what has been called a "perceptual template" which occupies a critical position in song ontogeny. The young bird's utterances progressively "crystallize" into the form of song which is characteristic of the species, and this process depends in some measure on an intact auditory system (Konishi, 1965*a,b*). In some species, e.g., the white-crowned sparrow, early auditory exposure to the species song of a given geographical area appears to modify the perceptual template in such a way that the final pattern, "fitted" to the template, becomes characteristic of that area (Marler and Tamura, 1964). In others, such "imitation" can extend to other environmental sounds.

Furthermore, in some species, although a richer auditory environment seems to result in the development of a more diverse repertoire, the influence is not necessarily of the specific sort involved in imitation but seems rather to be unspecific: in Marler's terms (Marler *et al.*, 1962), the presence of singers, even of other species, may stimulate *invention* or *improvisation*. It appears likely that in many cases the bird's provisional utterances progressively modify its own "perceptual template." The term "perceptual template" may prove to have been an unfortunate choice, implying as it does some sort of rather static, simultaneous matching, as of the steric configurations of lock and key. Perhaps the analogy of a combination lock is more appropriate: a temporal pattern of instructions for positions of the parts provides the "key."[20]

Among the thrushes, only the blackbird (Messmer and Messmer, 1956) and the robin (Konishi, 1965a) have been subjected to experimental studies of song ontogeny. In these species, the evidence for "perceptual templates" is rather convincing; it seems almost certain that the elements in the repertoire represent complex skills acquired through some sort of matching of the results of motor performance by auditory feedback against sets of perceptual criteria. Furthermore, in the European blackbird at least, the order of singing may be experimentally altered by broadcasting songs to the bird over a loudspeaker; the specific song which is played influences the choice of the next song type sung by the subject (Todt, 1970). It is therefore likely that in this species of thrush, the *order* of songs in the absence of experimental intervention is also in part dependent on the results of auditory feedback. Finally, it seems most likely that the same "sets of perceptual criteria" involved in acquisition of song or pattern types are also the ones involved in their ordering during production of the finished song. Why not? I have not gone so far as to quarrel with William of Ockham on this score; I have no wish *needlessly* to "multiply entities."

Now, it is not essential that such a set of perceptual criteria be homomorphic in some way with a particular sound pattern, but the evidence from studies of imitation in birds (including the blackbird; Tretzel, 1967) strongly suggests that this is at least sometimes the case. The detailed pattern of frequency and amplitude against time is often imitated rather precisely, which would be difficult if the perceptual criteria did not in fact constitute a very similar pattern. It seems that such a set, fully developed, constitutes a pattern-sensitive *gate*, allowing to pass only certain ordered combinations of input frequencies; further, it seems that during development its specificity may increase.

How many such "gates," considered as *functional* units, must a bird possess? I assume one for each unit of recombination in its primary song. In the white-crowned sparrow and chaffinch, the number is one or two; within *Hylocichla*, different individuals have repertoires of from four to 12 or so primary patterns in the Swainson's thrush, to 15 or so in the hermit

---

[20]All parts of a temporal pattern could in principle be simultaneously compared with a static template, however, by use of delay circuits, for example (Reiss, 1964). In fact, combination locks do not "open" progressively but only after a final configuration of the mechanism is arrived at. It seems we cannot escape the notions supplied by our machines.

thrush, to 20 or more in the wood thrush. But in other thrushes such as the song thrush (*Turdus philomelos*) there may be as many as several hundred units of recombination. If there must be literally hundreds of such functional gates, we face a difficult *anatomical* question: is each gate embodied in an anatomically separable unit? And if, as in these turdine thrushes, rather free but still lawful recombination may occur, what have we to say about the pattern of connections between all these gates?

I believe it is more parsimonious to assume that there is just one anatomical gate, which allows to pass at least as many patterns as there are units of recombination in the repertoire. I will now proceed to a "highly specific hypothesis" as to the nature of this gate, its anatomy and functioning.

B. A Highly Specific Model of the Perceptual Template

Imagine that there exists in the bird's head, somewhere in the no-man's-land between perception and action, a group of elements in which the *functional* arrangement along at least one dimension is according to the frequency of sound to which the elements are maximally sensitive. For convenience, let us assume that frequency is represented *tonotopically*, that is, that the elements are spatially ordered monotonically according to their frequency of maximal responsiveness. Potash (1970) has described such tonotopic organization in the avian nucleus lateralis, pars dorsalis; interestingly, he found this nucleus to be surrounded by a vocalization motor area, the nucleus intercollicularis, so that we seem to be well within the no-man's-land.

Now imagine that the spatial distribution of activity upon reception of a note of a particular frequency so biases the subsequent responsiveness of the elements in the array that notes of only certain frequencies will be able to activate their respective portions of the array during the ensuing epoch, whichever of these is sounded will further limit the responsiveness of the array during the next ensuing epoch, etc. We now have a device which will be maximally sensitive to only particular temporal patterns of sound, in a contingent manner: each succeeding note actualized may further limit the potential range of patterns to which the array is sensitive.[21] We now have, in fact, a generator of perceptual templates, a gate which passess certain patterns and does not pass others. If, depending on details of phylogeny and ontogeny, only a few patterns come to be "passable," the bird's motor output (in some poorly understood way) comes to be restricted to a correspondingly small repertoire of patterns.

In the case of the *Hylocichla* thrushes, we can make more specific guesses

[21]This is the familiar and often tragic ontogenetic pattern: an organism's range of potential is always greatest at birth.

as to the nature of the gating array. First, we note that songs are sung to particular *scales* composed of relatively similar arrays of musical intervals: thus we may hypothesize that passage is similarly restricted to regions on the array corresponding to the degrees of the scale, once the first note of the pattern has determined which scale it shall be.

How could such a gating be obtained? Let me return for a moment to the example of phyllotaxy patterns in plants. There is disagreement as to the exact nature of the effect, but all workers seem to agree that the position of a new primordium is determined by some sort of inhibition from already established primordia. This inhibition is presumed to be either of a chemical, diffusional nature (Turing, 1952; Richards, 1948) or of the nature of a mechanical "stacking" limitation (van Iterson, 1907; Snow, 1955). Turing has proposed a reaction-diffusion model of chemical processes occurring along a ring, which within certain ranges of parameter is unstable and leads to the development of regularly spaced concentration peaks and troughs. The mathematics involved is similar to that used by Reichardt and Mac-Ginitie (1962), following Hartline and Ratliff (1958), to model the lateral inhibitory network found in such receptor systems as the lateral eye of *Limulus*. Such a lateral inhibitory network model is a (linear) extension of Harmon's model of a reciprocally inhibitory pair of neuromimes, in which inhibition by an element of its neighbors is made a decreasing function of distance.

Without an extensive mathematical treatment, it is easiest to demonstrate how instability may arise if we consider first an alternative representation of Reiss' or Harmon's model, as in Fig. 5C. Seen from the point of view of one unit, the effect on it of the reciprocal inhibitory connections is one of positive feedback: a minus times a minus makes a plus. In fact, with a larger lateral inhibitory array (Fig. 5D), the whole remainder of the network may be lumped as a positive feedback loop to the unit in question (Fig. 5C again). Under wide ranges of conditions, such loops are unstable and result in "runaway" behavior, but in this case when the output of the unit in question is sufficiently large to silence all its neighbors, further increase in its output will have no further effect via the positive feedback loops and output will be stabilized at a high level.

Depending on the nature of the function relating the lateral inhibitory coefficients to distance, a single disturbing input at a particular point will result in one or more peaks of excitation (or troughs of inhibition) distributed at regular intervals along the array. If the array is circular, the number of peaks will be integral-valued (Turing, 1952). It seems, in short, that we have here a natural candidate for the generator of perceptual templates, at least to the extent that these take the form of musical scales.

In Turing's model, a slight increase in complexity (addition of a third "morphogen" component) may result in the appearance of *traveling waves* of concentration, one or more peaks of concentration which move around the circular array at a speed which is a function of the dynamics of the system as a whole. In a circular array of reciprocally inhibitory neural elements, fatigue or adaptation may play a similar role, so that the *succession* of musical scalelike perceptual templates takes the form of a rhomboidal patterning similar to Fig. 4C. It should be remembered that the mathematical conjugate of such a pattern, the troughs between the peaks, forms something like a double series of traveling waves going in opposite directions.

C. The Functioning of Gating Arrays

Such a "highly specific model" is certainly not specific enough to account for the details of the primary pattern, song, and sequence organization in these thrushes, but it is a start. The approximately pentatonic nature of the scales may arise during ontogeny through frequent hearing of octaves, fifths, and fourths, the intervals associated with the first few harmonics of any fundamental frequency, but it may also have a genetic basis. (The relationship of the octave in particular may have arisen in phylogeny.) The real problems seem to lie in obtaining motor patterns which progressively fit the developing "templates"; these problems are the general ones of the organization of acquired skills. Somehow during ontogeny the passing of an appropriate pattern by the gate evidently serves to reinforce, stabilize, "increase the probability of," the spatiotemporal distribution of motor activity which gave rise to the pattern, and yet during the morning song in these species it is rare for the same primary pattern to be followed by itself, and therefore passage through the gate seems also to inhibit immediate recurrence of the pattern.

I suggest the following TOTE-like general scheme. A pattern of excitation on the "gating array" is transmitted to the corresponding "motor array" and the corresponding sound pattern is generated. When the sound feedback arrives, it serves either to add to the residual pattern on the gating array or to cancel it, depending mainly on the time constants and coefficients of self-inhibition (adaptation) of the units of the array; if these time constants and coefficients are small, the pattern may "pass" and the song will be repeated. If they are large, the pattern will be canceled and a new, more or less conjugate pattern will become established and passed to the "motor array." In the song sparrow (Nice, 1943) or meadowlark (Fish *et al.*, 1962), a "gating pattern" may remain dominant for many songs, resulting in long bouts of the same song before self-inhibition is able to gain ascendency over the self-reinforcing effects of lateral inhibition, and a distribution in space and time characteristic of a new song type appears. In the robin (*Turdus migratorius*) and song thrush (*T. philomelos*), a pair of recombination units is often alternated in a bout, *ABAB* . . . . It is not clear whether such a pattern requires an addition to the model; probably I will only know when I can actually build such a gating array. But I am impressed that such "trilling" can occur at all levels of the temporal hierarchy: within a primary pattern, between recombination units, between different "song cycles" in Swainson's thrush. Does the rule of parsimony not suggest that similar features at different levels of organization may arise in the same way? If so, then why not on the same gating array?

I believe that the answer may lie in the presence of different sorts of units, possibly whole reciprocally inhibitory networks with different parameters, interleaved on the same gating array. Such an arrangement would go far toward accounting for the species-specific and individual-specific characteristics of song, in the following way. One such lateral inhibitory gating array, you will agree, would be capable of gating or generating a huge variety of patterns. With two such, the variety would be practically infinite. But if the two were interconnected in an inhibitory way, and they were merely "somewhat" different in their parameters, the number of acceptable patterns might be much reduced, and by and large they would be much more complex than those that a single array could produce. I have in mind a process analogous to the generation of various interference effects such as moiré patterns and complex Chladni figures, and the "superposition" and "magnet" effects of von Holst (1948).

D. Self-Tuning: Limit Cycle Behavior

Whether or not such an arrangement would be too sensitive, too susceptible to internal disruption and outside interference to account for song stability, can only be determined by experiment with an appropriately complex model, which I hope to be able to do soon. I suspect rather that the system as a whole will under certain parameter ranges be found to have a "self-tuning" property: in plant phyllotaxy, minor deviations in the position or time of insertion of the next primordium are corrected; major ones "flip" the system to a new and stable pattern. There are amazingly few occurrences of deviations which are propagated in this manner. (There is one notable exception to this: the distichous arrangement of successive primordia at 180° angles found in many families may gradually slip into the Fibonacci system as growth slows down or the rate of production of primordia increases. But the results are the same; the transient from one to the other system is simply longer.)

Such "self-tuning" arrays are nonlinear and energetically nonconservative in their functioning and may be said to show a characteristic kind of "limit-cycle" property. For example, if ("instantaneous") frequency of firing of one of the two neuromimes in Reiss' (1962) analog model is plotted against the frequency of firing of the other, the points will fall more or less into one of a number of cycles or loops, more or less boomerang-shaped, representing equilibrium *cycles*, not points, for the system. Slight deviations will be corrected; major ones will flip the system into another loop. As the number of elements is increased, such "phase-space" representation rapidly becomes impossible, for the number of dimensions of the phase space is at least as large as the number of variables, i.e., elements, in the system. Furthermore, in the case of Fibonacci-like systems, the cycle is never quite the same again: for any two elements arbitrarily selected, like a Lissajou figure relating the mutual course of amplitude of two wave-trains differing slightly and nonharmonically in frequency, the phase-space loop changes its shape each time around. But the limit-cycle property of correcting deviations toward an equilibrium cycle remains; a representation similar to Fig. 6A is perhaps more appropriate. In the examples of *Hylocichla* song showing cyclic organization, the limit cycle may be quite long but does eventually repeat.

E. Hierarchical Organization vs. Distributed Control

The TOTE hierarchy model of Miller *et al.* is an effort to bridge the gap between perception and action which has so long plagued psychologists. The gating array model presented above similarly attempts to bridge this gap, simply by eliminating it. Each element in the array, or the array as a whole, may be said equally to be "testing" and "operating." I claim that the array lies exactly along the frontier between perception and action. True, there does remain the little problem of translating the spatiotemporal distribution of activity in the array into a spatiotemporal distribution of motor activity; no doubt also the *particular* spatiotemporal distribution of motor activity will depend on activity of other parts of the system—the cerebellum, for example. But my point is, there should never have been a *gap*. I suggest, very tentatively, that arrays governing patterning of the other things the bird does may be organized in a similar fashion and that all such arrays may usefully be considered as being *on the same level*; it is likely that whole groups of such units may be interconnected by reciprocal inhibition. If this should prove to be the case, the organization of connection would still be in a way hierarchical but lacking a hierarch. Control or dominance would shift from one to another part of the distributed perceptual–motor interface according to internal or external necessity, a possibility more or less dismissed by Miller *et al.*

Now, it is true that I have no firm evidence that there is auditory perceptual control over song performance in thrushes of the genus *Hylocichla*.

It is also true that although certain species, such as the white-crowned sparrow, require an intact auditory system for the *development* of their characteristic song, they can *maintain* the developed song pattern following deafening without much degradation (Konishi, 1965b). Accordingly, I will hedge my bets by speculating that the perceptual gating array is able to *generate* patterns without peripheral input. Konishi (1965a) has deafened several robins (*T. migratorius*) prior to song development; he finds that the syllables (recombination units) are markedly different from those of the wild intact bird. Furthermore, in several deafened birds a syllable of a particular type tends to be given in long bouts, *AAAA, BBBB, . . .* , a pattern which is very rare in wild robins except at the very beginning of a period of singing (personal experience). This would be expected, according to the model, if the auditory input regulating song switching in the intact bird were removed.[22]

Finally, I should distinguish between those features of the model which I am prepared to sacrifice, singly or in groups, and those which I will relinquish only with reluctance. The latter list includes the notion of representation in the nervous system of a particular pattern of behavior by a spatiotemporal distribution of activity rather than by a center; more specifically, the notion of control distributed along an array of interacting elements with its implication of complex limit cycle organization; the absence of any need to have "test" and "operate" phases in different locations; and finally perhaps the absence of a hierarch. I believe these notions taken together define a kind of consistent holistic approach to the analysis of behavior and its underlying neural organization.

VII. . . . BUT IS IT ART?

To recapitulate, in the introduction I suggested that among the virtues of bird song were its dimensional simplicity and resulting ease of quantitative treatment, and its closed-loop nature and relative isolation from environmental disturbance. These enable us to see further virtues in the thrush song patterns. They are sung with beautiful precision; by the application of one of several simple culling rules, they may be made to fall unequivocally into one or another of a limited number of equivalent and mutually exclusive categories; long sequences of these elements may be knit together by diagonal patterning in a precise and rational order, which may be parsed unequivocally in a hierarchical manner. All of these features seem to be related, but the "syndrome" seems to be unremarked in the ethological literature. In Section

[22]Any TOTE-type organization, however, should produce the same result.

III, I suggested that similar structuring might be found in other taxa and thus that it should be taken seriously.

Next, I contrasted such complex determinate sequences with similar-appearing but "less orderly" Markovian progressions in which the sequence is generated in some measure by unrelated extrinsic ("chance") events. I meant by this comparison to suggest that where the former are found to be likely, the latter are hardly suitable as models. Linear control theory likewise seemed inappropriate for models of processes generated by nervous systems. But aspects of the "syndrome" began to appear as soon as we considered simple networks of nonlinear elements resembling neurons in their functioning. Analysis is soon beggared entirely by the complexity and versatility of behavior of such networks, and their study perforce becomes holistic.

But the possibility remained that the parts of behaving systems form hierarchies of connection, as in several popular recent theories. In this case, subsystems might be sufficiently isolated from one another to render their separate study meaningful. This may be the case with such phylogenetically old automatisms as the eyeblink reflex, but a brief examination of the vertebrate nervous system dispelled this hope immediately for the complex interactions that mainly interest the ethologist, and we are denied this final refuge from a holistic ethology.

The ground was thus prepared for the introduction of the notion of "distributed" as contrasted with "hierarchical control," in the concrete form of a perceptual gating array that seemed to go far in accounting for the patterning "syndrome" of the thrushes. In this model, the "perceptual template," long supposed to play an essential role in song development, becomes a kind of complex spatiotemporal resonance, one of a number of modes of operation of a tonotopically ordered array of mutually inhibitory elements.

Now, the notion of "control," usually somewhat vague even in speaking of the most strictly defined hierarchy of connection, becomes meaningless when extended to something distributed throughout a large part of the CNS at any one time. Wolfgang Schleidt terminated a discussion of Miller *et al.* (1960) one time in 1967 simply by asking "Can control have babies?" The answer seems to be "Never." By "distributed control" of a particular behavior, I mean merely the corresponding spatiotemporal pattern of nervous activity. Nor should this latter phrase be taken to refer only to the firing pattern: I assert that the nonfiring pattern, of "waves of silence," is equally important to a properly holistic approach.[23]

---

[23]Have you heard it said that the nerve impulse is the carrier of information in the nervous system? Is it the "1"s or the "0"s, or is it not their pattern, which performs the same function in a digital computer? After all, the all-or-nothing impulse is not free to vary at all, whereas the interval is free to vary at least in duration!

So also with the patterns of animal behavior, and in conclusion I must now take up the hitherto neglected matter of the intervals between thrush songs. In contrast to the highly determinate appearance of the songs themselves and of their ordering, their timing seems to be generated by a more or less indeterminate (extrinsic, unrelated) process. All attempts to relate interval duration to nature of preceding or following song, or to recurrence number, or to a number of other variables of sequencing have ended more or less in failure. All that is known for sure is that, as Wolfgang Schleidt took pains to discover, the distribution of interval length is approximately log-normal, or perhaps the tail of the distribution is a wee bit closer to a negative exponential.

Yet, although when timed with a stopwatch an individual's intervals vary enormously and unpredictably, sometimes when I am "just listening" I find that I am predicting the "entrances" of several individuals in a chorus with great success. It is my opinion that a particular bird, a member of a net of "distributed control," inserts his contribution at a time appropriate to the overall pattern of the chorus, within the limits imposed by his own internal constraints, and that I, an intruder upon the scene, am sufficiently similar in physiology to the bird that I am able to enter as an observer into this net.

To the extent that this is true, there is considerable merit in regarding animal communication in terms of unitary distributed systems; this is beyond my competence at the moment. But I must mention that when I "enter the net" I am overcome by a proper religious feeling, quite holistic.

VIII. REFERENCES

Ashby, W. R. (1956). *An Introduction to Cybernetics,* Wiley, New York.
Bohm, D. (1968a). Some remarks on the notion of order. In Waddington, C. H. (ed.), *Towards a Theoretical Biology,* University of Edinburgh Press, Edinburgh, pp. 18–40.
Bohm, D. (1968b). Further remarks on order. In Waddington, C. H. (ed.), *Towards a Theoretical Biology,* University of Edinburgh Press, Edinburgh, pp. 41–61.
Borror, D. J., and Gunn, W. W. H. (undated). Songs of thrushes, wrens and mockingbirds of America (record). In *Sounds of Nature,* Vol. 8, Federation of Ontario Naturalists, Houghton Mifflin, Boston.
Cane, V. (1961). Some ways of describing behaviour. In Thorpe, W. H., and Zangwill, O. L. (eds.), *Current Problems in Animal Behaviour,* Cambridge University Press, Cambridge, England, pp. 361–388.
Cox, D. R., and Lewis, P. A. W. (1966). *The Statistical Analysis of Series of Events,* Methuen, London.
Craig, W. (1943). The song of the wood pewee *Myiochanes virens* Linnaeus: A study of bird music. *N.Y. State Mus. Bull.* No. 334.
Crossley-Holland, P. (1960). I. Non-western music. In Robertson, A., and Stevens, D. (eds.), *The Pelican History of Music,* Penguin Books, Baltimore, pp. 13–135.
Dilger, W. C. (1956). Adaptive modifications and ecological isolating mechanisms in the thrush genera *Catharus* and *Hylocichla. Wilson Bull.* **68(3):**171–199.
Fish, W. R., Nelson, K., and Isaac, D. (1962). The temporal patterning of meadowlark song. *Am. Zoologist* **2(3):**ap. 54.

Greenewalt, C. H. (1968). *Bird Song: Acoustics and Physiology*, Smithsonian Institution Press, Washington.

Harmon, L. D. (1961). Studies with artificial neurons, I: Properties and functions of an artificial neuron. *Kybernetik* **1(3)**:89–117.

Harmon, L. D. (1964a). Problems in neural modelling. In Reiss, R. F. (ed.), *Neural Theory and Modelling*, Stanford University Press, Stanford, Calif., pp. 9–30.

Harmon, L. D. (1964b). Neuromimes: Action of a reciprocally inhibitory pair. *Science* **146**:1323–1325.

Hartline, H. K., and Ratliff, F. (1858). Spatial summation of inhibitory influences in the eye of *Limulus*, and the mutual interaction of receptor units. *J. Gen. Physiol.* **41(5)**: 1049–1066.

Huggins, W. H. (1957). Signal-flow graphs and random signals. *Proc. I. R. E.* **45**:74–86.

Kavanau, J. L. (1967). Behavior of captive white-footed mice. *Science* **155**:1623–1639.

Konishi, M. (1965a). Effects of deafening on song development in American robins and black-headed grosbeaks. *Z. Tierpsychol.* **22(5)**:584–599.

Konishi, M. (1965b). The role of auditory feedback in the control of vocalization in the white-crowned sparrow. *Z. Tierpsychol.* **22(7)**:770–783.

Marler, P., and Tamura, M. (1964). Culturally transmitted patterns of vocal behavior in a sparrow. *Science* **146**:1483–1486.

Marler, P., Kreith, M., and Tamura, M. (1962). Song development in hand-raised Oregon juncos. *Auk* **79**:12–30.

Messmer, E., and Messmer, I. (1956). Die Entwicklung der Lautaüsserungen und einiger Verhaltensweisen der Amsel (*Turdus merula merula* L.) unter natürlichen Bedingungen und nach Einzelaufzucht in schalldichten Raümen. *Z. Tierpsychol.* **13**:341–441.

Miller, G. A., Galanter, E., and Pribram, K. H. (1960). *Plans and the Structure of Behavior*, Holt, Rinehart and Winston, New York.

Milsum, J. H. (1966). *Biological Control System Analysis*, McGraw-Hill, New York.

Newell, A., Shaw, J. C., and Simon, H. A. (1959). A general problem-solving program for a computer. *Computers and Automation* **8(7)**:10–17.

Nice, M. M. (1943). Studies in the life history of the song sparrow. II. The behavior of the song sparrow and other passerines. *Trans. Linn. Soc. N.Y.* **6**:1–238.

Potash, L. M. (1970). Neuroanatomical regions relevant to production and analysis of vocalization within the avian *Torus semicircularis*. *Experientia* **26**:1104–1105.

Reichardt, W., and MacGinitie, G. (1962). Zur Theorie der lateralen Inhibition. *Kybernetik* **I**:155–165.

Reiss, R. F. (1962). A theory and simulation of rhythmic behavior due to reciprocal inhibition in small nerve nets. *Proc. Spring Joint Computer Conf., A. F. I. P. S.*

Reiss, R. F. (1964). A theory of resonant networks. In Reiss, R. E. (ed.), *Neural Theory and Modelling*, Stanford University Press, Stanford, Calif., pp. 105–137.

Richards, F. J. (1948). The geometry of phyllotaxis and its origin. *Symp. Soc. Exptl. Biol.* **2**:217–248.

Ruch, T. C. (1965). The cerebral cortex: Its structure and motor functions. In Ruch, T. C., and Patton, H. O. (eds.), *Physiology and Biophysics*, Saunders, Philadelphia, pp. 252–279.

Snow, R. (1955). Problems of phyllotaxis and leaf determination. *Endeavour*, October 1955, pp. 190–199.

Stein, R. C. (1956). A comparative study of "advertising song" in the *Hylocichla* thrushes. *Auk* **73**:503–512.

Thielke, H., and Thielke, G. (1960). Akustisches Lernen verschieden alter schallisolierter Amseln (*Turdus merula* L.) und die Entwicklungerlernter Motive ohne und mit künstlichen Einfluss von testosteron. *Z. Tierpsychol.* **17**:211–244.

Thompson, D. W. (1961). *On Growth and Form*, Abridgement ed. J. T. Bonner, Cambridge University Press, Cambridge, England.

Tinbergen, N. (1951). *The Study of Instinct*, Oxford University Press, New York.

Todt, D. (1968). Zur Steuerung unregelmässiger Verhaltensabläufe—Ergebnisse einer Analyse des Gesangs der Amsel (*Turdus merula*). *Kybernetik 1968*, Oldenburg, Munich.

Todt, D. (1970). Gesang und gesangliche Korrespondenz der Amsel. *Naturwissenschaften* **57**:61–66.

Toffler, A. (1970). *Future Shock*, Random House, New York.

Tretzel, E. (1967). Imitation und Transposition menschlicher Pfiffe durch Amseln (*Turdus m. merula* L.). Ein weiterer Nachweis relativen Lernens und akustischer Abstraktion bei Vögeln. *Z. Tierpsychol.* **24**:136–161.

Turing, A. M. (1952). The chemical basis of morphogenesis. *Phil. Trans. Roy. Soc. Ser. B* **237**:37–72.

von Holst, E. (1948). Von der Mathematik der nervösen Ordnungsleistung. *Experientia* **4**(10):374–381.

von Holst, E., and von St. Paul, U. (1963). On the functional organization of drives. *Anim. Behav.* **11**:1–20, translated from *Naturwissenschaften* **18**:409–422.

van Iterson, G. (1907). *Studien über Blattstellungen*, Fischer, Jena.

Wilson, D. M. (1964). Relative refractoriness and patterned discharge of locust flight motor neurons. *J. Exptl. Biol.* **41**:191–205.

Woolsey, C. N. (1958). Organization of somatic sensory and motor areas of the cerebral cortex. In Harlow, H. F., and Woolsey, C. N. (eds.), *Biological and Biochemical Bases of Behavior*, University of Wisconsin Press, Madison.

INDEX